Contemporary Quantum Mechanics in Practice

This helpful and pedagogical book offers problems and solutions in quantum mechanics from areas of current research rarely addressed in introductory courses or textbooks. It is based on the authors' own experience of teaching undergraduate and graduate courses in quantum mechanics and adapts problems from contemporary research publications to be accessible to students. Each section introduces key quantum mechanical concepts, which are followed by exercises that grow progressively more challenging throughout the chapter. The step-by-step solutions provide detailed mathematical derivations and explore their application to wider research topics. This is an indispensable resource for undergraduate and graduate students alike, expanding the range of topics usually covered in the classroom, as well as for instructors and early-career researchers in quantum mechanics, quantum computation and communication, and quantum information.

Lilia M. Woods is Professor at the Department of Physics, University of South Florida, where she has taught many courses in the past two decades. An academician with a stellar publication record, her research focus is on several aspects of quantum mechanics, nanomaterials, and device-modeling.

Pablo Rodríguez López is an assistant professor at Universidad Rey Juan Carlos de Madrid and is a part of the GISC. Since finishing his PhD at the Complutense University of Madrid, followed by postdoctoral experience at the University of Loughborough, the National Centre for Scientific Research of France (CNRS) and Spain (CSIC), and the University of South Florida, he has been a collaborator of Lilia M. Woods and has authored many high-impact publications.

Contemporary Quantum Mechanics in Practice

Problems and Solutions

Lilia M. Woods
University of South Florida

Pablo Rodríguez López
Universidad Rey Juan Carlos de Madrid

CAMBRIDGE
UNIVERSITY PRESS

CAMBRIDGE
UNIVERSITY PRESS

Shaftesbury Road, Cambridge CB2 8EA, United Kingdom

One Liberty Plaza, 20th Floor, New York, NY 10006, USA

477 Williamstown Road, Port Melbourne, VIC 3207, Australia

314–321, 3rd Floor, Plot 3, Splendor Forum, Jasola District Centre, New Delhi – 110025, India

103 Penang Road, #05-06/07, Visioncrest Commercial, Singapore 238467

Cambridge University Press is part of Cambridge University Press & Assessment, a department of the University of Cambridge.

We share the University's mission to contribute to society through the pursuit of education, learning and research at the highest international levels of excellence.

www.cambridge.org
Information on this title: www.cambridge.org/9781009355407

DOI: 10.1017/9781009355414

First published 2024

A catalogue record for this publication is available from the British Library.

Library of Congress Cataloging-in-Publication Data
Names: Woods, Lilia, author. | Rodríguez López, Pablo, author.
Title: Contemporary quantum mechanics in practice : problems and solutions / Lilia M. Woods, University of South Florida, Pablo Rodríguez López, Universidad Rey Juan Carlos de Madrid.
Description: Cambridge, United Kingdom ; New York, NY : Cambridge University Press, 2024. | Includes bibliographical references and index.
Identifiers: LCCN 2023053891 | ISBN 9781009355407 (hardback) | ISBN 9781009355445 (paperback) | ISBN 9781009355414 (ebook)
Subjects: LCSH: Quantum theory – Textbooks. | Quantum theory – Mathematics – Textbooks.
Classification: LCC QC174.12 .W66 2024 | DDC 530.12–dc23/eng/20231221
LC record available at https://lccn.loc.gov/2023053891

ISBN 978-1-009-35540-7 Hardback
ISBN 978-1-009-35544-5 Paperback

Contents

1 Introduction

Quantum mechanics is one of the fundamental subjects in physics. It is a key component for any undergraduate and graduate curricula, and it is required for students to obtain a good degree of understanding in order to conduct meaningful research. From a student's immediate perspective, however, this goal may be difficult to achieve without taking fundamental concepts and applying them in practice problems.

As part of our teaching experiences as professors, we have had students constantly asking for more practice problems. The consensus is that having a variety of available exercises illustrating key concepts *more than one time in different scenarios* can give students that extra knowledge. This skill is strongly desired, especially in quantum mechanics as being one of the pillars in physics. On the other hand, from a practical point of view of teaching the course and having to give multiple exams during the semester, it is useful if a diverse set of problems reflecting modern developments in quantum mechanics is available.

There are many books on quantum mechanics available in the literature. In addition to textbooks adapted by universities, professors develop custom-made notes emphasizing a particular topic or set of topics, and when it comes to practice problems, the majority of books cover rather standard topics, such as 1D Schrödinger equations, perturbation theory, addition of angular momenta, the simple harmonic oscillator and hydrogen atom Schrödinger equations, and scattering. However, new research areas are emerging that rely on many concepts that are less represented in this list (which is not meant to be exhaustive, of course).

Here are some examples. Recent discoveries of many new materials with topologically nontrivial nature have elevated the importance of symmetry and geometry in quantum mechanics. There is a great need to better understand Berry phases and related properties in free space and periodic environments. Manipulating functions of operators and various transformations in this context also becomes important. Our experience in the classroom, as students and professors now, shows that geometrical phases are given rather peripheral attention and are not much studied beyond the standard Aharonov–Bohm effect. But since topology with its underlined symmetry and group theory is gaining prominence in various areas, we feel that a larger body of introductory exercises is needed. After all, future researchers can benefit tremendously if these concepts are taught in quantum mechanics classes with the opportunity for extra practice.

Quantum information is forming into a separate scientific field, which uses Dirac notation with theories relying heavily on the density operator. Describing a given system with a density matrix is in fact more general than the wave function formalism, for which many problems already exist throughout the literature. However, the statistical nature of quantum particles, processes, and measurements is much better captured using the density operator. This formalism is also necessary for the probabilistic nature of entropy, pure and entangled states, and measurements for quantum information science. Although the density operator with some of its basic properties is discussed in standard textbooks, there is a shortage of practice problems at a more advanced level.

Perhaps the most challenging problems in quantum mechanics involve identical particles. It is not easy to understand how to apply symmetrization/antisymmetrization procedures to composite systems that contain two, three, four, seventeen, or an infinite number of fermions or bosons. Most books focus on the quantum mechanical states of two identical fermions and bosons making a connection with addition of angular momenta. For this purpose, the Clebsch–Gordan coefficient table is conveniently introduced. Our experience shows, however, that students cannot easily translate this knowledge when they are dealing with three or four identical particles and their angular momenta have to be added. The problem of handling a large but finite number of fermions becomes especially important in atoms for which the electron shell structures are needed. Even though this is well known in chemistry, here we tried to reinforce that the electron shells in atoms from the periodic table are a consequence of the nature of identical particles. When dealing with quantum mechanical states that represent essentially an infinite number of identical fermions or bosons (such as in macroscopic materials), second quantization becomes necessary, which is seldom discussed in standard courses. Here we try to give problems showing how second quantization is performed in relatively simple Hamiltonians by giving the reader opportunity for extra practice and to make broader connections in the context of the statistical nature of quantum mechanics. The problems of second quantization are also quite necessary for students to see how current research can become textbook material. Much of the research in the past several years has been graphene-driven, thus it would be beneficial to see how its Hamiltonian quantization occurs. Starting with "simpler" problems and delving into statistical quantum mechanical concepts, the problems we work out in detail give an exemplary connection between textbook quantum mechanics and quantum field theory.

Understanding relativistic effects in quantum mechanics has become especially important in light of many recently discovered new materials, in which the Dirac equation plays a prominent role. The most important example perhaps is graphene, whose properties can be described and understood in terms of relativistic quantum mechanics, where the speed of light is substituted by an effective velocity of massless particles. Scattering of relativistic particles as well as topologically nontrivial edge states also require knowledge of the Dirac equation and its variations. In addition to these new developments, relativistic problems are always left at the end of quantum mechanics courses and students are shortchanged. The problems in the last chapter of the book give different variations of the Dirac equation, which gives an opportunity for practice in the context of contemporary research situations.

These emerging fields of quantum research have created new challenges for teaching and practice. Universities are currently creating programs and curricula to answer these demands and train the future workforce with diverse background, so students are better prepared for cutting-edge research and industry of tomorrow. The collection of problems in this book emphasizes less-represented topics in the mainstream literature. At the beginning of each chapter, a brief introduction of basic concepts is provided. The reader will notice that this is not a textbook format; rather, it is a brief summary of some important formulas and relations with concise explanations (a more in-depth discussion can be found in textbooks). The problems in each chapter are of varying difficulty and we tried to arrange them by starting with simpler examples. Some problems are created from scratch, while others are revised from the research literature in order to make them suitable for practice and instruction.

We certainly hope that the reader will find this book useful. Technical details and connections that a student or a beginning researcher were expected to make on their own are now shown explicitly, which is quite beneficial in the interest of saving time and building skills. We also hope readers will find the book enjoyable. Even though we have tried very hard to make all notation uniform throughout, it is possible that improvements are necessary. Some typos are also likely to occur (not many, we hope) simply because of the nature of the content. We would appreciate it if readers communicate with us in case they find such unintentional errors. It would also be great if readers send us variations of the current problems or new ones based on their research to include in a new edition of this book in the future.

2 Operators, Transformations, and Symmetry

In physics we often deal with functions of operators, and in this chapter, we give many problems to practice this. The key concept to remember is that what stands behind $f(\widehat{A})$ ($\widehat{A}$ – an operator) is the Taylor series $f(\widehat{A}) = \sum_{n=0}^{\infty} \frac{f^{(n)}(0)}{n!}\widehat{A}^n$ with $f^{(n)}(0)$ being the nth derivative of the function f evaluated at 0. Actually, such a meaning is embedded in the translation and rotation operators given in what follows. Here we have given a collection of problems by including concepts related to magnetic translations, coherent and squeezed states, for example. Understanding symmetry also relies on functions of operators, as evident from the many problems on rotations, parity, and time-reversal operations. Several problems are also given to apply the Bloch theorem in the context of symmetry, which is especially relevant in condensed matter physics.

For the benefit of the reader, we give some basic concepts and definitions, which are critical for being successful in solving the problems in this chapter. Some useful formulas that appear often and can shorten the work process are also given here.

Translation Operator: $\widehat{T}(\boldsymbol{a}) = e^{-\frac{i}{\hbar}\hat{\boldsymbol{p}}\cdot\boldsymbol{a}}$

The basic property of the translation operator is to shift the argument of the Schrödinger wave function by $\boldsymbol{a} \rightarrow \widehat{T}(\boldsymbol{a})\psi(\boldsymbol{x}) = \psi(\boldsymbol{x}+\boldsymbol{a})$.

Rotation Operator: $\widehat{R}_{\widehat{\boldsymbol{n}}}(\phi) = e^{-i\widehat{\boldsymbol{J}}\cdot\widehat{\boldsymbol{n}}\phi/\hbar}$

The basic property of the rotation operator is to rotate a quantum mechanical state about a given axis $\widehat{\boldsymbol{n}}$ by an angle ϕ such that $\psi(\boldsymbol{x}) \rightarrow \psi(\boldsymbol{x}_R) = \psi(\widehat{R}^{+}_{\widehat{\boldsymbol{n}}}(\phi)\boldsymbol{x}\widehat{R}_{\widehat{\boldsymbol{n}}}(\phi))$.

Parity Operator: $\widehat{\pi}^{+}\widehat{r}\widehat{\pi} = -\widehat{r}$

The basic property of the parity operator is to perform a reflection of the displacement operator.

Time-Reversal Operator: $\widehat{\theta} = \widehat{U}\widehat{K}$

The time-reversal operator is an antiunitary operator, as it is a product of a unitary operator $\widehat{U}$ and a complex conjugation operation $\widehat{K}$. Its property is to reverse the particle trajectory direction $\widehat{\theta}\widehat{\boldsymbol{p}}\widehat{\theta}^{-1} = -\widehat{\boldsymbol{p}}$.

Transforming a Quantum Mechanical State under Operation $\widehat{D}$: $\psi_D(\boldsymbol{r}) = \widehat{D}\psi(\boldsymbol{r})$

Transforming an Operator $\widehat{A}$ under Hermitian Operation $\widehat{D}$: $\widehat{A}_D = \widehat{D}^{+}\widehat{A}\widehat{D}$

Transforming an Operator $\widehat{A}$ under Anti-Hermitian Operation $\widehat{M}$: $\widehat{A}_M = \widehat{M}\widehat{A}\widehat{M}^{-1}$

Operator $\widehat{A}$ Invariant under $\widehat{D}$ Operation: $\widehat{A} = \widehat{D}^{+}\widehat{A}\widehat{D} \rightarrow [\widehat{A},\widehat{D}] = 0$

Operator $\widehat{A}$ Invariant under $\widehat{M}$ Operation: $\widehat{A} = \widehat{M}\widehat{A}\widehat{M}^{-1} \rightarrow [\widehat{A},\widehat{M}] = 0$

Useful Formulas:

$e^{\widehat{A}+\widehat{B}} = e^{\widehat{A}}e^{\widehat{B}}e^{-\frac{1}{2}[\widehat{A},\widehat{B}]}$, where $[\widehat{A},[\widehat{A},\widehat{B}]] = [\widehat{B},[\widehat{A},\widehat{B}]] = 0$

$[e^{\lambda\widehat{A}},\widehat{B}] = \lambda\widehat{C}e^{\lambda\widehat{B}}$, where $[\widehat{A},\widehat{B}] = \widehat{C}$ and $[\widehat{A},\widehat{C}] = [\widehat{B},\widehat{C}] = 0$

$e^{\widehat{B}}e^{\widehat{A}} = e^{\widehat{A}}e^{\widehat{B}}e^{-[\widehat{A},\widehat{B}]}$, where $[\widehat{A},[\widehat{A},\widehat{B}]] = [\widehat{B},[\widehat{A},\widehat{B}]] = 0$

$e^{\widehat{A}}\widehat{B}e^{-\widehat{A}} = \widehat{B} + [\widehat{A},\widehat{B}] + \frac{1}{2!}[\widehat{A},[\widehat{A},\widehat{B}]] + \frac{1}{3!}[\widehat{A},[\widehat{A},[\widehat{A},\widehat{B}]]] + \ldots$

Problem 2.1

The Galilean transformation $\boldsymbol{x}_{\boldsymbol{v}} = \boldsymbol{x} - \boldsymbol{v}t$ (and $t_{\boldsymbol{v}} = t$) gives the space-time relation between two inertial frames, as known from classical physics. Here $\boldsymbol{x},\boldsymbol{x}_{\boldsymbol{v}}$ are vectors in space for the two reference frames, $\boldsymbol{v}$ is the constant relative velocity between the two frames, and t denotes time.

Compare the Schrödinger equation for the quantum mechanical wave functions $\psi(\boldsymbol{x},t)$ and $\psi(\boldsymbol{x}_{\boldsymbol{v}},t_{\boldsymbol{v}})$. Construct a unitary transformation that transforms $\psi(\boldsymbol{r},t)$ into $\psi(\boldsymbol{x}_{\boldsymbol{v}},t_{\boldsymbol{v}})$, such that the Schrödinger equation is invariant.

The problem is essentially asking us to write the Schrödinger equation in both reference frames corresponding to $\boldsymbol{x},\boldsymbol{x}_{\boldsymbol{v}}$ spatial variables and compare them to find the relation for the invariance of the Schrödinger equation. It has been considered explicitly in Brandsen and Joachain (2000).

We begin with the Schrödinger equation in the $(\boldsymbol{r},t)$ reference frame,

$$i\hbar\frac{\partial\psi(\boldsymbol{x},t)}{\partial t} = \left(-\frac{\hbar^2\nabla_x^2}{2m} + V(\boldsymbol{x},t)\right)\psi(\boldsymbol{x},t).$$

By writing $\boldsymbol{x} = \boldsymbol{x}_{\boldsymbol{v}} + \boldsymbol{v}t_{\boldsymbol{v}}$ and $t = t_{\boldsymbol{v}}$, we express $\psi(\boldsymbol{x},t) = \psi(\boldsymbol{x}_{\boldsymbol{v}} + \boldsymbol{v}t_{\boldsymbol{v}},t_{\boldsymbol{v}})$. From the chain rule $\frac{\partial\psi(x_\nu(y'_\mu))}{\partial y'_\mu} = \frac{\partial x_\nu}{\partial y'_\mu}\frac{\partial\psi(x_\nu)}{\partial x_\nu}\Big|_{x_\nu\to x_\nu(y'_\mu)}$, it is then straightforward to see that

$$\begin{aligned}
\frac{\partial\psi(\boldsymbol{x},t)}{\partial t} &= \frac{\partial\psi(\boldsymbol{x}_{\boldsymbol{v}}+\boldsymbol{v}t_{\boldsymbol{v}},t_{\boldsymbol{v}})}{\partial t} = \frac{\partial\boldsymbol{x}_{\boldsymbol{v}}}{\partial t}\frac{\partial\psi(\boldsymbol{x}_{\boldsymbol{v}}+\boldsymbol{v}t_{\boldsymbol{v}},t_{\boldsymbol{v}})}{\partial\boldsymbol{x}_{\boldsymbol{v}}} + \frac{\partial t_{\boldsymbol{v}}}{\partial t}\frac{\partial\psi(\boldsymbol{x}_{\boldsymbol{v}}+\boldsymbol{v}t_{\boldsymbol{v}},t_{\boldsymbol{v}})}{\partial t_{\boldsymbol{v}}}\\
&= \frac{\partial(\boldsymbol{x}-\boldsymbol{v}t)}{\partial t}\frac{\partial\psi(\boldsymbol{x}_{\boldsymbol{v}}+\boldsymbol{v}t_{\boldsymbol{v}},t_{\boldsymbol{v}})}{\partial\boldsymbol{x}_{\boldsymbol{v}}} + \frac{\partial t}{\partial t}\frac{\partial\psi(\boldsymbol{x}_{\boldsymbol{v}}+\boldsymbol{v}t_{\boldsymbol{v}},t_{\boldsymbol{v}})}{\partial t_{\boldsymbol{v}}}\\
&= -\boldsymbol{v}\cdot\frac{\partial\psi(\boldsymbol{x}_{\boldsymbol{v}}+\boldsymbol{v}t_{\boldsymbol{v}},t_{\boldsymbol{v}})}{\partial\boldsymbol{x}_{\boldsymbol{v}}} + \frac{\partial\psi(\boldsymbol{x}_{\boldsymbol{v}}+\boldsymbol{v}t_{\boldsymbol{v}},t_{\boldsymbol{v}})}{\partial t_{\boldsymbol{v}}}\\
&= \frac{\partial\psi(\boldsymbol{x}_{\boldsymbol{v}}+\boldsymbol{v}t_{\boldsymbol{v}},t_{\boldsymbol{v}})}{\partial t_{\boldsymbol{v}}} - \boldsymbol{v}\cdot\nabla_{\boldsymbol{x}_{\boldsymbol{v}}}\psi(\boldsymbol{x}_{\boldsymbol{v}}+\boldsymbol{v}t_{\boldsymbol{v}},t_{\boldsymbol{v}}),\\
\nabla_{\boldsymbol{x}}\psi(\boldsymbol{x},t) &= \frac{\partial\psi(\boldsymbol{x}_{\boldsymbol{v}}+\boldsymbol{v}t_{\boldsymbol{v}},t_{\boldsymbol{v}})}{\partial\boldsymbol{x}} = \frac{\partial\boldsymbol{x}_{\boldsymbol{v}}}{\partial\boldsymbol{x}}\frac{\partial\psi(\boldsymbol{x}_{\boldsymbol{v}}+\boldsymbol{v}t_{\boldsymbol{v}},t_{\boldsymbol{v}})}{\partial\boldsymbol{x}_{\boldsymbol{v}}} + \frac{\partial t_{\boldsymbol{v}}}{\partial\boldsymbol{x}}\frac{\partial\psi(\boldsymbol{x}_{\boldsymbol{v}}+\boldsymbol{v}t_{\boldsymbol{v}},t_{\boldsymbol{v}})}{\partial t_{\boldsymbol{v}}}\\
&= \frac{\partial(\boldsymbol{x}-\boldsymbol{v}t)}{\partial\boldsymbol{x}}\frac{\partial\psi(\boldsymbol{x},t)}{\partial\boldsymbol{x}_{\boldsymbol{v}}} + \frac{\partial t}{\partial\boldsymbol{x}}\frac{\partial\psi(\boldsymbol{x},t)}{\partial t} = \frac{\partial\psi(\boldsymbol{x},t)}{\partial\boldsymbol{x}_{\boldsymbol{v}}} + 0 = \nabla_{\boldsymbol{x}_{\boldsymbol{v}}}\psi(\boldsymbol{x}_{\boldsymbol{v}}+\boldsymbol{v}t_{\boldsymbol{v}},t_{\boldsymbol{v}}),\\
\nabla_{\boldsymbol{x}}^2\psi(\boldsymbol{x},t) &= \nabla_{\boldsymbol{x}}\cdot\nabla_{\boldsymbol{x}}\psi(\boldsymbol{x},t) = \nabla_{\boldsymbol{x}_{\boldsymbol{v}}}\cdot\nabla_{\boldsymbol{x}_{\boldsymbol{v}}}\psi(\boldsymbol{x}_{\boldsymbol{v}}+\boldsymbol{v}t_{\boldsymbol{v}},t_{\boldsymbol{v}}) = \nabla_{\boldsymbol{x}_{\boldsymbol{v}}}^2\psi(\boldsymbol{x}_{\boldsymbol{v}}+\boldsymbol{v}t_{\boldsymbol{v}},t_{\boldsymbol{v}}).
\end{aligned}$$

Substituting everything in the Schrödinger equation allows its expression in the $(\boldsymbol{x}_{\boldsymbol{v}},t_{\boldsymbol{v}})$ coordinate system:

$$i\hbar\frac{\partial\psi(x_v+vt_v,t_v)}{\partial t_v}-i\hbar\, v\cdot\nabla_{x_v}\psi(x_v+vt_v,t_v)$$
$$=-\frac{\hbar^2}{2m}\nabla^2_{x_v}\psi(x_v+vt_v,t_v)+V(x_v+vt_v,t_v)\psi(x_v+vt_v,t_v),$$
$$i\hbar\frac{\partial\psi(x_v+vt_v,t_v)}{\partial t_v}=\left(-\frac{\hbar^2\nabla^2_{x_v}}{2m}+V(x_v+vt_v,t_v)\right)\psi(x_v+vt_v,t_v)+i\hbar\, v\cdot\nabla_{x_v}\psi(x_v+vt_v,t_v).$$

Clearly, due to the appearance of the term $i\hbar v\cdot\nabla_{x_v}\psi(x_v+vt_v,t_v)$, the Schrödinger equation is not satisfied if the wavefunction is not transformed. In quantum mechanics, a symmetry transformation (like the Galilean transformation G) is represented by a unitary operator $\widehat{U}_v$; then, operators are transformed as

$$\widehat{A}_G(x,t)=\widehat{A}(Gx,Gt)=\widehat{U}_v\widehat{A}(x_v,t_v)\widehat{U}_v^+,$$

while the wavefunctions are transformed as

$$\psi_G(x,t)=\psi(Gx,Gt)=\psi(x_v+vt_v,t_v)=\widehat{U}_v\psi(x_v,t_v).$$

For the unitary transformation of the Galilean transformation, we define

$$\psi_v(x,t)=\psi(x_v+vt_v,t_v)=e^{\frac{i}{\hbar}\left(mv\cdot x_v+\frac{mv^2}{2}t_v\right)}\psi(x_v,t_v)=\widehat{U}_v\psi(x_v,t_v).$$

One also easily finds

$$\frac{\partial}{\partial t_v}\widehat{U}_v=\frac{\partial}{\partial t_v}e^{\frac{i}{\hbar}\left(mv\cdot x_v+\frac{mv^2}{2}t_v\right)}=\frac{i}{\hbar}\frac{mv^2}{2}e^{\frac{i}{\hbar}\left(mv\cdot x_v+\frac{mv^2}{2}t_v\right)}=\frac{i}{\hbar}\frac{mv^2}{2}\widehat{U}_v,$$
$$\nabla_{x_v}\widehat{U}_v=\nabla_{x_v}e^{\frac{i}{\hbar}\left(mv\cdot x_v+\frac{mv^2}{2}t_v\right)}=\frac{i}{\hbar}mv\, e^{\frac{i}{\hbar}\left(mv\cdot x_v+\frac{mv^2}{2}t_v\right)}=\frac{i}{\hbar}mv\,\widehat{U}_v.$$

Then the Schrödinger equation in the new Galilean transformed system becomes

$$\frac{\partial\psi_v(x,t)}{\partial t}=\frac{\partial\psi(x_v+vt_v,t_v)}{\partial t_v}-v\cdot\nabla_{x_v}\psi(x_v+vt_v,t_v)=\frac{\partial\widehat{U}_v\psi(x_v,t_v)}{\partial t_v}-v\cdot\nabla_{x_v}\widehat{U}_v\psi(x_v,t_v)$$
$$=\widehat{U}_v\frac{\partial\psi(x_v,t_v)}{\partial t_v}+\frac{\partial\widehat{U}_v}{\partial t_v}\psi(x_v,t_v)-v\cdot\left(\nabla_{x_v}\widehat{U}_v\right)\psi(x_v,t_v)-\widehat{U}_v v\cdot\nabla_{x_v}\psi(x_v,t_v)$$
$$=\widehat{U}_v\frac{\partial\psi(x_v,t_v)}{\partial t_v}+\frac{i}{\hbar}\frac{mv^2}{2}\widehat{U}_v\psi(x_v,t_v)-v\cdot\left(\frac{i}{\hbar}mv\widehat{U}_v\right)\psi(x_v,t_v)-\widehat{U}_v v\cdot\nabla_{x_v}\psi(x_v,t_v)$$
$$=\widehat{U}_v\left(\frac{\partial\psi(x_v,t_v)}{\partial t_v}-v\cdot\nabla_{x_v}\psi(x_v,t_v)-\frac{i}{\hbar}\frac{mv^2}{2}\psi(x_v,t_v)\right),$$

$$\nabla_x\psi_v(x,t)=\nabla_{x_v}\psi(x_v+vt_v,t_v)=\nabla_{x_v}\widehat{U}_v\psi(x_v,t_v)=\left(\nabla_{x_v}\widehat{U}_v\right)\psi(x_v,t_v)+\widehat{U}_v\nabla_{x_v}\psi(x_v,t_v)$$
$$=\left(\frac{i}{\hbar}mv\widehat{U}_v\right)\psi(x_v,t_v)+\widehat{U}_v\nabla_{x_v}\psi(x_v,t_v)=\widehat{U}_v\left(\frac{i}{\hbar}mv\psi(x_v,t_v)+\nabla_{x_v}\psi(x_v,t_v)\right),$$
$$\nabla_x^2\psi_v(x,t)=\frac{i}{\hbar}mv\cdot\nabla_{x_v}\left[\widehat{U}_v\psi(x_v,t_v)\right]+\nabla_{x_v}\cdot\left[\widehat{U}_v\nabla_{x_v}\psi(x_v,t_v)\right]$$
$$=\frac{i}{\hbar}mv\cdot\left[\nabla_{x_v}\widehat{U}_v\right]\psi(x_v,t_v)+\frac{i}{\hbar}m\widehat{U}_v v\cdot\nabla_{x_v}\psi(x_v,t_v)+\left[\nabla_{x_v}\widehat{U}_v\right]\cdot\nabla_{x_v}\psi(x_v,t_v)$$

$$+\widehat{U}_v\left[\nabla_{x_v}\cdot\nabla_{x_v}\psi(x_v,t_v)\right]$$

$$=\widehat{U}_v\left(\frac{-1}{\hbar^2}m^2v^2\psi(x_v,t_v)+2\frac{i}{\hbar}m\,v\cdot\nabla_{x_v}\psi(x_v,t_v)+\nabla^2_{x_v}\psi(x_v,t_v)\right).$$

The potential is transformed as an operator,

$$V_v(r,t)=V(x_v+vt_v,t_v)=\widehat{U}_vV(x_v,t_v)\widehat{U}_v^{+}.$$

Substituting everything in the Schrödinger equation again allows its expression in the (x_v,t_v) coordinate system:

$$i\hbar\frac{\partial\psi_v(x,t)}{\partial t}=\left(-\frac{\hbar^2\nabla_x^2}{2m}+V_v(x,t)\right)\psi_v(x,t),$$

$$i\hbar\widehat{U}_v\left(\frac{\partial\psi(x_v,t_v)}{\partial t_v}-v.\nabla_{x_v}\psi(x_v,t_v)-\frac{i}{\hbar}\frac{mv^2}{2}\psi(x_v,t_v)\right)$$

$$=-\frac{\hbar^2}{2m}\widehat{U}_v\left(\frac{-1}{\hbar^2}m^2v^2\psi(x_v,t_v)+2\frac{i}{\hbar}m\,v\cdot\nabla_{x_v}\psi(x_v,t_v)+\nabla^2_{x_v}\psi(x_v,t_v)\right)$$

$$+\widehat{U}_vV(x_v,t_v)\widehat{U}_v^{+}\widehat{U}_v\psi(x_v,t_v).$$

$$i\hbar\frac{\partial\psi(x_v,t_v)}{\partial t_v}-i\hbar v\cdot\nabla_{x_v}\psi(x_v,t_v)+\frac{mv^2}{2}\psi(x_v,t_v)$$

$$=\frac{mv^2}{2}\psi(x_v,t_v)-i\hbar\,v\cdot\nabla_{x_v}\psi(x_v,t_v)-\frac{\hbar^2}{2m}\nabla^2_{x_v}\psi(x_v,t_v)+V(x_v,t_v)\psi(x_v,t_v).$$

$$i\hbar\frac{\partial\psi(x_v,t_v)}{\partial t_v}=\left(-\frac{\hbar^2\nabla^2_{x_v}}{2m}+V(x_v,t_v)\right)\psi(x_v,t_v),$$

which is precisely the Schrödinger equation in the (x_v,t_v) reference frame.

> *Food for thought*: This problem is actually quite important in showing that, for an isolated system of particles, the Schrödinger equation is invariant. In the case of a two-particle system, one can write the Galilean transformation $R_v=R-vt$, where the center of mass R_v is related to some initial coordinate R and $v=\frac{dx}{dt}$ is the velocity for the relative coordinate x. The unitary transformation ensuring that the Schrödinger equation is invariant in both reference frames is $\psi(R_v+vt,t)=e^{\frac{i}{\hbar}\left(mv\cdot R_v+\frac{mv^2}{2}t\right)}\psi(R,t)$, where $M=m_1+m_2$ is the mass for the mass center of the two particles.

Problem 2.2

Consider a particle of mass m constrained to move in a periodic potential $U(x+a)=U(x)$ with period a on a line with length $L\to\infty$, such that the wave function $\psi(x)$ has periodic boundary conditions at the end points of the line. The translation operator for this periodic environment is then given as $\widehat{T}(a)=e^{i\frac{a\widehat{p}}{\hbar}}$, where $\widehat{p}$ is the momentum operator. What are the eigenstates and eigenvalues of $\widehat{T}(a)$?

This problem is a variation of the Bloch theorem, whose importance for condensed matter physics can never be understated.

The periodicity of the wave function at $L = Na$ ($N \to \infty$) shows that $\psi(x+L) = \psi(x)$. Thus, the wave function can be written as $\psi(x) = u_k(x)e^{ikx}$, where $u_k(x+a) = u_k(x)$ are the periodic Bloch functions. Also, using $\psi(x+L) = \psi(x)$, we see that

$$\psi(x+L) = u_k(x+Na)e^{ik(x+Na)} = u_k(x)e^{ikx} = \psi(x),$$

thus $e^{ikNa} = 1$, $k = \frac{2n\pi}{Na}$, with $n \in \mathbb{Z}$. Let us then consider,

$$\widehat{T}(a)\psi(x) = \psi(x+a) = u_k(x+a)e^{ik(x+a)} = u_k(x)e^{ikx}e^{ika},$$
$$\widehat{T}(a)\psi(x) = \psi(x+a) = \lambda\psi(x) = \lambda u_k(x)e^{ikx}.$$

Comparing the preceding relations, we see that $\psi(x)$ is an eigenstate of $\widehat{T}(a)$ with an eigenvalue $\lambda = e^{ika}$, where $k = \frac{2n\pi}{Na}$, with $n \in \mathbb{Z}$.

Problem 2.3

An electron with charge $(-q)$ and mass m moves in a periodic crystal potential with an applied constant external magnetic field $\boldsymbol{B}$. The periodicity is such that, for an arbitrary real space vector $\boldsymbol{r} = \boldsymbol{\rho} + \boldsymbol{R}$, where $\boldsymbol{\rho}$ changes inside the unit cell of the crystal, we have that $V(\boldsymbol{r}) = V(\boldsymbol{\rho}+\boldsymbol{R}) = V(\boldsymbol{\rho})$. Show that the Hamiltonian of the electron in the periodic environment can be written as

$$\widehat{H}(\widehat{\rho},\widehat{\boldsymbol{R}}) = \widehat{H}_0(\widehat{\rho}) + \frac{\left(\widehat{\boldsymbol{p}}_{\boldsymbol{R}} - q\boldsymbol{A}(\widehat{\boldsymbol{R}})\right)^2}{2m} + \frac{\left(\widehat{\boldsymbol{p}}_\rho - q\boldsymbol{A}(\widehat{\rho})\right)\cdot\left(\widehat{\boldsymbol{p}}_{\boldsymbol{R}} - q\boldsymbol{A}(\widehat{\boldsymbol{R}})\right)}{m}$$
$$\widehat{H}_0(\widehat{\rho}) = \frac{\left(\widehat{\boldsymbol{p}}_\rho - q\boldsymbol{A}(\widehat{\rho})\right)^2}{2m} + V(\widehat{\rho}) - \frac{q\hbar}{2m}(\widehat{\sigma}\cdot\boldsymbol{B}),$$

where $\boldsymbol{\sigma} = (\sigma_1, \sigma_2, \sigma_3)$ are the Pauli matrices.

We begin with the Hamiltonian in real space:

$$\widehat{H}(\widehat{\boldsymbol{r}}) = \frac{\left(\widehat{\boldsymbol{p}}_{\boldsymbol{r}} - q\boldsymbol{A}(\widehat{\boldsymbol{r}})\right)}{2m} + V(\widehat{\boldsymbol{r}}) - \frac{q\hbar}{2m}(\widehat{\sigma}\cdot\boldsymbol{B}).$$

It is straightforward to show that $\boldsymbol{A}(\widehat{\boldsymbol{r}}) = \boldsymbol{A}(\widehat{\rho}) + \boldsymbol{A}(\widehat{\boldsymbol{R}})$ and $\widehat{\boldsymbol{p}}_{\boldsymbol{r}} = \widehat{\boldsymbol{p}}_\rho + \widehat{\boldsymbol{p}}_{\boldsymbol{R}}$, as we have done here:

$$\widehat{\boldsymbol{p}}_{\boldsymbol{r}}\psi(r,t) = -i\hbar\nabla_{\boldsymbol{r}}\psi(\boldsymbol{\rho}+\boldsymbol{R},t)$$
$$= -i\hbar\nabla_\rho\psi(\boldsymbol{\rho}+\boldsymbol{R},t) - i\hbar\nabla_{\boldsymbol{R}}\psi(\boldsymbol{\rho}+\boldsymbol{R},t) = \left(\widehat{\boldsymbol{p}}_\rho + \widehat{\boldsymbol{p}}_{\boldsymbol{R}}\right)\psi(\boldsymbol{\rho}+\boldsymbol{R},t),$$
$$\boldsymbol{A}(\widehat{\boldsymbol{r}}) = \frac{1}{2}(\boldsymbol{B}\times\widehat{\boldsymbol{r}}) = \frac{1}{2}\left(\boldsymbol{B}\times\left[\widehat{\rho}+\widehat{\boldsymbol{R}}\right]\right) = \frac{1}{2}(\boldsymbol{B}\times\widehat{\rho}) + \frac{1}{2}\left(\boldsymbol{B}\times\widehat{\boldsymbol{R}}\right) = \boldsymbol{A}(\widehat{\rho}) + \boldsymbol{A}(\widehat{\boldsymbol{R}}).$$

Therefore, we have

$$\widehat{H}(\widehat{\boldsymbol{r}}) = \widehat{H}\left(\widehat{\rho},\widehat{\boldsymbol{R}}\right) = \frac{\left(\widehat{\boldsymbol{p}}_\rho + \widehat{\boldsymbol{p}}_{\boldsymbol{R}} - q\boldsymbol{A}(\widehat{\rho}) - q\boldsymbol{A}(\widehat{\boldsymbol{R}})\right)^2}{2m} + V(\widehat{\rho}+\widehat{\boldsymbol{R}}) - \frac{q\hbar}{2m}(\widehat{\sigma}\cdot\boldsymbol{B})$$

$$= \frac{\left(\widehat{p}_\rho - qA(\widehat{\rho})\right)^2}{2m} + \frac{\left(\widehat{p}_R - qA(\widehat{R})\right)^2}{2m} + 2\frac{\left(\widehat{p}_\rho - qA(\widehat{\rho})\right)\cdot\left(\widehat{p}_R - qA(\widehat{R})\right)}{2m}$$

$$+ V(\widehat{\rho}) - \frac{q\hbar}{2m}(\widehat{\sigma}\cdot \boldsymbol{B}) = \widehat{H}_0(\widehat{\rho}) + \frac{\left(\widehat{p}_R - qA(\widehat{R})\right)^2}{2m} + \frac{\left(\widehat{p}_\rho - qA(\widehat{\rho})\right)\cdot\left(\widehat{p}_R - qA(\widehat{R})\right)}{m}.$$

Food for thought: The equation $i\hbar\partial_t\psi(\boldsymbol{r},t) = \widehat{H}(\widehat{\boldsymbol{r}})\psi(\boldsymbol{r},t)$ with $\widehat{H}(\widehat{\boldsymbol{r}}) = \frac{(\widehat{p}_r - qA(\widehat{r}))^2}{2m} + V(\widehat{\boldsymbol{r}}) - \frac{q\hbar}{2m}(\widehat{\sigma}\cdot \boldsymbol{B})$ for a two-component spinor $\psi(\boldsymbol{r},t)$ is called **the Pauli equation** in quantum mechanics and corresponds to the motion of an electron in the presence of an external electromagnetic field (Nowakowski, 1999).

Problem 2.4

The Hamiltonian for a given particle $\widehat{H}_0 = K(\widehat{\boldsymbol{p}}) + U(\widehat{\boldsymbol{r}})$ has a periodic potential with period $\boldsymbol{a}$, such that $U(\widehat{\boldsymbol{r}}+\boldsymbol{a}) = U(\widehat{\boldsymbol{r}})$, and $K(\widehat{\boldsymbol{p}}) = \frac{\widehat{\boldsymbol{p}}^2}{2m}$ is the kinetic energy with $\widehat{\boldsymbol{p}}$ being the momentum operator. In this case, we know that the translation operator $\widehat{T}_0(\boldsymbol{a}) = e^{\frac{i\boldsymbol{a}\cdot\widehat{\boldsymbol{p}}}{\hbar}}$ commutes with $\widehat{H}_0$, as evident from Bloch's theorem.

Suppose an external magnetic field $\boldsymbol{B} = (0,0,B)$ along the z-direction is applied to this system. The Hamiltonian then becomes

$$\widehat{H} = \frac{(\widehat{\boldsymbol{p}} - e\boldsymbol{A}(\widehat{\boldsymbol{r}}))^2}{2m} + U(\widehat{\boldsymbol{r}}),$$

where $\boldsymbol{A}(\widehat{\boldsymbol{r}}) = \frac{1}{2}(-By, Bx, 0)$ is the vector potential and e is the charge of the particle. What is the translation operator that commutes with $\widehat{H}$ in this case?

This problem alerts us that the translation operator in a periodic potential when an external magnetic field is applied is different from the familiar $\widehat{T}_0(\boldsymbol{a}) = e^{\frac{i\boldsymbol{a}.\widehat{\boldsymbol{p}}}{\hbar}}$ operator. We immediately see that $\left[\widehat{T}_0(\boldsymbol{a}), \widehat{H}\right] \neq 0$ simply because $\left[\widehat{T}_0(\boldsymbol{a}), \boldsymbol{A}(\widehat{\boldsymbol{r}})\right] \neq 0$. Thus, $\widehat{T}_0(\boldsymbol{a})$ must be modified so that $\left[\widehat{T}(\boldsymbol{a}), \widehat{H}\right] = 0$.

We note that when $\boldsymbol{B}$ is introduced, then the momentum is rescaled by the vector potential as given in the Hamiltonian $\widehat{H}$. Let's assume that something similar happens in the translation operator and construct $\widehat{T}(\boldsymbol{a}) = e^{\frac{i}{\hbar}(\widehat{\boldsymbol{p}}+e\boldsymbol{A}(\widehat{\boldsymbol{r}}))\cdot\boldsymbol{a}}$. Now let's consider

$$\left[\widehat{T}(\boldsymbol{a}), (K(\widehat{\boldsymbol{p}} - e\boldsymbol{A}(\widehat{\boldsymbol{r}})) + U(\widehat{\boldsymbol{r}}))\right] = \left[\widehat{T}(\boldsymbol{a}), K(\widehat{\boldsymbol{p}} - e\boldsymbol{A}(\widehat{\boldsymbol{r}}))\right] + \left[\widehat{T}(\boldsymbol{a}), U(\widehat{\boldsymbol{r}})\right].$$

However, since $\left[(\widehat{\boldsymbol{p}} - e\boldsymbol{A}(\widehat{\boldsymbol{r}}))_i, (\widehat{\boldsymbol{p}} + e\boldsymbol{A}(\widehat{\boldsymbol{r}}))_j\right] = i\hbar e\left(\frac{\partial A_j}{\partial x_i} - \frac{\partial A_i}{\partial x_j}\right)$, where $i,j = x,y,z$, we realize that for the $\boldsymbol{A}(\widehat{\boldsymbol{r}}) = \frac{1}{2}(-By, Bx, 0)$ gauge, one has $\left[(\widehat{\boldsymbol{p}} - e\boldsymbol{A}(\widehat{\boldsymbol{r}}))_i, (\widehat{\boldsymbol{p}} + e\boldsymbol{A}(\widehat{\boldsymbol{r}}))_j\right] = 0$ for all i, j. This means that

$$\left[\widehat{T}(\boldsymbol{a}), K(\widehat{\boldsymbol{p}} - e\boldsymbol{A}(\widehat{\boldsymbol{r}}))\right] = 0.$$

On the other hand, by using $e^{\frac{i}{\hbar}(\widehat{p}+e\mathbf{A}(\widehat{r}))\cdot a} = e^{\frac{i}{\hbar}\widehat{p}\cdot a} e^{\frac{ie}{\hbar}\mathbf{A}(\widehat{r})\cdot a}$,

$$\left[\widehat{T}(\boldsymbol{a}), U(\widehat{\boldsymbol{r}})\right] = e^{\frac{i}{\hbar}(\widehat{p}+eA(\widehat{r}))\cdot a} U(\widehat{r}) - U(\widehat{r}) e^{\frac{i}{\hbar}(\widehat{p}+eA(\widehat{r}))\cdot a}$$
$$= U(\widehat{r}+a) e^{\frac{i}{\hbar}(\widehat{p}+eA(\widehat{r}))\cdot a} - U(\widehat{r}) e^{\frac{i}{\hbar}(\widehat{p}+eA(\widehat{r}))\cdot a}$$
$$= U(\widehat{r}) e^{\frac{i}{\hbar}(\widehat{p}+eA(\widehat{r}))\cdot a} - U(\widehat{r}) e^{\frac{i}{\hbar}(\widehat{p}+eA(\widehat{r}))\cdot a} = 0.$$

Food for thought: The so-constructed translation operator $\widehat{T}(a) = e^{\frac{i}{\hbar}(\widehat{p}+eA(\widehat{r}))\cdot a}$ is called a **magnetic translation operator**. The concept of a magnetic translation operator was considered by Brown (1964) and Zak (1964), which turns out to be quite useful, especially in the construction of Wannier function basis sets for periodic environments in the presence of magnetic fields.

Problem 2.5

Under the application of a magnetic field $\boldsymbol{B} = (0,0,B)$ along the z-direction, the Hamiltonian is $\widehat{H} = \frac{(\widehat{p}-eA(\widehat{r}))^2}{2m} + U(\widehat{\boldsymbol{r}})$, where the potential is a periodic function such that $U(\boldsymbol{x}+\boldsymbol{a}) = U(\boldsymbol{x})$. For the gauge $\boldsymbol{A}(\boldsymbol{r}) = \frac{1}{2}(\boldsymbol{B}\times\boldsymbol{r}) = \frac{1}{2}(-By, Bx, 0)$, show that the product of two magnetic translation operators satisfies

$$\widehat{T}(\boldsymbol{a}_1)\widehat{T}(\boldsymbol{a}_2) = \widehat{T}(\boldsymbol{a}_2)\widehat{T}(\boldsymbol{a}_1)\, e^{\frac{ie}{\hbar}\boldsymbol{B}\cdot(\boldsymbol{a}_1\times\boldsymbol{a}_2)} = \widehat{T}(\boldsymbol{a}_1+\boldsymbol{a}_2) e^{i\boldsymbol{\beta}\cdot(\boldsymbol{a}_1\times\boldsymbol{a}_2)/2},$$

where $\widehat{T}(\boldsymbol{a}) = e^{\frac{i}{\hbar}(\widehat{p}+eA(\widehat{r}))\cdot a}$ and $\boldsymbol{\beta} = e\boldsymbol{B}/\hbar$.

This problem builds on our understanding of magnetic translation operators (introduced in the previous problem) by considering the successive application of two such translations with different substitute with vectors $\boldsymbol{a}_{1,2}$.

Let's first see that

$$\widehat{T}(\boldsymbol{a}_1)\psi(\boldsymbol{x}) = e^{\frac{i}{\hbar}(\widehat{p}+eA(\widehat{r}))\cdot a_1}\psi(\boldsymbol{x}) = e^{\frac{ie}{\hbar}A(\widehat{r})\cdot a_1}\psi(\boldsymbol{x}+\boldsymbol{a}_1) = e^{\frac{ie}{2\hbar}(B\times\widehat{r})\cdot a_1}\widehat{T}_0(\boldsymbol{a}_1)\psi(\boldsymbol{x}),$$

where we make the definition of $\widehat{T}_0(\boldsymbol{a}) = e^{\frac{i}{\hbar}\widehat{p}\cdot a}$. From the preceding result, $\widehat{T}(\boldsymbol{a}_1) = e^{\frac{ie}{2\hbar}(B\times\widehat{r})\cdot a_1}\widehat{T}_0(\boldsymbol{a}_1) = e^{\frac{-i}{2}(\beta\times a_1)\cdot\widehat{r}}\widehat{T}_0(\boldsymbol{a}_1)$, thus we have

$$\widehat{T}(\boldsymbol{a}_1)\widehat{T}(\boldsymbol{a}_2) = e^{\frac{-i}{2}[\beta\times(a_1+a_2)]\cdot\widehat{r}}\widehat{T}_0(\boldsymbol{a}_1+\boldsymbol{a}_2)\, e^{\frac{i}{2}(a_1\times a_2)\cdot\beta}$$
$$= \widehat{T}_0(\boldsymbol{a}_1+\boldsymbol{a}_2)\, e^{\frac{-i}{2}[\beta\times(a_1+a_2)]\cdot\widehat{r}} e^{\frac{i}{2}(a_1\times a_2)\cdot\beta}$$
$$= \widehat{T}_0(\boldsymbol{a}_1+\boldsymbol{a}_2)\, e^{i\frac{e}{\hbar}A(\widehat{r})\cdot(a_1+a_2)} e^{\frac{i}{2}(a_1\times a_2)\cdot\beta} = \widehat{T}(\boldsymbol{a}_1+\boldsymbol{a}_2)\, e^{\frac{i}{2}(a_1\times a_2)\cdot\beta},$$
$$\widehat{T}(\boldsymbol{a}_1)\widehat{T}(\boldsymbol{a}_2) = \widehat{T}(\boldsymbol{a}_1+\boldsymbol{a}_2)\, e^{\frac{i}{2}(a_1\times a_2)\cdot\beta}.$$

In the preceding, we have utilized cross-dot properties in the vector potential $\frac{e}{\hbar}A(\widehat{r}) = \frac{1}{2}[\boldsymbol{\beta}\times\widehat{\boldsymbol{r}}]$, such that

$$\frac{-i}{2}[\boldsymbol{\beta}\times(\boldsymbol{a}_1+\boldsymbol{a}_2)]\cdot\widehat{\boldsymbol{r}} = \frac{i}{2}[\boldsymbol{\beta}\times\widehat{\boldsymbol{r}}]\cdot(\boldsymbol{a}_1+\boldsymbol{a}_2) = i\frac{e}{\hbar}A(\widehat{\boldsymbol{r}})\cdot(\boldsymbol{a}_1+\boldsymbol{a}_2).$$

By exchanging $\boldsymbol{a}_1$ and $\boldsymbol{a}_2$,

$$\widehat{T}(\boldsymbol{a}_2)\widehat{T}(\boldsymbol{a}_1) = \widehat{T}(\boldsymbol{a}_2+\boldsymbol{a}_1)e^{\frac{i}{2}(\boldsymbol{a}_2\times\boldsymbol{a}_1)\cdot\boldsymbol{\beta}} = \widehat{T}(\boldsymbol{a}_1+\boldsymbol{a}_2)e^{-\frac{i}{2}(\boldsymbol{a}_1\times\boldsymbol{a}_2)\cdot\boldsymbol{\beta}}.$$

Therefore, we have

$$\widehat{T}(\boldsymbol{a}_1+\boldsymbol{a}_2) = \widehat{T}(\boldsymbol{a}_2)\widehat{T}(\boldsymbol{a}_1)e^{\frac{i}{2}(\boldsymbol{a}_1\times\boldsymbol{a}_2)\cdot\boldsymbol{\beta}}.$$

By relating the two underlined equations,

$$\begin{aligned}\widehat{T}(\boldsymbol{a}_1)\widehat{T}(\boldsymbol{a}_2) &= \widehat{T}(\boldsymbol{a}_1+\boldsymbol{a}_2)e^{\frac{i}{2}(\boldsymbol{a}_1\times\boldsymbol{a}_2)\cdot\boldsymbol{\beta}}\\ &= \widehat{T}(\boldsymbol{a}_2)\widehat{T}(\boldsymbol{a}_1)e^{\frac{i}{2}(\boldsymbol{a}_1\times\boldsymbol{a}_2)\cdot\boldsymbol{\beta}}e^{\frac{i}{2}(\boldsymbol{a}_1\times\boldsymbol{a}_2)\cdot\boldsymbol{\beta}}\\ &= \widehat{T}(\boldsymbol{a}_2)\widehat{T}(\boldsymbol{a}_1)e^{i(\boldsymbol{a}_1\times\boldsymbol{a}_2)\cdot\boldsymbol{\beta}}.\end{aligned}$$

Therefore, one finds

$$\widehat{T}(\boldsymbol{a}_1)\widehat{T}(\boldsymbol{a}_2) = \widehat{T}(\boldsymbol{a}_1+\boldsymbol{a}_2)e^{i\boldsymbol{\beta}\cdot(\boldsymbol{a}_1\times\boldsymbol{a}_2)}.$$

Problem 2.6

Under the application of a magnetic field $\boldsymbol{B} = (0,0,B)$ along the z-direction, the Hamiltonian is $\widehat{H} = \frac{(\widehat{\boldsymbol{p}}-e\boldsymbol{A}(\widehat{\boldsymbol{r}}))^2}{2m} + U(\widehat{\boldsymbol{r}})$, where the potential is a periodic function such that $U(\boldsymbol{x}+\boldsymbol{a}) = U(\boldsymbol{x})$. For the gauge $\boldsymbol{A} = \frac{1}{2}(\boldsymbol{B}\times\boldsymbol{r}) = \frac{1}{2}(-By, Bx, 0)$, show that when $\boldsymbol{R}_1+\boldsymbol{R}_2+\boldsymbol{R}_3 = 0$, then $\widehat{T}(\boldsymbol{R}_1)\widehat{T}(\boldsymbol{R}_2)\widehat{T}(\boldsymbol{R}_3) = e^{2\pi i\frac{\phi}{\phi_0}}$, where $\phi_0 = \frac{\pi\hbar}{e}$ and $\phi = \boldsymbol{B}\cdot(\boldsymbol{R}_2\times\boldsymbol{R}_3)$.

This problem of three successive translations is closely related with the previous one, for which two successive translations were considered in the presence of a magnetic field.

Let's use the results from the previous problem for two successive magnetic translations:

$$\widehat{T}(\boldsymbol{a}_1)\widehat{T}(\boldsymbol{a}_2) = \widehat{T}(\boldsymbol{a}_1+\boldsymbol{a}_2)e^{\frac{i}{2}(\boldsymbol{a}_1\times\boldsymbol{a}_2)\cdot\boldsymbol{\beta}} = \widehat{T}(\boldsymbol{a}_2)\widehat{T}(\boldsymbol{a}_1)e^{i(\boldsymbol{a}_1\times\boldsymbol{a}_2)\cdot\boldsymbol{\beta}}.$$

Let's try now performing three such successive translations:

$$\begin{aligned}\widehat{T}(\boldsymbol{R}_1)\widehat{T}(\boldsymbol{R}_2)\widehat{T}(\boldsymbol{R}_3) &= \widehat{T}(\boldsymbol{R}_1)\widehat{T}(\boldsymbol{R}_2+\boldsymbol{R}_3)e^{\frac{i}{2}\boldsymbol{\beta}\cdot(\boldsymbol{R}_2\times\boldsymbol{R}_3)}\\ &= \widehat{T}(\boldsymbol{R}_1+\boldsymbol{R}_2+\boldsymbol{R}_3)e^{\frac{i}{2}\boldsymbol{\beta}\cdot[\boldsymbol{R}_1\times(\boldsymbol{R}_2+\boldsymbol{R}_3)]}e^{\frac{i}{2}\boldsymbol{\beta}\cdot(\boldsymbol{R}_2\times\boldsymbol{R}_3)}.\end{aligned}$$

Let us further use that $\boldsymbol{R}_2+\boldsymbol{R}_3 = -\boldsymbol{R}_1$ and $\boldsymbol{\beta} = e\boldsymbol{B}/\hbar$:

$$\begin{aligned}\widehat{T}(\boldsymbol{R}_1)\widehat{T}(\boldsymbol{R}_2)\widehat{T}(\boldsymbol{R}_3) &= \widehat{T}(\boldsymbol{R}_1-\boldsymbol{R}_1)e^{\frac{i}{2}\boldsymbol{\beta}\cdot[\boldsymbol{R}_1\times(-\boldsymbol{R}_1)]}e^{\frac{i}{2}\boldsymbol{\beta}\cdot(\boldsymbol{R}_2\times\boldsymbol{R}_3)}\\ &= \widehat{T}(\boldsymbol{0})e^{-\frac{i}{2}\boldsymbol{\beta}\cdot[\boldsymbol{R}_1\times\boldsymbol{R}_1]}e^{\frac{i}{2}\boldsymbol{\beta}\cdot(\boldsymbol{R}_2\times\boldsymbol{R}_3)} = e^{\frac{i}{2}\boldsymbol{\beta}\cdot(\boldsymbol{R}_2\times\boldsymbol{R}_3)}.\end{aligned}$$

Then, using the definitions $\phi_0 = \frac{\pi\hbar}{e}$ and $\phi = \boldsymbol{B}\cdot(\boldsymbol{R}_2\times\boldsymbol{R}_3)$:

$$\widehat{T}(\boldsymbol{R}_1)\widehat{T}(\boldsymbol{R}_2)\widehat{T}(\boldsymbol{R}_3) = e^{\frac{ie}{2\hbar}\boldsymbol{B}\cdot(\boldsymbol{R}_2\times\boldsymbol{R}_3)} = e^{2\pi i\frac{\phi}{\phi_0}}.$$

Problem 2.7

Consider a 2D lattice whose periodic potential satisfies $U(\boldsymbol{x}+N_1\boldsymbol{a}_1+N_2\boldsymbol{a}_2) = U(\boldsymbol{x})$, where $N_{1,2}$ are integers and $\boldsymbol{a}_{1,2}$ are the unit cell primitive vectors. When an external magnetic field is applied, this natural periodicity is not fulfilled any more.

Show that for $\boldsymbol{B} = (0,0,B)$, the magnetic translation operators each associated with $\boldsymbol{a}_1$ and $\boldsymbol{a}_2$ will commute with each other only when the flux $\Phi = \boldsymbol{B}\cdot(\boldsymbol{a}_1 \times \boldsymbol{a}_2)$ satisfies the conditions $\frac{e\Phi}{h} = \frac{m}{N_1 N_2}$, where m is an integer.

From the previous problems on magnetic translation operations, we realize that this is just another variation of Problem 2.5 for the product of two translations, but now things are considered in the context of periodicity.

Given the periodic potential, we infer that the wave function also contains the same type of periodicity (as implied by Bloch's theorem). Thus, we must have

$$\widehat{T}(N_1\boldsymbol{a}_1)\widehat{T}(N_2\boldsymbol{a}_2)\psi(\boldsymbol{x}) = \psi(\boldsymbol{x} + N_1\boldsymbol{a}_1 + N_2\boldsymbol{a}_2) = \psi(\boldsymbol{x}).$$

Using the previously obtained result in Problem 2.4 that $\widehat{T}(\boldsymbol{a}_1)\widehat{T}(\boldsymbol{a}_2) = \widehat{T}(\boldsymbol{a}_1 + \boldsymbol{a}_2)e^{i\beta\cdot(\boldsymbol{a}_1\times\boldsymbol{a}_2)}$, we get

$$\begin{aligned}\widehat{T}(N_1\boldsymbol{a}_1)\widehat{T}(N_2\boldsymbol{a}_2)\psi(\boldsymbol{x}) &= \widehat{T}(N_2\boldsymbol{a}_2)\widehat{T}(N_1\boldsymbol{a}_1)e^{\frac{ie}{\hbar}\boldsymbol{B}\cdot(N_1\boldsymbol{a}_1\times N_2\boldsymbol{a}_2)}\psi(\boldsymbol{x})\\ &= e^{\frac{ieN_1N_2}{\hbar}\boldsymbol{B}\cdot(\boldsymbol{a}_1\times\boldsymbol{a}_2)}\widehat{T}(N_2\boldsymbol{a}_2)\widehat{T}(N_1\boldsymbol{a}_1)\psi(\boldsymbol{x})\\ &= e^{\frac{ieN_1N_2}{\hbar}\boldsymbol{B}\cdot(\boldsymbol{a}_1\times\boldsymbol{a}_2)}\psi(\boldsymbol{x} + N_1\boldsymbol{a}_1 + N_2\boldsymbol{a}_2)\\ &= e^{\frac{ieN_1N_2}{\hbar}\boldsymbol{B}\cdot(\boldsymbol{a}_1\times\boldsymbol{a}_2)}\psi(\boldsymbol{x}).\end{aligned}$$

Comparing the last two relations, we find $e^{\frac{ieN_1N_2}{\hbar}\boldsymbol{B}\cdot(\boldsymbol{a}_1\times\boldsymbol{a}_2)} = 1 \rightarrow e^{\frac{ieN_1N_2}{\hbar}\boldsymbol{B}\cdot(\boldsymbol{a}_1\times\boldsymbol{a}_2)} = e^{i2\pi m}$, where m is an integer. Thus,

$$\frac{ieN_1N_2}{\hbar}\boldsymbol{B}\cdot(\boldsymbol{a}_1\times\boldsymbol{a}_2) = \frac{ieN_1N_2}{\hbar}\Phi = i2\pi m,$$

where $\Phi = \boldsymbol{B}\cdot(\boldsymbol{a}_1\times\boldsymbol{a}_2)$. After some simplifications, we arrive at

$$\frac{e}{2\pi\hbar}\Phi = \frac{e}{h}\Phi = \frac{m}{N_1N_2}.$$

Problem 2.8

Consider now the Landau gauge with $A = -By\widehat{\boldsymbol{u}}_x$. What is the magnetic translation vector in this case? Can you derive a property for $\widehat{T}(\boldsymbol{a}_1)\widehat{T}(\boldsymbol{a}_2)$ similar to the one found in Problem 2.5?

This problem gives us an opportunity to practice the derivation of two successive magnetic translations (similar to Problem 2.5), but for a different gauge for the vector potential.

Let's begin with the definition $\widehat{T}(\boldsymbol{a}) = e^{\frac{i}{\hbar}(\widehat{\boldsymbol{p}}+eA(\widehat{\boldsymbol{r}}))\cdot\boldsymbol{a}} = e^{\frac{i}{\hbar}\widehat{\boldsymbol{p}}\cdot\boldsymbol{a}}e^{\frac{i}{\hbar}eA(\widehat{\boldsymbol{r}})\cdot\boldsymbol{a}} = \widehat{T}_0(\boldsymbol{a})e^{\frac{i}{\hbar}eA(\widehat{\boldsymbol{r}})\cdot\boldsymbol{a}} = e^{\frac{i}{\hbar}eA(\widehat{\boldsymbol{r}})\cdot\boldsymbol{a}}\widehat{T}_0(\boldsymbol{a})$, in which $A(x)\cdot\boldsymbol{a} = -By\widehat{\boldsymbol{u}}_x\cdot\boldsymbol{a} = -Bya_x$. Therefore,

$$\widehat{T}(\boldsymbol{a}) = e^{-\frac{i}{\hbar}eBya_x}\widehat{T}_0(\boldsymbol{a}) = \widehat{T}_0(\boldsymbol{a})e^{-\frac{i}{\hbar}eBya_x}.$$

The product of two magnetic translations becomes

$$\widehat{T}(\boldsymbol{a}_1)\widehat{T}(\boldsymbol{a}_2) = e^{-\frac{i}{\hbar}eBya_{1,x}}\widehat{T}_0(\boldsymbol{a}_1)e^{-\frac{i}{\hbar}eBya_{2,x}}\widehat{T}_0(\boldsymbol{a}_2).$$

Exchanging $\widehat{T}_0(\boldsymbol{a}_1)e^{-\frac{i}{\hbar}eBya_{2,x}} = e^{-\frac{i}{\hbar}eBa_{2,x}(y+a_{1,y})}\widehat{T}_0(\boldsymbol{a}_1)$, we arrive at

$$\begin{aligned}\widehat{T}(\boldsymbol{a}_1)\widehat{T}(\boldsymbol{a}_2) &= e^{-\frac{i}{\hbar}eBya_{1,x}}e^{-\frac{i}{\hbar}eBa_{2,x}(y+a_{1,y})}\widehat{T}_0(\boldsymbol{a}_1)\widehat{T}_0(\boldsymbol{a}_2)\\ &= e^{-\frac{i}{\hbar}eBya_{1,x}}e^{-\frac{i}{\hbar}eBya_{2,x}}e^{-\frac{i}{\hbar}eB(a_{1,y}a_{2,x})}\widehat{T}_0(\boldsymbol{a}_1+\boldsymbol{a}_2)\\ &= e^{-\frac{i}{\hbar}eBy(a_{1,x}+a_{2,x})}\widehat{T}_0(\boldsymbol{a}_1+\boldsymbol{a}_2)e^{-\frac{i}{\hbar}eB(a_{1,y}a_{2,x})} = \widehat{T}(\boldsymbol{a}_1+\boldsymbol{a}_2)e^{-\frac{i}{\hbar}eB(a_{1,y}a_{2,x})}.\end{aligned}$$

Exchanging $\boldsymbol{a}_1$ and $\boldsymbol{a}_2$ in the preceding,

$$\begin{aligned}\widehat{T}(\boldsymbol{a}_2)\widehat{T}(\boldsymbol{a}_1) &= \widehat{T}(\boldsymbol{a}_2+\boldsymbol{a}_1)\,e^{-\frac{i}{\hbar}eB(a_{2,y}\,a_{1,x})},\\ \widehat{T}(\boldsymbol{a}_1+\boldsymbol{a}_2) &= \widehat{T}(\boldsymbol{a}_2)\widehat{T}(\boldsymbol{a}_1)e^{\frac{i}{\hbar}eB(a_{2,y}\,a_{1,x})}.\end{aligned}$$

Thus, we obtain

$$\begin{aligned}\widehat{T}(\boldsymbol{a}_1)\widehat{T}(\boldsymbol{a}_2) &= \widehat{T}(\boldsymbol{a}_1+\boldsymbol{a}_2)e^{-\frac{i}{\hbar}eB(a_{1,y}a_{2,x})} = \widehat{T}(\boldsymbol{a}_2)\widehat{T}(\boldsymbol{a}_1)e^{\frac{i}{\hbar}eB(a_{2,y}a_{1,x})}e^{-\frac{i}{\hbar}eB(a_{1,y}a_{2,x})}\\ &= \widehat{T}(\boldsymbol{a}_2)\widehat{T}(\boldsymbol{a}_1)e^{\frac{i}{\hbar}eB(a_{1,x}a_{2,y}-a_{1,y}a_{2,x})} = \widehat{T}(\boldsymbol{a}_2)\widehat{T}(\boldsymbol{a}_1)e^{i\frac{e}{\hbar}\boldsymbol{B}\cdot(\boldsymbol{a}_1\times\boldsymbol{a}_2)}\\ &= \widehat{T}(\boldsymbol{a}_2)\widehat{T}(\boldsymbol{a}_1)e^{i\boldsymbol{\beta}\cdot(\boldsymbol{a}_1\times\boldsymbol{a}_2)},\end{aligned}$$

where we have used $\boldsymbol{\beta} = \frac{e\boldsymbol{B}}{\hbar}$. This is exactly the same result as found in Problem 2.5.

Problem 2.9
Consider the following transformation, $\widehat{D}(\alpha) = e^{\alpha\widehat{a}^+ - \alpha^*\widehat{a}}$, where $\widehat{a}, \widehat{a}^+$ are the annihilation and creation operators for the simple harmonic oscillator and $\alpha = |\alpha|e^{i\psi}$ is a complex number. Show that

$$\begin{aligned}\widehat{D}(\alpha) &= e^{-\frac{1}{2}|\alpha|^2}e^{\alpha\widehat{a}^+}e^{-\alpha^*\widehat{a}},\\ \widehat{D}^+(\alpha)\widehat{a}\widehat{D}(\alpha) &= \widehat{a}+\alpha,\\ \widehat{D}^+(\alpha)\widehat{a}^+\widehat{D}(\alpha) &= \widehat{a}^+ + \alpha^*,\\ \widehat{D}(\alpha+\beta) &= \widehat{D}(\alpha)\widehat{D}(\beta)e^{-iIm(\alpha\beta^*)}.\end{aligned}$$

Because of the second and third relations, $\widehat{D}(\alpha)$ is called a *displacement operator*. This displacement operator is often applied to the **coherent state of the simple harmonic oscillator**, as we will see in the problems that follow.

For the *first relation*, let's do the following:

$$\widehat{D}(\alpha) = e^{\alpha\widehat{a}^+ - \alpha^*\widehat{a}} = e^{\alpha\widehat{a}^+}e^{-\alpha^*\widehat{a}}e^{-\frac{1}{2}[\alpha\widehat{a}^+, -\alpha^*\widehat{a}]} = e^{-\frac{1}{2}|\alpha|^2}e^{\alpha\widehat{a}^+}e^{-\alpha^*\widehat{a}},$$

where we have used that $[\widehat{a}, \widehat{a}^+] = 1$ and the relations in the useful formulas.

For the *second relation*, we use the following property: $e^{\widehat{A}}\widehat{B}e^{-\widehat{A}} = \widehat{B} + [\widehat{A}, \widehat{B}] + \frac{1}{2!}\left[\widehat{A}, [\widehat{A}, \widehat{B}]\right] + \frac{1}{3!}\left[\widehat{A}, \left[\widehat{A}, \left[\widehat{A}, \widehat{B}\right]\right]\right] + \ldots$. We further note that $\widehat{D}^+(\alpha) = \widehat{D}(-\alpha) = [\widehat{D}(\alpha)]^{-1}$. Therefore,

$$\begin{aligned}\widehat{D}^+(\alpha)\widehat{a}\widehat{D}(\alpha) &= \widehat{a} + \left[\left(-\alpha\widehat{a}^+ + \alpha^*\widehat{a}\right), \widehat{a}\right] + \frac{1}{2!}\left[\left(-\alpha\widehat{a}^+ + \alpha^*\widehat{a}\right), \left[\left(-\alpha\widehat{a}^+ + \alpha^*\widehat{a}\right), \widehat{a}\right]\right] + \ldots\\ &= \widehat{a} + \alpha + 0 + \ldots,\end{aligned}$$

where we have used again that $[\widehat{a}, \widehat{a}^+] = 1$.

The *third relation* can be proven similarly as the second one.

The *fourth relation* can be shown by using

$$\widehat{D}(\alpha+\beta) = e^{\alpha\widehat{a}^{+}-\alpha^{*}\widehat{a}+\beta\widehat{a}^{+}-\beta^{*}\widehat{a}} = e^{\alpha\widehat{a}^{+}-\alpha^{*}\widehat{a}}e^{\beta\widehat{a}^{+}-\beta^{*}\widehat{a}}e^{-\frac{1}{2}[(\alpha\widehat{a}^{+}-\alpha^{*}\widehat{a}),(\beta\widehat{a}^{+}-\beta^{*}\widehat{a})]}$$
$$= \widehat{D}(\alpha)\widehat{D}(\beta)e^{-\frac{1}{2}(\alpha\beta^{*}-\alpha^{*}\beta)} = \widehat{D}(\alpha)\widehat{D}(\beta)e^{-iIm(\alpha\beta^{*})}.$$

Problem 2.10

A coherent state for the simple harmonic oscillator is defined as $|\alpha\rangle = \widehat{D}(\alpha)|0\rangle$, where $\widehat{D}(\alpha) = e^{\alpha\widehat{a}^{+}-\alpha^{*}\widehat{a}}$ is the displacement operator for the harmonic oscillator, $|0\rangle$ is its vacuum state, and $\widehat{a}, \widehat{a}^{+}$ are the annihilation and creation operators. Show that the coherent state can be written alternatively as $|\alpha\rangle = e^{-\frac{1}{2}|\alpha|^2}\sum_n \frac{\alpha^n}{\sqrt{n!}}|n\rangle$. Demonstrate that $\widehat{a}|\alpha\rangle = \alpha|\alpha\rangle$.

We start with

$$|\alpha\rangle = \widehat{D}(\alpha)|0\rangle = e^{\alpha\widehat{a}^{+}-\alpha^{*}\widehat{a}}|0\rangle = e^{-\frac{1}{2}|\alpha|^2}e^{\alpha\widehat{a}^{+}}e^{-\alpha^{*}\widehat{a}}|0\rangle = e^{-\frac{1}{2}|\alpha|^2}e^{\alpha\widehat{a}^{+}}|0\rangle,$$

since $e^{-\alpha^{*}\widehat{a}}|0\rangle = \left(1-\alpha^{*}\widehat{a}+\frac{1}{2}\alpha^{2,*}\widehat{a}^2+\ldots\right)|0\rangle = |0\rangle$.

Also, $e^{\alpha\widehat{a}^{+}}|0\rangle = \sum_n \frac{\alpha_n}{n!}(\widehat{a}^{+})^n|0\rangle = \sum_n \frac{\alpha^n}{n!}\sqrt{n!}|n\rangle = \sum_n \frac{\alpha^n}{\sqrt{n!}}|n\rangle$. Therefore, $|\alpha\rangle = e^{-\frac{1}{2}|\alpha|^2}\sum_n \frac{\alpha^n}{\sqrt{n!}}|n\rangle$.

For the second quantity, realizing that $\widehat{D}(\alpha)\widehat{D}^{+}(\alpha) = 1$, we can obtain

$$\widehat{D}(\alpha)\widehat{D}^{+}(\alpha)\widehat{a}|\alpha\rangle = \widehat{D}(\alpha)\widehat{D}^{+}(\alpha)\widehat{a}\widehat{D}(\alpha)|0\rangle = \widehat{D}(\alpha)(\widehat{a}+\alpha)|0\rangle = \alpha\widehat{D}(\alpha)|0\rangle = \alpha|\alpha\rangle.$$

> *Food for thought*: This is a problem found in many textbooks, and here we are re-visiting it again in the context of the displacement operator.

Problem 2.11

Show that $\langle(\Delta\widehat{x})^2\rangle_\alpha = \langle\widehat{x}^2\rangle_\alpha - \langle\widehat{x}\rangle_\alpha^2 = \frac{\hbar}{2m\omega}$ and $\langle(\Delta\widehat{p})^2\rangle_\alpha = \langle\widehat{p}^2\rangle_\alpha - \langle\widehat{p}\rangle_\alpha^2 = \frac{\hbar m\omega}{2}$, where the expectation values associated with the displacement and momentum operators of the simple harmonic oscillator in 1D are evaluated with respect to the coherent state $|\alpha\rangle = \widehat{D}(\alpha)|0\rangle$.

We remember that $\widehat{x} = \sqrt{\frac{\hbar}{2m\omega}}(\widehat{a}+\widehat{a}^{+})$and $\widehat{p} = i\sqrt{\frac{\hbar m\omega}{2}}(\widehat{a}-\widehat{a}^{+})$. Therefore, using the results from Problem 2.9, we find that

$$\langle\alpha|(\widehat{a}+\widehat{a}^{+})|\alpha\rangle = \langle 0|\widehat{D}^{+}(\alpha)(\widehat{a}+\widehat{a}^{+})\widehat{D}(\alpha)|0\rangle = \langle 0|(\widehat{a}+\widehat{a}^{+}+\alpha+\alpha^{*})|0\rangle = \alpha+\alpha^{*},$$
$$\langle\alpha|(\widehat{a}-\widehat{a}^{+})|\alpha\rangle = \langle 0|\widehat{D}^{+}(\alpha)(\widehat{a}-\widehat{a}^{+})\widehat{D}(\alpha)|0\rangle = \langle 0|(\widehat{a}-\widehat{a}^{+}+\alpha-\alpha^{*})|0\rangle = \alpha-\alpha^{*},$$
$$\langle\alpha|(\widehat{a}+\widehat{a}^{+})(\widehat{a}+\widehat{a}^{+})|\alpha\rangle = (\alpha+\alpha^{*})^2+1,$$
$$\langle\alpha|(\widehat{a}-\widehat{a}^{+})(\widehat{a}-\widehat{a}^{+})|\alpha\rangle = (\alpha-\alpha^{*})^2-1.$$

By simple substitution we further find that

$$\langle(\Delta\widehat{x})^2\rangle_\alpha = \frac{\hbar}{2m\omega}; \langle(\Delta\widehat{p})^2\rangle_\alpha = \frac{\hbar m\omega}{2}, \text{ and } \langle(\Delta\widehat{x})^2\rangle_\alpha\langle(\Delta\widehat{p})^2\rangle_\alpha = \frac{\hbar^2}{4}.$$

Therefore, the coherent state satisfies the minimum uncertainty relation.

Problem 2.12

Show that the expectation value for the displacement operator of a simple harmonic oscillator in 1D with respect to its time-dependent coherent state $|\alpha(t)\rangle$ is $\langle\widehat{x}\rangle_\alpha(t) = \sqrt{2}x_0|\alpha|\cos(\omega t - \phi)$. Here $x_0 = \sqrt{\frac{\hbar}{m\omega}}$ and we have used that $\alpha = |\alpha|e^{i\varphi}$.

To solve this problem, we realize that we need to evaluate the expectation value of the displacement operator $\widehat{x}$ of the harmonic oscillator with respect to the time-dependent wave function $\phi_\alpha(x,t)$ corresponding to the coherent state α.

We begin with the representation for the coherent state at time $t=0$ from Problem 2.10: - $|\alpha\rangle = e^{-\frac{1}{2}|\alpha|^2}\sum_n \frac{\alpha^n}{\sqrt{n!}}|n\rangle$. Thus, the wave function at $t=0$ corresponding to this state is

$$\phi_\alpha(x) = \langle x|\alpha\rangle = e^{-\frac{1}{2}|\alpha|^2}\sum_n \frac{\alpha_n}{\sqrt{n!}}\langle x|n\rangle = e^{-\frac{1}{2}|\alpha|^2}\sum_n \frac{\alpha^n}{\sqrt{n!}}\psi_n(x),$$

where the eigenfunctions of the simple harmonic oscillator are $\psi_n(x) = \frac{1}{\sqrt{2^n n!\sqrt{\pi}x_0}} H_n(\xi)e^{-\frac{\xi^2}{2}}$, $\xi = \frac{x}{x_0}$, $x_0 = \sqrt{\frac{\hbar}{m\omega}}$. Since the Hamiltonian for the harmonic oscillator does not depend explicitly on time, the time dependence of $\psi_n(x)$ can be expressed as

$$\psi_n(x,t) = e^{-\frac{i\widehat{H}t}{\hbar}}\psi_n(x,0) = e^{-i\omega\left(\frac{1}{2}+n\right)t}\psi_n(x),$$

where we have used that $\psi_n(x,t) = \widehat{U}(t)\psi_n(x,0)$ with the evolution operator $\widehat{U}(t) = e^{-\frac{i\widehat{H}t}{\hbar}}$.

Therefore, the *time-dependent coherent state* is simply $\phi_\alpha(x,t) = e^{-\frac{1}{2}|\alpha|^2}\sum_n \frac{\alpha^n}{\sqrt{n!}}\psi_n(x,t)$ $= e^{-\frac{1}{2}|\alpha|^2}e^{-\frac{i\omega t}{2}}\sum_n \frac{\alpha^n e^{-in\omega t}}{\sqrt{n!}}\psi_n(x)$. Making the substitution $\alpha(t) = \alpha e^{-i\omega t}$, we write

$$\phi_\alpha(x,t) = e^{-\frac{1}{2}|\alpha|^2}e^{-\frac{i\omega t}{2}}\sum_n \frac{\alpha^n(t)}{\sqrt{n!}}\psi_n(x) = \phi_{\alpha(t)}(x)e^{-\frac{i\omega t}{2}}.$$

To find $\langle\widehat{x}\rangle_\alpha(t)$, we write

$$\langle\widehat{x}\rangle_\alpha(t) = \langle\phi_{\alpha(t)}(x)|\widehat{x}|\phi_{\alpha(t)}(x)\rangle = \sqrt{\frac{\hbar}{2m\omega}}\langle\phi_{\alpha(t)}(x)|(\widehat{a}+\widehat{a}^+)|\phi_{\alpha(t)}(x)\rangle$$
$$= \sqrt{\frac{\hbar}{2m\omega}}(\alpha(t)+\alpha^*(t)) = \sqrt{2}x_0 Re(\alpha(t)) = \sqrt{2}x_0|\alpha|\cos(\omega t - \phi).$$

Food for thought: We note that the relation $\langle \widehat{x} \rangle_\alpha(t) = \sqrt{2} x_0 |\alpha| \cos(\omega t - \phi)$ indicates that the expectation value of the displacement operator oscillates with respect to the time-dependent quantum mechanical. This outcome is actually quite different when compared with the expectation value of the displacement operator found with respect to the time-dependent quantum mechanical states. *Can you show this quantitatively?*

Problem 2.13

Define the following transformation, $\widehat{S}(\xi) = e^{\frac{\xi^*}{2}\widehat{a}^{+,2} - \frac{\xi}{2}\widehat{a}^2}$, where $\widehat{a}$, $\widehat{a}^+$ are the annihilation and creation operators for the 1D simple harmonic oscillator and $\xi = re^{i\theta}$ is a complex number (r is real). Show that

$$\widehat{S}^+(\xi)\widehat{a}S(\widehat{\xi}) = \widehat{a}\cosh(r) + \widehat{a}^+ e^{-i\theta}\sinh(r),$$
$$\widehat{S}^+(\xi)\widehat{a}^+ S(\widehat{\xi}) = \widehat{a}^+\cosh(r) + \widehat{a}e^{i\theta}\sinh(r).$$

Using the preceding properties, find $\widehat{S}^+(\xi)\widehat{x}S(\widehat{\xi})$ and $\widehat{S}^+(\xi)\widehat{p}S(\widehat{\xi})$ when $\xi = r$ is real.

To show the *first relation*, we use the property $e^{\widehat{A}}\widehat{B}e^{-\widehat{A}} = \widehat{B} + \left[\widehat{A},\widehat{B}\right] + \frac{1}{2!}\left[\widehat{A},[\widehat{A},\widehat{B}]\right] + \frac{1}{3!}\left[\widehat{A},\left[\widehat{A},\left[\widehat{A},\widehat{B}\right]\right]\right] + \ldots$, where $\widehat{A} = \frac{\xi}{2}\widehat{a}^2 - \frac{\xi^*}{2}\widehat{a}^{+,2}$ and $\widehat{B} = \widehat{a}$. Applying it requires the evaluation of several commutators,

$$\left[\widehat{A},\widehat{a}\right] = \left[\frac{\xi}{2}\widehat{a}^2 - \frac{\xi^*}{2}\widehat{a}^{+,2},\widehat{a}\right] = \frac{\xi}{2}\left[\widehat{a}^2,\widehat{a}\right] - \frac{\xi^*}{2}\left[\widehat{a}^{+,2},\widehat{a}\right] = \xi^*\widehat{a}^+,$$

where we have used that $\left[\widehat{a}^2,\widehat{a}\right] = 0$ and $\left[\widehat{a}^{+,2},\widehat{a}\right] = -2\widehat{a}^+$ Similarly, using $\left[\widehat{a}^{+,2},\widehat{a}^+\right] = 0$ and $\left[\widehat{a}^2,\widehat{a}^+\right] = 2\widehat{a}$,

$$\left[\widehat{A},\left[\widehat{A},\widehat{a}\right]\right] = \left[\widehat{A},\xi^*\widehat{a}^+\right] = \xi^*\left[\widehat{A},\widehat{a}^+\right] = \xi^*\frac{\xi}{2}\left[\widehat{a}^2,\widehat{a}^+\right] - \xi^*\frac{\xi^*}{2}\left[\widehat{a}^{+,2},\widehat{a}^+\right] = |\xi|^2\widehat{a}.$$

The rest of the commutators are found easily, giving the desired result:

$$\widehat{S}^+(\xi)\widehat{a}S(\widehat{\xi}) = \widehat{a} - \xi^*\widehat{a}^+ + \frac{1}{2!}|\xi|^2\widehat{a} - \frac{1}{3!}\xi^*|\xi|^2\widehat{a} + \frac{1}{4!}|\xi|^4\widehat{a} - \frac{1}{5!}\xi^*|\xi|^4\widehat{a} + \ldots$$
$$= \widehat{a}\cosh(r) + \widehat{a}^+ e^{-i\theta}\sinh(r).$$

The *second relation* can be shown analogously, and it is left for extra practice.

For the *last part* of the problem, we remember that $\widehat{a} = \sqrt{\frac{m\omega}{2\hbar}}\widehat{x} + i\sqrt{\frac{1}{2m\hbar\omega}}\widehat{p}$; $\widehat{a}^+ = \sqrt{\frac{m\omega}{2\hbar}}\widehat{x} - i\sqrt{\frac{1}{2m\hbar\omega}}\widehat{p}$. Then, for the case of real $\xi = r$,

$$\widehat{S}^+(r)(\widehat{a}+\widehat{a}^+)S(r) = (\widehat{a}+\widehat{a}^+)\cosh(r) + (\widehat{a}+\widehat{a}^+)\sinh(r),$$
$$\widehat{S}^+(r)\widehat{x}S(r) = \widehat{x}[\cosh(r)+\sinh(r)] = e^r\widehat{x},$$
$$\widehat{S}^+(r)(\widehat{a}-\widehat{a}^+)S(r) = (\widehat{a}-\widehat{a}^+)\cosh(r) - (\widehat{a}-\widehat{a}^+)\sinh(r),$$
$$\widehat{S}^+(r)\widehat{p}S(r) = \widehat{p}[\cosh(r)-\sinh(r)] = e^{-r}\widehat{p}.$$

Food for thought: The so-defined $\widehat{S}(\xi)$ is called "a squeezed operator," defined by Stoler (1970) to deal with squeezed states whose property is to give the minimum uncertainty principle for the expectation values of the displacement and momentum operators of the simple harmonic oscillator. More on this in the next problem.

Problem 2.14

In quantum optics one often has to consider squeezed states, which are defined as $|\xi\rangle = \widehat{S}(\xi)|0\rangle$, where $\widehat{S}(\xi) = e^{\frac{\xi^*}{2}\widehat{a}^{+,2} - \frac{\xi}{2}\widehat{a}^2}$ and $|0\rangle$ is the vacuum state of the simple harmonic oscillator with $\widehat{a}^+$, $\widehat{a}$ being the creation and annihilation operators.

By taking that $\xi = r$ is a real number, find the variance of the displacement and momentum operators $\Delta\widehat{x}_\xi$, $\Delta\widehat{p}_\xi$, and then check the validity of the uncertainty principle.

We start with

$$\left(\Delta\widehat{x}_\xi\right)^2 = \langle\xi|\widehat{x}^2|\xi\rangle - \langle\xi|\widehat{x}|\xi\rangle^2 = \langle 0|\widehat{S}^+(\xi)\widehat{x}^2 S(\xi)|0\rangle - \langle 0|\widehat{S}^+(\xi)\widehat{x}S(\xi)|0\rangle^2$$
$$= \langle 0|e^{2r}\widehat{x}^2|0\rangle - \langle 0|e^{r}\widehat{x}|0\rangle^2 = e^{2r}\langle 0|\widehat{x}^2|0\rangle - e^{2r}\langle 0|\widehat{x}|0\rangle^2 = e^{2r}\frac{\hbar}{2mw}.$$

In the preceding, we have used the results from the previous problem and $\widehat{S}^+(\xi) = e^{\frac{\xi}{2}\widehat{a}^2 - \frac{\xi^*}{2}\widehat{a}^{+,2}} = \widehat{S}^{-1}(\xi)$. Similarly, one can find that

$$\left(\Delta\widehat{p}_\xi\right)^2 = e^{-2r}\frac{\hbar m\omega}{2}.$$

Given the preceding results, we find that

$$\Delta\widehat{x}_\xi\Delta\widehat{p}_\xi = \frac{\hbar}{2}.$$

We note that $\Delta\widehat{x}_\xi \to 0$ and $\Delta\widehat{p}_\xi \to \infty$ as $\xi = r \to \infty$; the minimum uncertainty relation is always fulfilled.

Problem 2.15

In quantum optics one often has to consider *squeezed coherent states*, which are defined as $|\alpha,\xi\rangle = \widehat{S}(\xi)|\alpha\rangle = \widehat{S}(\xi)\widehat{D}(\alpha)|0\rangle$, where $\widehat{S}(\xi) = e^{\frac{\xi^*}{2}\widehat{a}^{+,2} - \frac{\xi}{2}\widehat{a}^2}$ and $\widehat{D}(\alpha) = e^{\alpha\widehat{a}^+ - \alpha^*\widehat{a}}$ for the 1D simple harmonic oscillator, whose vacuum state is $|0\rangle$ and $\widehat{a}^+, \widehat{a}$ are the creation and annihilation operators.

Find equivalent expressions for the following relations in the case of α, ξ being real parameters:

$$\widehat{S}^+(\xi)\widehat{D}^+(\alpha)\widehat{x}\widehat{D}(\alpha)S(\xi), \quad \widehat{S}^+(\xi)\widehat{D}^+(\alpha)\widehat{p}\widehat{D}(\alpha)S(\xi),$$
$$\langle\alpha,\xi|\widehat{x}|\alpha,\xi\rangle, \quad \langle\alpha,\xi|\widehat{p}|\alpha,\xi\rangle.$$

Using the results from the previous problems, we write

$$\widehat{S}^{+}(\xi)\widehat{D}^{+}(\alpha)\widehat{x}\widehat{D}(\alpha)\widehat{S}(\xi) = \widehat{S}^{+}(\xi)\left(\widehat{x}+2\alpha\sqrt{\frac{\hbar}{2m\omega}}\right)\widehat{S}(\xi)$$
$$= \widehat{S}^{+}(\xi)\widehat{x}\widehat{S}(\xi)+2\alpha\sqrt{\frac{\hbar}{2m\omega}} = e^{\xi}\widehat{x}+2\alpha\sqrt{\frac{\hbar}{2m\omega}}.$$

Similarly,

$$\widehat{S}^{+}(\xi)\widehat{D}^{+}(\alpha)\widehat{p}\widehat{D}(\alpha)\widehat{S}(\xi) = \widehat{S}^{+}(\xi)\widehat{p}\widehat{S}(\xi) = e^{-\xi}\widehat{p}.$$

For the last two relations,

$$\langle\alpha,\xi|\widehat{x}|\alpha,\xi\rangle = \langle 0|\left(e^{\xi}\widehat{x}+2\alpha\sqrt{\frac{\hbar}{2m\omega}}\right)|0\rangle = 2\alpha\sqrt{\frac{\hbar}{2m\omega}},$$
$$\langle\alpha,\xi|\widehat{p}|\alpha,\xi\rangle = \langle 0|e^{-\xi}\widehat{p}|0\rangle = 0.$$

Food for thought: The transformed displacement operator consists of a squeezed term and a displacement constant, while the transformed momentum operator consists of a squeezed contribution only.

Problem 2.16

Consider the Hamiltonian $\widehat{H} = \frac{\omega_1}{2}\widehat{\sigma}_z + \omega_2\widehat{N} + \lambda(\widehat{a}+\widehat{a}^{+})\widehat{\sigma}_x$, where $\omega_1, \omega_2, \lambda$ are real constants, $\widehat{N} = \widehat{a}^{+}\widehat{a}$ is the number operator for the simple harmonic oscillator $\widehat{a}^{+}, \widehat{a}$ are the creation and annihilation operators), and $\sigma_{x,y,z}$ are the Pauli matrices.

a) Find an expression for the transformed Hamiltonian $\widehat{H}_{12} = \widehat{U}_2\widehat{U}_1\widehat{H}\widehat{U}_1^{+}\widehat{U}_2^{+}$ within first order of $\zeta_1 = \frac{\lambda}{\omega_1+\omega_2}$ and $\zeta_2 = \frac{\lambda}{\omega_1-\omega_2}$ for the unitary operators $\widehat{U}_1 = e^{\zeta_1(\widehat{a}^{+}\widehat{\sigma}_{+}-\widehat{a}\widehat{\sigma}_{-})}$ and $\widehat{U}_2 = e^{\zeta_2(\widehat{a}\widehat{\sigma}_{+}-\widehat{a}^{+}\widehat{\sigma}_{-})}$

b) Find the expectation value of $\widehat{H}$ with respect to the nth simple harmonic state whose spin is in the positive z-direction.

This problem gives an opportunity to exercise manipulations of creation and annihilation operators for the simple harmonic oscillator and spin $1/2$ operators.

a) A useful formula here is the expression $e^{\widehat{A}}\widehat{B}e^{-\widehat{A}} = \widehat{B} + \left[\widehat{A},\widehat{B}\right] + \frac{1}{2!}\left[\widehat{A},\left[\widehat{A},\widehat{B}\right]\right] + \frac{1}{3!}\left[\widehat{A},\left[\widehat{A},\left[\widehat{A},\widehat{B}\right]\right]\right] + \ldots$. The transformed Hamiltonian becomes

$$\widehat{H}_{12} = \widehat{U}_2\widehat{U}_1\widehat{H}\widehat{U}_1^{+}\widehat{U}_2^{+} = \widehat{H} + \left[\zeta_1(\widehat{a}^{+}\widehat{\sigma}_{+}-\widehat{a}\widehat{\sigma}_{-}),\widehat{H}\right] + \left[\zeta_2(\widehat{a}\widehat{\sigma}_{+}-\widehat{a}^{+}\widehat{\sigma}_{-}),\widehat{H}\right] + \ldots.$$

We have kept only the first three terms, which turn out to be enough to obtain the contribution to first order with respect to ζ_1 and ζ_2. Working out the commutators yields

$$\left[\zeta_1(\widehat{a}^{+}\widehat{\sigma}_{+}-\widehat{a}\widehat{\sigma}_{-}),\widehat{H}\right] = \omega_2\zeta_1\left(\widehat{a}\widehat{\sigma}_{-}+\widehat{a}^{+}\widehat{\sigma}_{+}\right)$$
$$-\lambda\zeta_1\left[(\widehat{\sigma}_z-\widehat{\sigma}_0)+\left(\widehat{a}\widehat{a}+2\widehat{a}^{+}\widehat{a}+\widehat{a}^{+}\widehat{a}^{+}\right)\widehat{\sigma}_z\right],$$

$$\left[\zeta_2(\hat{a}\hat{\sigma}_+ - \hat{a}^+\hat{\sigma}_-), \hat{H}\right] = -\omega_2\zeta_2\left(\hat{a}\hat{\sigma}_+ + \hat{a}^+\hat{\sigma}_-\right)$$
$$-\lambda\zeta_2\left[(\hat{\sigma}_z + \hat{\sigma}_0) + \left(\hat{a}\hat{a} + 2\hat{a}^+\hat{a} + \hat{a}^+\hat{a}^+\right)\hat{\sigma}_z\right],$$

where $\hat{\sigma}_0$ is the 2×2 identity matrix.

Grouping everything together and using $\zeta_1 + \zeta_2 = \frac{2\lambda\omega_1}{\omega_1^2 - \omega_2^2}$, $\zeta_1 - \zeta_2 = \frac{-2\lambda\omega_2}{\omega_1^2 - \omega_2^2}$, $\hat{N} = \hat{a}^+\hat{a}$, and $\hat{\sigma}_\pm = \frac{1}{2}\left(\hat{\sigma}_x \pm i\hat{\sigma}_y\right)$, we obtain

$$\hat{H}_{12} = \frac{\omega_1}{2}\hat{\sigma}_z + \omega_2\hat{N} + \lambda\left(\hat{a} + \hat{a}^+\right)\hat{\sigma}_x + \left[\zeta_1(\hat{a}^+\hat{\sigma}_+ - \hat{a}\hat{\sigma}_-), \hat{H}\right]$$
$$+\left[\zeta_2(\hat{a}\hat{\sigma}_+ - \hat{a}^+\hat{\sigma}_-), \hat{H}\right] + \ldots,$$

$$\hat{H}_{12} = \omega_1\left(\frac{1}{2} + \frac{2\lambda^2}{\omega_2^2 - \omega_1^2}(1 + 2\hat{N} + \hat{a}\hat{a} + \hat{a}^+\hat{a}^+)\right)\hat{\sigma}_z + \omega_2\left(\hat{N} + \frac{2\lambda^2}{\omega_2^2 - \omega_1^2}\right)\hat{\sigma}_0$$
$$+\lambda\left(1 + \frac{\omega_2^2}{\omega_2^2 - \omega_1^2}\right)(\hat{a} + \hat{a}^+)\hat{\sigma}_x + i\frac{\lambda\omega_1\omega_2}{\omega_2^2 - \omega_1^2}(\hat{a} - \hat{a}^+)\hat{\sigma}_y.$$

b) The expectation value for $\hat{H}$ for the nth-particle state whose spin is $1/2$ in the positive z-axis is

$$\langle\hat{H}_{12}\rangle_n = \left\langle n, \frac{1}{2}\middle|\hat{H}_{12}\middle|n, \frac{1}{2}\right\rangle = \omega_1\left(\frac{1}{2} + \frac{2\lambda^2}{\omega_2^2 - \omega_1^2}(1 + 2n)\right) + \omega_2\left(n + \frac{2\lambda^2}{\omega_2^2 - \omega_1^2}\right),$$

where we have used that $\langle n|\hat{a}|n\rangle = \langle n|\hat{a}^+|n\rangle = \langle n|\hat{a}\hat{a}|n\rangle = \langle n|\hat{a}^+\hat{a}^+|n\rangle = 0$, $\langle n|\hat{N}|n\rangle = n$, and $\left\langle\frac{1}{2}\middle|\hat{\sigma}_z\middle|\frac{1}{2}\right\rangle = 1$.

Problem 2.17

Find an equivalent expression for the operator $\hat{\Pi} = e^{i\lambda\hat{\pi}}$, where $\hat{\pi}$ is the parity operator and λ is a real constant.

To find such an expression, we use the fact that functions of operators can be written using their Taylor series equivalent.

$$\hat{\Pi} = e^{i\lambda\hat{\pi}} = \sum_n \frac{(i\lambda)^n}{n!}\hat{\pi}^n.$$

Remembering the basic property for the parity operator that $\hat{\pi}^2 = \hat{1}$, we can easily recombine this series in even and odd components:

$$\hat{\Pi} = \sum_n\left(1 - \frac{\lambda^2}{2!} + \frac{\lambda^4}{4!} - \ldots\right)\hat{1} + i\left(\lambda - \frac{\lambda^3}{3!} + \frac{\lambda^5}{5!}\ldots\right)\hat{\pi} = \cos(\lambda)\hat{1} + i\sin(\lambda)\hat{\pi}.$$

Problem 2.18

Consider the bound motion of a quantum mechanical particle in 1D under an even potential $V(x) = V(-x)$. Demonstrate that the eigenfunctions of the stationary Schrödinger equation for discrete bound states have a well-defined parity.

This is actually a standard problem that can be seen in many textbooks. Due to its importance, here we present it again. What is asked in the problem is to show that each eigenfunction of the basic 1D Schrödinger equation is either even or odd. A sensible way to begin is by writing the basic 1D Schrödinger equation

$$\left[-\frac{\hbar^2}{2m}\frac{d^2}{dx^2}+V(x)\right]\psi(x)=E\psi(x),$$

where $\psi(x)$ is the eigenfunction whose eigenenergy is E. However, since $V(x)=V(-x)$, we can also write

$$\left[-\frac{\hbar^2}{2m}\frac{d^2}{dx^2}+V(-x)\right]\psi(x)=E\psi(x).$$

After executing the change of variable $x\to -x$ in the preceding equation, one arrives at

$$\left[-\frac{\hbar^2}{2m}\frac{d^2}{dx^2}+V(x)\right]\psi(-x)=E\psi(-x).$$

Therefore, $\psi(x)$ and $\psi(-x)$ are solutions to the stationary Schrödinger equation with the same energy E. In 1D motion, however, the energy spectrum of bound states is always nondegenerate (**Can you explain *why?***), meaning that $\psi(x)$ and $\psi(-x)$ must be linearly dependent. Also, both functions must be normalized:

$$\int_{-\infty}^{\infty}|\psi(x)|^2dx=\int_{-\infty}^{\infty}|\psi(-x)|^2dx=1.$$

Because of the nondegenerate 1D spectrum and the normalization condition for the eigenfunctions, one concludes that the wave function must satisfy $\psi(x)=\psi(-x)$ or $\psi(x)=-\psi(-x)$.

> *Food for thought:* Can you give examples of specific 1D potentials for which their eigenfunctions have definite parity?

Problem 2.19

Show that the eigenstates for the Hamiltonian of the simple harmonic oscillator potential have a definite parity.

Show that the parity of the eigenstates for even n have an even parity and the eigenstates with odd n have an odd parity.

Obviously, this problem is related to the previous one, since the potential for the simple harmonic oscillator is $V(x)=\frac{1}{2}m\omega^2x^2=V(-x)$. The Hamiltonian is $\widehat{H}=\frac{\widehat{p}^2}{2m}+\frac{1}{2}m\omega^2\widehat{x}^2$ and it commutes with the parity, $\left[\widehat{H},\widehat{\pi}\right]=0$. Thus, according to the previous problem, the eigenstates must have definite parity.

To demonstrate the parity property of the eigenfunctions, we use the relations for the annihilation and creation operators $\widehat{a}=\sqrt{\frac{m\omega}{2\hbar}}\left(\widehat{x}+i\frac{\widehat{p}}{m\omega}\right)$; $a^+=\sqrt{\frac{m\omega}{2\hbar}}\left(\widehat{x}-i\frac{\widehat{p}}{m\omega}\right)$. Remembering that $\widehat{\pi}^+\widehat{x}\widehat{\pi}=-\widehat{x}$ and $\widehat{\pi}^+\widehat{p}\widehat{\pi}=-\widehat{p}$, one easily shows

$$\widehat{\pi}^+\widehat{a}^+\widehat{\pi}=-\widehat{a}^+;\quad \widehat{\pi}^+\widehat{a}\widehat{\pi}=-\widehat{a}.$$

Thus, one finds

$$\widehat{\pi}|n+1\rangle = \widehat{\pi}\frac{\widehat{a}^+}{\sqrt{n+1}}|n\rangle = -\widehat{a}^+\frac{\widehat{\pi}}{\sqrt{n+1}}|n\rangle,$$

where we have used the properties of the creation and annihilation operators $\widehat{a}^+|n\rangle = \sqrt{n+1}|n+1\rangle$; $\widehat{a}|n\rangle = \sqrt{n}|n-1\rangle$. On the other hand, since the simple harmonic oscillator eigenstates $|n\rangle$ have definite parity, we have

$$\widehat{\pi}|n\rangle = \eta_n|n\rangle; \qquad \widehat{\pi}|n+1\rangle = \eta_{n+1}|n+1\rangle.$$

Comparing the preceding relations, we arrive at

$$\eta_{n+1} = -\eta_n,$$

meaning that the parities of the successive eigenstates alternate. Noting that the ground state $\psi_0(x) = \langle x|0\rangle \sim e^{-\frac{m\omega x^2}{2\hbar}}$ is even with $\eta_0 = (+1)$. Thus, $\eta_n = (-1)^n$. Another way to realize this is by using

$$\widehat{\pi}^+|n\rangle\widehat{\pi} = \widehat{\pi}^+\frac{(a^+)^n}{\sqrt{n!}}|0\rangle\widehat{\pi} = (-1)^n\frac{(\widehat{a}^+)^n}{\sqrt{n!}}\widehat{\pi}^+|0\rangle\widehat{\pi} = (-1)^n\frac{(\widehat{a}^+)^n}{\sqrt{n!}}|0\rangle\eta_0 = (-1)^n\eta_0|n\rangle = \eta_n|n\rangle.$$

Again, since $\eta_n = (-1)^n$, we conclude that the eigenstates for even n have an even parity, while the eigenstates for odd n have an odd parity.

Problem 2.20

Consider the Hamiltonian for the simple harmonic oscillator and the definition for the parity operator in the continuum basis of the position operator $\widehat{\pi} = \int dx|x\rangle\langle -x|$.

Find an expression for the parity operator in terms of the number operator $\widehat{N} = \widehat{a}^+\widehat{a}$, where $\widehat{a}^+, \widehat{a}$ are the creation and annihilation operators respectively.

Using bra-ket notation for the wave function of the simple harmonic oscillator $\psi_n(x) = \langle x|n\rangle$, we can write

$$\widehat{\pi}\psi_n(x) = \widehat{\pi}\langle x|n\rangle = \langle x|\widehat{\pi}|n\rangle = \int_{-\infty}^{\infty} dx_0\langle x|x_0\rangle\langle -x_0|n\rangle = \langle -x|n\rangle = \psi_n(-x),$$

where the definition for the parity $\widehat{\pi} = \int dx|x\rangle\langle -x|$ was used. However, we know that $\psi_n(-x) = \begin{cases} -\psi_n(x), & \text{for n} - \text{odd} \\ +\psi_n(x), & \text{for n} - \text{even} \end{cases}$. Since $\widehat{N}|n\rangle = n|n\rangle$, we arrive at

$$\widehat{\pi}|n\rangle = (-1)^n|n\rangle = e^{i\pi\widehat{N}}|n\rangle \rightarrow \widehat{\pi} = e^{i\pi\widehat{N}}.$$

Problem 2.21

Consider a Hamiltonian $\widehat{H}$ with eigenenergies and eigenstates given by the characteristic equation $\widehat{H}|\psi_n\rangle = E_n|\psi_n\rangle$. Let $\widehat{H}$ commute with the parity operator, $\left[\widehat{H}, \widehat{\pi}\right] = 0$.

Show explicitly that $|\psi_n\rangle$ is also an eigenstate of the parity operator $\widehat{\pi}$ and find its eigenvalues.

We will show that $|\psi_n\rangle$ is an eigenstate of $\widehat{\pi}$ by construction. Let us take

$$|\widetilde{\psi}_n\rangle = \frac{1}{2}(1 \pm \widehat{\pi})|\psi_n\rangle.$$

It is easy to see that $\widehat{H}|\widetilde{\psi}_n\rangle = E_n|\widetilde{\psi}_n\rangle$. Check:

$$\widehat{H}|\widetilde{\psi}_n\rangle = \frac{1}{2}(\widehat{H} \pm \widehat{H}\widehat{\pi})|\psi_n\rangle = \frac{1}{2}(E_n|\psi_n\rangle \pm \widehat{\pi}\widehat{H}|\psi_n\rangle) = \frac{1}{2}(E_n|\psi_n\rangle \pm E_n\widehat{\pi}|\psi_n\rangle) = E_n|\widetilde{\psi}_n\rangle.$$

It is also easy to see that $\widehat{\pi}|\widetilde{\psi}_n\rangle = \pm|\widetilde{\psi}_n\rangle$. Check:

$$\widehat{\pi}|\widetilde{\psi}_n\rangle = \frac{1}{2}\left(\widehat{\pi} \pm \widehat{\pi}^2\right)|\psi_n\rangle = \frac{1}{2}(\widehat{\pi} \pm 1)|\psi_n\rangle = \pm\frac{1}{2}(1 \pm \widehat{\pi})|\psi_n\rangle = \pm|\widetilde{\psi}_n\rangle.$$

Food for thought: This problem was considered explicitly in Sakurai and Napolitano (2017). Can you give examples of specific Hamiltonians that commute with the parity operator?

Problem 2.22

The electric dipole moment operator of a quantum mechanical particle is defined as $\widehat{\mathbf{D}} = q\widehat{\mathbf{x}}$, where q is the charge and $\widehat{\mathbf{x}}$ is the displacement operator. If in the absence of any external electric field $\widehat{\mathbf{D}} \neq \mathbf{0}$, then the particle is said to have a *spontaneous* dipole moment. Suppose the Hamiltonian $\widehat{H}$ for this particle has nondegenerate spectrum and is invariant under parity. Show that the expectation value of the spontaneous dipole moment of the particle is zero.

Since $\widehat{H}$ is invariant under parity, then $\widehat{\pi}^{+}\widehat{H}\widehat{\pi} = \widehat{H}$ and $\left[\widehat{\pi}, \widehat{H}\right] = 0$.

Consider the nondegenerate spectrum of the Hamiltonian $\widehat{H}|\psi_n\rangle = E_n|\psi_n\rangle$ and do the following:

$$\left[\widehat{\pi}, \widehat{H}\right]|\psi_n\rangle = 0 \rightarrow \widehat{H}|\psi_n'\rangle = E|\psi_n'\rangle, \quad \text{where} \quad |\psi_n'\rangle = \widehat{\pi}|\psi_n\rangle.$$

Also, we realize that

$$\langle\psi_n|\widehat{\mathbf{D}}|\psi_n\rangle = \langle\psi_n|\widehat{\pi}^{+}\widehat{\pi}\widehat{\mathbf{D}}\widehat{\pi}^{+}\widehat{\pi}|\psi_n\rangle = \langle\psi_n'|\widehat{\pi}\widehat{\mathbf{D}}\widehat{\pi}^{+}|\psi_n'\rangle = -\langle\psi_n'|\widehat{\mathbf{D}}|\psi_n'\rangle,$$

where we have used that $\widehat{\pi}^{+}\widehat{\pi} = \widehat{1}$ and the fact that $\widehat{\pi}\widehat{\mathbf{D}}\widehat{\pi}^{+} = q\widehat{\pi}\widehat{\mathbf{x}}\widehat{\pi}^{+} = -q\widehat{\mathbf{x}} = -\widehat{\mathbf{D}}$. However, since the Hamiltonian and parity operators commute, then the eigenfunctions of the Hamiltonian have definite parity, thus $|\psi_n'\rangle = \widehat{\pi}|\psi_n\rangle = \pm|\psi_n\rangle$. Therefore,

$$\langle\psi_n|\widehat{\mathbf{D}}|\psi_n\rangle = -\langle\psi_n'|\widehat{\mathbf{D}}|\psi_n'\rangle = -\langle\psi_n|\widehat{\mathbf{D}}|\psi_n\rangle,$$

meaning that the expectation value of the spontaneous dipole moment is zero, $\langle\psi_n|\widehat{\mathbf{D}}|\psi_n\rangle = \mathbf{0}$.

Why is it necessary to require the spectrum of $\widehat{H}$ to be nondegenerate? Is it possible that for such a Hamiltonian for which $\widehat{\pi}^{+}\widehat{H}\widehat{\pi} = \widehat{H}$, the expectation value of $\widehat{\mathbf{D}}$ is not zero?

Suppose $\widehat{H}|\psi_n^e\rangle = E_n|\psi_n^e\rangle$ and $\widehat{H}|\psi_n^o\rangle = E_n|\psi_n^o\rangle$ (e = even parity, o = odd parity). Then for the linear combination $|\tilde{\psi}_n\rangle = C_e|\psi_n^e\rangle + C_o|\psi_n^0\rangle$, one has $|\tilde{\psi}_n'\rangle = \widehat{\pi}|\tilde{\psi}_n\rangle = C_e|\psi_n^e\rangle - C_o|\psi_n^0\rangle$. Therefore $|\tilde{\psi}_n\rangle$ does not have a definite parity and the expectation value of the spontaneous dipole moment can be nonzero.

> *Food for thought:* This situation reflects the fact that in systems that lack inversion symmetry, the average dipole moment can be nonzero.

Problem 2.23

a) Using the definition $(\langle\phi|\widehat{A})|\psi\rangle = \langle\phi|(\widehat{A}|\psi\rangle)$ of a linear operator $\widehat{A}$, write the corresponding definition for an *antilinear operator* $\widehat{B}$. Also, using the definition $\langle\phi|\widehat{A}^+|\psi\rangle = (\langle\phi|\widehat{A}|\psi\rangle)^*$ of an adjoint operator, write the definition of an *adjoint antilinear operator* $\widehat{B}$.
b) Show that transition probabilities between two states $|\psi\rangle$ and $|\phi\rangle$ are conserved under the antiunitary operator transformation $\widehat{B}$.

This problem gives a straightforward exercise for the basic properties of antiunitary operators.

a) The corresponding definition of an antilinear operator is $(\langle\phi|\widehat{B})|\psi\rangle = [\langle\phi|(\widehat{B}|\psi\rangle)]^*$.
   The corresponding definition of the adjoint antilinear operator is $\langle\phi|(\widehat{B}^+|\psi\rangle) = [(\langle\phi|\widehat{B})|\psi\rangle]^* = \langle\phi|(\widehat{B}|\psi\rangle)$.
   The antilinear operator is also antiunitary if it satisfies $\widehat{B}^+\widehat{B} = \widehat{B}\widehat{B}^+ = \widehat{1}$.
b) Let $|\psi'\rangle = \widehat{B}|\psi\rangle$ and $|\phi'\rangle = \widehat{B}|\phi\rangle$, then

$$\langle\phi'|\psi'\rangle = \left(\langle\phi|\widehat{B}^+\right)\left(\widehat{B}|\psi\rangle\right) = [\langle\phi|(\widehat{B}^+\widehat{B}|\psi\rangle)]^* = \langle\phi|\psi\rangle^*.$$

   Therefore, $|\langle\phi|\psi\rangle|^2$ remains unchanged.
   We conclude that transition probabilities are not altered by $\widehat{B}$.

Problem 2.24

Let us consider the time-reversal operator of spin $1/2$. Using the general definition of the time-reversal operator $\widehat{\Theta} = \widehat{U}\widehat{K}$, where $\widehat{U}$ is a suitable unitary operator and $\widehat{K}$ denotes complex conjugation, construct the explicit form $\widehat{\Theta}$. Specifically, from the basic transformation for the spin $1/2$ operator under time reversal and the fact that any 2×2 unitary matrix can be given as $\widehat{U} = \alpha\sigma_1 + \beta\sigma_2 + \gamma\sigma_3 + \delta\sigma_0$ ($\alpha, \beta, \gamma, \delta$ are constants), show that the time-reversal operator for a spin $1/2$ particle can be written as $\widehat{\Theta} = \eta e^{i\sigma_2\pi/2}\widehat{K}$. Note that there is an accumulated phase η upon time-reversal operation, which often is chosen to be 1. Feel free to take $\eta = 1$ in your solution.

We use the following property: $\widehat{\Theta}\widehat{\boldsymbol{S}}\widehat{\Theta}^{-1} = -\widehat{\boldsymbol{S}}$.

Thus, since $\widehat{\boldsymbol{S}} = \frac{\hbar}{2}\boldsymbol{\sigma}$, we find that

$$-\sigma_1 = \widehat{\Theta}\sigma_1\widehat{\Theta}^{-1} = (\widehat{U}\widehat{K})\sigma_1(\widehat{U}\widehat{K})^{-1} = \widehat{U}\widehat{K}\sigma_1\widehat{K}\widehat{U}^{-1} = \widehat{U}\sigma_1\widehat{K}^2\widehat{U}^{-1} = \widehat{U}\sigma_1\widehat{U}^{-1},$$

where we have used that σ_1 is real, thus $\widehat{K}\sigma_1 = \sigma_1\widehat{K}$ and that $\widehat{K}^2 = 1$. From here one finds that the anticommutator

$$\{\widehat{U}, \sigma_1\} = \widehat{U}\sigma_1 + \sigma_1\widehat{U} = 0.$$

Noting that σ_2 is imaginary and $\widehat{K}\sigma_2 = -\sigma_2\widehat{K}$, we find that

$$-\sigma_2 = \widehat{\Theta}\sigma_2\widehat{\Theta}^{-1} = (\widehat{U}\widehat{K})\sigma_2(\widehat{U}\widehat{K})^{-1} = \widehat{U}\widehat{K}\sigma_2\widehat{K}\widehat{U}^{-1} = -\widehat{U}\sigma_2\widehat{K}^2\widehat{U}^{-1} = -\widehat{U}\sigma_2\widehat{U}^{-1},$$

thus,

$$\left[\widehat{U}, \sigma_2\right] = 0.$$

Noting that σ_3 is real and $\widehat{K}\sigma_3 = \sigma_3\widehat{K}$, we find that

$$-\sigma_3 = \widehat{\Theta}\sigma_3\widehat{\Theta}^{-1} = \widehat{U}\widehat{K}\sigma_3(\widehat{U}\widehat{K})^{-1} = \widehat{U}\widehat{K}\sigma_3\widehat{K}\widehat{U}^{-1} = \widehat{U}\sigma_3\widehat{K}^2\widehat{U}^{-1} = \widehat{U}\sigma_3\widehat{U}^{-1},$$

thus,

$$\left\{\widehat{U}, \sigma_3\right\} = 0.$$

In summary, $\widehat{U}$ anticommutes with σ_1, σ_3 and it commutes with σ_2. Use each result explicitly,

$$\begin{aligned}\left\{\widehat{U}, \sigma_1\right\} &= 0 = \alpha\{\sigma_1, \sigma_1\} + \beta\{\sigma_2, \sigma_1\} + \gamma\{\sigma_3, \sigma_1\} + \delta\{\sigma_0, \sigma_1\} \\ &= 2\alpha\sigma_1^2 + 2\delta\sigma_1 = 2\alpha\sigma_0 + 2\delta\sigma_1 = 2\begin{pmatrix} \alpha & \delta \\ \delta & \alpha \end{pmatrix}, \\ \alpha = \delta = 0 \quad &\rightarrow \quad \widehat{U} = \beta\sigma_2 + \gamma\sigma_3.\end{aligned}$$

Also, $\left\{\widehat{U}, \sigma_3\right\} = 0 = \beta\{\sigma_2, \sigma_3\} + \gamma\{\sigma_3, \sigma_3\} = 2\gamma\sigma_3^2 = 2\gamma\sigma_0 = 2\begin{pmatrix} \gamma & 0 \\ 0 & \gamma \end{pmatrix}$, thus $\gamma = 0$,

$$\widehat{U} = \beta\sigma_2.$$

Additionally, $\widehat{U}\widehat{U}^{+} = \widehat{U}^{+}\widehat{U} = \widehat{1}$; thus, by taking $\beta = i$,

$$\widehat{U} = i\sigma_2 = e^{\frac{i\pi\sigma_2}{2}} = \sigma_0\cos\left(\frac{\pi}{2}\right) + i\sigma_2\sin\left(\frac{\pi}{2}\right) \quad \rightarrow \quad \widehat{\Theta} = e^{i\sigma_2\pi/2}\widehat{K}.$$

We note that, from $\widehat{U}\widehat{U}^{+} = \widehat{U}^{+}\widehat{U} = \widehat{1}$, we could take $\beta = i\eta$, with $|\eta| = 1$ being an arbitrary phase. Then $\widehat{\Theta} = \eta e^{i\sigma_2\pi/2}\widehat{K}$.

Problem 2.25

Let a spin $1/2$ particle be in a spinor state $\chi = \frac{1}{\sqrt{|\alpha|^2+|\beta|^2}}\binom{\alpha}{\beta}$, where this vector is expressed in the standard eigenbasis representation in which σ_3 is diagonal (σ_i, $i = 1, 2, 3$ are the Pauli matrices). What is the time-reversed spinor state?

For spin 1/2 particle the time-reversal operator is $\widehat{\Theta} = i\eta\sigma_2\widehat{K}$, where $\sigma_2 = \begin{pmatrix} 0 & -i \\ i & 0 \end{pmatrix}$ and $\widehat{K}$ is the complex conjugation operation. One easily finds that

$$\chi_\Theta = \widehat{\Theta}\chi = i\eta\sigma_2\widehat{K}\frac{1}{\sqrt{|\alpha|^2+|\beta|^2}}\begin{pmatrix} \alpha \\ \beta \end{pmatrix} = \frac{\eta}{\sqrt{|\alpha|^2+|\beta|^2}}\begin{pmatrix} 0 & 1 \\ -1 & 0 \end{pmatrix}\begin{pmatrix} \alpha^* \\ \beta^* \end{pmatrix}$$
$$= \frac{\eta}{\sqrt{|\alpha|^2+|\beta|^2}}\begin{pmatrix} \beta^* \\ -\alpha^* \end{pmatrix}.$$

Problem 2.26

Consider a particle with angular momentum $J = 1$, whose spinor is given as $\chi = \frac{1}{\sqrt{|\alpha|^2+|\beta|^2+|\gamma|^2}}\begin{pmatrix} \alpha \\ \beta \\ \gamma \end{pmatrix}$ in the eigenbasis representation of $\widehat{J}_z$ being diagonal. What happens to this spinor under the operation of time reversal?

Clearly, we have to know the time-reversal operator for the angular momentum $\widehat{\boldsymbol{J}}$, whose matrix representation is

$$\widehat{J}_x = \frac{\hbar}{\sqrt{2}}\begin{pmatrix} 0 & 1 & 0 \\ 1 & 0 & 1 \\ 0 & 1 & 0 \end{pmatrix}; \widehat{J}_y = \frac{h}{\sqrt{2}}\begin{pmatrix} 0 & -i & 0 \\ i & 0 & -i \\ 0 & i & 0 \end{pmatrix}; \widehat{J}_z = \frac{\hbar}{\sqrt{2}}\begin{pmatrix} 1 & 0 & 0 \\ 0 & 0 & 0 \\ 0 & 0 & -1 \end{pmatrix}.$$

The time-reversal operator can be written as

$$\widehat{\Theta} = \eta e^{-i\widehat{J}_y\pi/\hbar}\widehat{K},$$

where $\widehat{K}$ denotes complex conjugation. Using the series representation of $e^{-i\widehat{J}_y\pi/\hbar}$ and the relation $\left(\frac{\widehat{J}_y}{\hbar}\right)^3 = \frac{\widehat{J}_y}{\hbar}$, we find that

$$e^{-i\widehat{J}_y\alpha/\hbar} = \widehat{1} - \frac{i\widehat{J}_y\alpha}{\hbar} + \frac{1}{2!}\left(\frac{i\widehat{J}_y\alpha}{\hbar}\right)^2 - \frac{1}{3!}\left(\frac{i\widehat{J}_y\alpha}{\hbar}\right)^3 + \frac{1}{4!}\left(\frac{i\widehat{J}_y\alpha}{\hbar}\right)^4 + \dots$$
$$= \widehat{1} - \left(\frac{\widehat{J}_y}{\hbar}\right)^2(1-\cos(\alpha)) - i\left(\frac{\widehat{J}_y}{\hbar}\right)\sin(\alpha).$$

Therefore,

$$\widehat{\Theta} = \eta e^{-i\widehat{J}_y\pi/\hbar}\widehat{K} = \eta\left[\widehat{1} - \left(\frac{\widehat{J}_y}{\hbar}\right)^2(1-\cos(\pi)) - i\left(\frac{\widehat{J}_y}{\hbar}\right)\sin(\pi)\right]\widehat{K} = \eta\left[1 - 2\left(\frac{\widehat{J}_y}{\hbar}\right)^2\right]\widehat{K}.$$

Then we see that

$$\chi_\Theta = \widehat{\Theta}\chi = \eta\left[\widehat{1} - 2\left(\frac{\widehat{J}_y}{\hbar}\right)^2\right]\widehat{K}\frac{1}{\sqrt{|\alpha|^2+|\beta|^2+|\gamma|^2}}\begin{pmatrix} \alpha \\ \beta \\ \gamma \end{pmatrix}$$

$$= \frac{\eta}{\sqrt{|\alpha|^2+|\beta|^2+|\gamma|^2}} \begin{pmatrix} 0 & 0 & 1 \\ 0 & -1 & 0 \\ 1 & 0 & 0 \end{pmatrix} \begin{pmatrix} \alpha^* \\ \beta^* \\ \gamma^* \end{pmatrix}$$

$$= \frac{\eta}{\sqrt{|\alpha|^2+|\beta|^2+|\gamma|^2}} \begin{pmatrix} \gamma^* \\ -\beta^* \\ \alpha^* \end{pmatrix}.$$

Problem 2.27

a) How does the commutator for the displacement and momentum operators $[\widehat{x}_k, \widehat{p}_\ell]$ transform under time reversal?

b) How does the commutator between two components of the angular momentum operator $[\widehat{J}_k, \widehat{J}_\ell]$ transform under time reversal?

---

a) We recall that $[\widehat{x}_k, \widehat{p}_\ell] = i\hbar\delta_{k\ell}$. One finds that

$$\widehat{\Theta}[\widehat{x}_k, \widehat{p}_\ell]\widehat{\Theta}^{-1} = [\widehat{x}_k, -\widehat{p}_\ell]\widehat{\Theta}\widehat{\Theta}^{-1} = -[\widehat{x}_k, \widehat{p}_\ell],$$

which is consistent with the right-hand side of the commutator,

$$\widehat{\Theta}(i\hbar\delta_{k\ell})\widehat{\Theta}^{-1} = -i\hbar\delta_{k\ell}\widehat{\Theta}\widehat{\Theta}^{-1} = -i\hbar\delta_{k\ell}.$$

b) We recall that $\left[\widehat{J}_k, \widehat{J}_\ell\right] = i\hbar\varepsilon_{k\ell m}\widehat{J}_m$. Thus,

$$\widehat{\Theta}\left[\widehat{J}_k, \widehat{J}_\ell\right]\widehat{\Theta}^{-1} = \left[-\widehat{J}_k, -\widehat{J}_\ell\right]\widehat{\Theta}\widehat{\Theta}^{-1} = \left[\widehat{J}_k, \widehat{J}_\ell\right],$$

which is consistent with the right-hand side of the commutator,

$$\widehat{\Theta}\left(i\hbar\varepsilon_{k\ell m}\widehat{J}_m\right)\widehat{\Theta}^{-1} = -i\hbar\varepsilon_{k\ell m}\left(-\widehat{J}_m\right)\widehat{\Theta}\widehat{\Theta}^{-1} = i\hbar\varepsilon_{k\ell m}\widehat{J}_m.$$

Problem 2.28

Consider the wave function for the quantum mechanical state $|\alpha\rangle$ in momentum representation.

What is the momentum-space wave function for this state under time reversal $\widehat{\Theta}$?

In bra-ket notation, the momentum wave function is $\phi_\alpha(\boldsymbol{p}) = \langle \boldsymbol{p}|\alpha\rangle$. We then write

$$\tilde{\phi}_\alpha(\boldsymbol{p}) = \langle \boldsymbol{p}|\widehat{\Theta}|\alpha\rangle = \langle \boldsymbol{p}|\widehat{\Theta}\int d\boldsymbol{p}_0|\boldsymbol{p}_0\rangle\langle \boldsymbol{p}_0|\alpha\rangle = \langle \boldsymbol{p}|\int d(-\boldsymbol{p}_0)|-\boldsymbol{p}_0\rangle\widehat{\Theta}\phi_\alpha(\boldsymbol{p}_0)$$

$$= -\langle \boldsymbol{p}|\int d\boldsymbol{p}_0|-\boldsymbol{p}_0\rangle\phi_\alpha^*(\boldsymbol{p}_0) = \langle \boldsymbol{p}|\int d\boldsymbol{p}_1|\boldsymbol{p}_1\rangle\phi_\alpha^*(-\boldsymbol{p}_1) = \int d\boldsymbol{p}_1\langle \boldsymbol{p}|\boldsymbol{p}_1\rangle\phi_\alpha^*(-\boldsymbol{p}_1)$$

$$= \phi_\alpha^*(-\boldsymbol{p}),$$

where we have used that $\langle \boldsymbol{p}|\boldsymbol{p}_1\rangle = \delta(\boldsymbol{p}-\boldsymbol{p}_1)$ and some obvious change of variables.

In a similar way, one can obtain the wave function in real space under time reversal. Try it!

Problem 2.29

Let a spin 1/2 particle be subjected to the following potential: $V(\hat{r}) = V_0(\hat{r}) + V_1(\hat{r})(\boldsymbol{\sigma}\cdot\hat{\boldsymbol{L}})$, where $V_0(\hat{r}), V_1(\hat{r})$ are real scalar functions, where r is the distance magnitude $\boldsymbol{\sigma}$ are the Pauli matrices, and $\boldsymbol{L}$ is the three-dimensional angular momentum.

a) Determine if the following quantities are conserved: total energy E, total angular momentum $\hat{\boldsymbol{J}} = \hat{\boldsymbol{L}} + \hat{\boldsymbol{S}}$, and $\hat{J}^2$.

b) Is the underlying Hamiltonian invariant under parity or time reversal?

---

a) The Hamiltonian is time-independent, thus energy E is conserved.

A general operator $\hat{A}$ is conserved, providing $\frac{d\hat{A}}{dt} = \frac{1}{i\hbar}\left[\hat{A},\hat{H}\right] + \frac{\partial\hat{A}}{\partial t} = 0$, where $\hat{H}$ is the Hamiltonian for the system. We note that $\hat{\boldsymbol{J}}$, $\hat{\boldsymbol{L}}$, $\hat{\boldsymbol{S}}$ are time independent, thus $\frac{\partial\hat{\boldsymbol{J}}}{\partial t} = \mathbf{0}$ and $\frac{\partial\hat{J}^2}{\partial t} = 0$. We further examine $[\hat{\boldsymbol{J}},\hat{H}]$ and find that, for all components $i = \{x,y,z\}$:

$$\left[\hat{J}_i, \boldsymbol{\sigma}\cdot\hat{\boldsymbol{L}}\right] = \frac{2}{\hbar}\left[\hat{J}_i, \hat{\boldsymbol{S}}\cdot\hat{\boldsymbol{L}}\right] = \frac{1}{\hbar}\left[\hat{J}_i, \left(\hat{J}^2 - \hat{S}^2 - \hat{L}^2\right)\right] = 0,$$

$$\left[\hat{J}^2, \boldsymbol{\sigma}\cdot\hat{\boldsymbol{L}}\right] = \frac{2}{\hbar}\left[\hat{\boldsymbol{J}}^2, \hat{\boldsymbol{S}}\cdot\hat{\boldsymbol{L}}\right] = \frac{1}{\hbar}\left[\hat{J}^2, \left(\hat{J}^2 - \hat{S}^2 - \hat{L}^2\right)\right] = 0.$$

Therefore, $\hat{\boldsymbol{J}} = \hat{\boldsymbol{L}} + \hat{\boldsymbol{S}}$ and $\hat{J}^2$ are conserved.

b) For the second question, we have

$$\hat{\pi}^{+}V(\hat{r})\hat{\pi} = \hat{\pi}^{+}\left(V_0(\hat{r}) + V_1(\hat{r})(\boldsymbol{\sigma}\cdot\hat{\boldsymbol{L}})\right)\hat{\pi} = V_0(\hat{r}) + V_1(\hat{r})\hat{\pi}^{+}\boldsymbol{\sigma}\cdot\hat{\boldsymbol{L}}\hat{\pi}$$
$$= V_0(\hat{r}) + V_1(\hat{r})(\boldsymbol{\sigma}\cdot\hat{\boldsymbol{L}}) = V(\hat{r}),$$
$$\hat{\Theta}V(\hat{r})\hat{\Theta}^{-1} = \hat{\Theta}\left(V_0(\hat{r}) + V_1(\hat{r})(\boldsymbol{\sigma}\cdot\hat{\boldsymbol{L}})\right)\hat{\Theta}^{-1} = V_0(\hat{r}) + V_1(\hat{r})\hat{\Theta}\boldsymbol{\sigma}\cdot\hat{\boldsymbol{L}}\hat{\Theta}^{-1}$$
$$= V_0(\hat{r}) + V_1(\hat{r})((-\boldsymbol{\sigma})\cdot(-\hat{\boldsymbol{L}})) = V(\hat{r}).$$

Thus, the Hamiltonian is invariant under both parity and time reversal.

> *Food for thought:* On your own, answer the same questions, but for a Hamiltonian whose potential is $V(\hat{r}) = V_0(\hat{r}) + V_1(\hat{r})\,(\boldsymbol{\sigma}\cdot\hat{\boldsymbol{u}}_r)$.

Problem 2.30

Is the Hamiltonian $\hat{H} = \frac{\hat{p}^2}{2m} + \alpha(\hat{x}^4 + \hat{y}^4 + \hat{z}^4)$ invariant under (a) parity; (b) the angular momentum $\hat{\boldsymbol{L}}$; (c) rotation about the x-axis by an angle $\pi/2$?

The Hamiltonian is invariant under a given operator transformation, if their commutator is zero. Thus,

a) $\left[\hat{H},\hat{\pi}\right] = 0$ since $\hat{H}$ has even powers of $\hat{\boldsymbol{p}}$ and $\hat{\boldsymbol{x}}$ operators. Thus, $\hat{H}$ is *invariant* under parity.

b) $\left[\widehat{H},\widehat{L}_i\right]=\left[\frac{\widehat{p}_j^2}{2m},\widehat{L}_i\right]+\left[\widehat{x}^4,\widehat{L}_i\right]+\left[\widehat{y}^4,\widehat{L}_i\right]+\left[\widehat{z}^4,\widehat{L}_i\right]=\left[\widehat{x}^4,\widehat{L}_i\right]+\left[\widehat{y}^4,\widehat{L}_i\right]+\left[\widehat{z}^4,\widehat{L}_i\right]$ since $\left[\frac{\widehat{p}_j^2}{2m},\widehat{L}_i\right]=0.$

By using $\widehat{L}_i=\varepsilon_{ijk}\widehat{r}_j\widehat{p}_k$ and $\widehat{p}_k=-i\hbar\nabla_k$, it is easy to obtain

$$\left[\widehat{r}_n^4,\widehat{L}_i\right]=4i\hbar\,\widehat{r}_j\widehat{r}_n^3\delta_{nk}\varepsilon_{ijk},$$

$$\left[\widehat{x}^4,\widehat{L}_x\right]+\left[\widehat{y}^4,\widehat{L}_x\right]+\left[\widehat{z}^4,\widehat{L}_x\right]=-4i\hbar\widehat{y}\,\widehat{z}(\widehat{y}^2-\widehat{z}^2),$$

$$\left[\widehat{x}^4,\widehat{L}_y\right]+\left[\widehat{y}^4,\widehat{L}_y\right]+\left[\widehat{z}^4,\widehat{L}_y\right]=4i\hbar\widehat{x}\,\widehat{z}\left(\widehat{x}^2-\widehat{z}^2\right),$$

$$\left[\widehat{x}^4,\widehat{L}_z\right]+\left[\widehat{y}^4,\widehat{L}_z\right]+\left[\widehat{z}^4,\widehat{L}_z\right]=-4i\hbar\widehat{x}\,\widehat{y}(\widehat{x}^2-\widehat{y}^2).$$

Therefore, the Hamiltonian is *not invariant* under the $\widehat{L}_i$ operation.

c) The rotation operator about the x-axis by an angle β can be written as $\widehat{D}_x(\beta)=e^{-\frac{i\widehat{J}_x\beta}{\hbar}}$. We have to determine $\left[\widehat{H},\widehat{D}_x\left(\frac{\pi}{2}\right)\right]=0$ or equivalently $\widehat{D}_x^+\left(\frac{\pi}{2}\right)\widehat{H}\widehat{D}_x\left(\frac{\pi}{2}\right)=\widehat{H}$. Thus,

$$\widehat{D}_x^+\left(\frac{\pi}{2}\right)\left(\frac{\widehat{p}_x^2}{2m}+\frac{\widehat{p}_y^2}{2m}+\frac{\widehat{p}_z^2}{2m}+\alpha\widehat{x}^4+\alpha\widehat{y}^4+\alpha\widehat{z}^4\right)\widehat{D}_x\left(\frac{\pi}{2}\right)$$

$$=\frac{\widehat{p}_x^2}{2m}+\frac{\widehat{p}_z^2}{2m}+\frac{\left(-\widehat{p}_y\right)^2}{2m}+\alpha\widehat{x}^4+\alpha\widehat{z}^4+\alpha(-\widehat{y})^4=\widehat{H}.$$

In the preceding equation, we have used the fact that the specified rotation is reflection about the x-axis, such that $x\to x$; $y\to z$; $z\to -y$; $p_x\to p_x$; $p_y\to p_z$; $p_z\to -p_y$. Thus, $\widehat{H}$ is invariant under the specified rotation.

Food for thought: You can repeat the same problem by considering a different type of rotation or giving a different type of Hamiltonian, for example $\widehat{H}=\frac{\widehat{p}^2}{2m}+\alpha\widehat{r}^4$, where $\widehat{r}=(\widehat{x},\widehat{y},\widehat{z})$.

Problem 2.31

Consider a Hamiltonian that is invariant under time reversal, that is $[\widehat{H},\widehat{\Theta}]=0$. Suppose that $\widehat{\Theta}^2=-\widehat{1}$. Show that in this case, the eigenstates of the Hamiltonian are at least double-degenerate. Such states are said to have Kramer's degeneracy.

The eigenstates and energies of the Hamiltonian are denoted as $\widehat{H}|\psi_n\rangle=E_n|\psi_n\rangle$.

Since $\widehat{H}$ and $\widehat{\Theta}$ commute, then $\widehat{H}\left(\widehat{\Theta}|\psi_n\rangle\right)=E_n$, thus $|\psi_n\rangle$ and $\left(\widehat{\Theta}|\psi_n\rangle\right)$ share the same eigenvalue.

For $|\psi_n\rangle$ and $\left(\widehat{\Theta}|\psi_n\rangle\right)$ to be eigenstates, one also needs to show that they are orthogonal. We see that, using that $\widehat{\Theta}$ is an antiunitary operator:

$$\langle\psi_n|(\widehat{\Theta}\psi_n)\rangle = \langle(\widehat{\Theta}\psi_n)|(\widehat{\Theta}^2\psi_n)\rangle^* = -\langle(\widehat{\Theta}\psi_n)|\psi_n\rangle^* = -\langle\psi_n|(\widehat{\Theta}\psi_n)\rangle \Rightarrow \langle\psi_n|(\widehat{\Theta}\psi_n)\rangle = 0.$$

Therefore, when $[\widehat{H},\widehat{\Theta}] = 0$ and $\widehat{\Theta}^2 = -\widehat{1}$ the eigenvalues of $\widehat{H}$ are doubly degenerate.

Food for thought: What kind of particles does the time-reversal outcome $\widehat{\Theta}^2 = -\widehat{1}$ correspond to? Is there Kramer's degeneracy for the case of $\widehat{\Theta}^2 = \widehat{1}$?

Problem 2.32

Construct the explicit form of the time-reversal operator for a spinless particle.

The key point here is to realize that since the particle has no spin, the relevant operators are the displacement and momentum operators, $\widehat{x}$ and $\widehat{p}$.

Taking the definition $\widehat{\Theta} = \widehat{U}\widehat{K}$, we find that since the components of the position operator are real,

$$\widehat{r} = \widehat{\Theta}\widehat{r}\widehat{\Theta}^{-1} = \widehat{U}\widehat{K}\widehat{r}\widehat{K}^{-1}\widehat{U}^{-1} = \widehat{U}\widehat{r}\widehat{U} \to \widehat{U} = e^{i\lambda}\widehat{I},$$

where $\widehat{I}$ is the identity matrix, λ is a real parameter, and we have used the fact that $\widehat{U} = \widehat{U}^{-1}$. On the other hand, for the components of the momentum operator,

$$\widehat{p} = \widehat{\Theta}\widehat{p}\widehat{\Theta}^{-1} = \widehat{U}\widehat{K}\widehat{p}\widehat{K}^{-1}\widehat{U}^{-1} = e^{i\lambda}\widehat{I}\widehat{K}\widehat{p}\widehat{K}^{-1}e^{-i\lambda}\widehat{I} = e^{i\lambda}\widehat{I}(-\widehat{p})e^{-i\lambda}\widehat{I} = -\widehat{p}.$$

By choosing $\lambda = 0$, the time-reversal operator for a spinless particle is

$$\widehat{\Theta} = \widehat{I}\widehat{K}.$$

By choosing $\eta = e^{i\lambda}$, the time-reversal operator for a spinless particle is

$$\widehat{\Theta} = \eta\widehat{I}\widehat{K}.$$

Problem 2.33

a) Consider a spin $1/2$ particle in the presence of an *external* electric field $\boldsymbol{E}_{ext}(\boldsymbol{x})$. Is there Kramer's degeneracy for this particle?
b) Consider a spin $1/2$ particle in the presence of an *external* magnetic field $\boldsymbol{B}_{ext}(\boldsymbol{x})$. Is there Kramer's degeneracy for this particle?
c) Consider a spin $1/2$ particle in the presence of *internal* electric field $\boldsymbol{E}$ and magnetic field $\boldsymbol{B}$. Is there Kramer's degeneracy for this particle?

---

a) In the presence of an external electric field, the Hamiltonian of the particle is

$$\widehat{H} = \frac{\widehat{p}^2}{2m} - e\phi_{ext}(\widehat{r}),$$

where $V_{ext}(\widehat{r}) = e\phi_{ext}(\widehat{r})$ is the potential associated with the field $\boldsymbol{E}_{ext}(\boldsymbol{x}) = -\nabla V_{ext}(\boldsymbol{x})$. After the time-reversal operation, we find that

$$\widehat{H}_\Theta = \widehat{\Theta}\widehat{H}\widehat{\Theta}^{-1} = \frac{(-\widehat{p})^2}{2m} - e\phi_{ext}(\widehat{r}) = \widehat{H},$$

where we have used the fact that the external field $\phi_{ext}(\hat{r})$ is not affected by $\hat{\Theta}$. Since $[\hat{H},\hat{\Theta}] = 0$ for the preceding Hamiltonian, we conclude that there is at least twofold degeneracy of the eigenstates of $\hat{H}$, as we have explicitly shown in an earlier problem.

b) In the presence of an external magnetic field, the Hamiltonian of the particle is

$$\hat{H} = \frac{(\hat{\boldsymbol{p}} - e\boldsymbol{A}_{ext}(\hat{\boldsymbol{r}}))^2}{2m} - \gamma\hat{\boldsymbol{S}}\cdot\boldsymbol{B}_{ext}(\hat{\boldsymbol{r}}),$$

where $\boldsymbol{A}_{ext}(\hat{\boldsymbol{r}})$ is the vector potential ($\boldsymbol{B}_{ext}(\hat{\boldsymbol{r}}) = \nabla\times\boldsymbol{A}_{ext}(\boldsymbol{x})$) and γ is the gyromagnetic coefficient. The transformed Hamiltonian becomes

$$\hat{H}_\Theta = \hat{\Theta}\hat{H}\hat{\Theta}^{-1} = \frac{(-\hat{\boldsymbol{p}} - e\boldsymbol{A}_{ext}(\hat{\boldsymbol{r}}))^2}{2m} - \gamma(-\hat{\boldsymbol{S}})\cdot\boldsymbol{B}_{ext}(\hat{\boldsymbol{r}}) \neq \hat{H}.$$

In the preceding equation, we have used that $\hat{\Theta}\hat{\boldsymbol{p}}\hat{\Theta}^{-1} = -\hat{\boldsymbol{p}}$, $\hat{\Theta}\hat{\mathrm{S}}\hat{\Theta}^{-1} = -\hat{\mathrm{S}}$, and the fact that the external field and associated vector potential are not affected by the $\hat{\Theta}$ operation. Therefore, the preceding Hamiltonian does not commute with $\hat{\Theta}$, meaning that the *external* magnetic field breaks the time-reversal symmetry and the Kramer's degeneracy is lifted.

c) The internal electric and magnetic fields are the electromagnetic fields created by the dynamical degrees of freedom of our system,

$$\hat{H} = \frac{\left(\hat{\boldsymbol{p}} - e\hat{\boldsymbol{A}}\right)^2}{2m} - e\hat{\phi} - \gamma\hat{\boldsymbol{S}}\cdot\hat{\boldsymbol{B}}.$$

In this case, we have to consider the properties of fields and operators under time reversal: $\hat{\Theta}\hat{\boldsymbol{A}}\hat{\Theta}^{-1} = -\hat{\boldsymbol{A}}$; $\hat{\Theta}\hat{\phi}\hat{\Theta}^{-1} = \hat{\phi}$; $\hat{\Theta}\hat{\boldsymbol{E}}\hat{\Theta}^{-1} = \hat{\boldsymbol{E}}$; $\hat{\Theta}\hat{\boldsymbol{B}}\hat{\Theta}^{-1} = -\hat{\boldsymbol{B}}$; $\hat{\Theta}\hat{\mathrm{S}}\hat{\Theta}^{-1} = -\hat{\mathrm{S}}$. After the time-reversal operation, we find that

$$\hat{H}_\Theta = \hat{\Theta}\hat{H}\hat{\Theta}^{-1} = \frac{\left((-\hat{\boldsymbol{p}}) - e(-\hat{\boldsymbol{A}})\right)^2}{2m} - e(+\hat{\phi}) - \gamma(-\hat{\boldsymbol{S}})\cdot(-\hat{\boldsymbol{B}}) = \hat{H}.$$

Thus, $[\hat{H},\hat{\Theta}] = 0$, and we conclude that there is at least a twofold degeneracy of the eigenstates of $\hat{H}$ in the presence of *internal* electromagnetic fields.

Problem 2.34

Consider Bloch electrons in a periodic system. What is the condition for the Hamiltonian of Bloch electrons to remain invariant under the time-reversal operation?

What are the Bloch electron Kramer's partners for this periodic system?

The time-reversal operation connects Kramer's pairs and it can be given in a matrix form whose components are expressed in terms of matrix elements between various Bloch states $w_{ns,ms'} = \langle u_{ns,\boldsymbol{k}}|\hat{\Theta}|u_{ms',-\boldsymbol{k}}\rangle$. Write explicitly the matrix for $\hat{\Theta}$ in this Bloch state representation (n, m are state indices; s, s' are the spin projections of the Bloch states, $\boldsymbol{k}$ is the wave vector).

This problem probes our knowledge of the Bloch theorem for periodic systems and basic properties of the time-reversal operator.

We remember that in a periodic system, the eigenvalue equation for the Hamiltonian is

$$\widehat{H}(\boldsymbol{k})|\psi_{n\boldsymbol{k}}\rangle = E_n(\boldsymbol{k})|\psi_{n\boldsymbol{k}}\rangle; \quad |\psi_{n\boldsymbol{k}}\rangle = e^{i\boldsymbol{k}\cdot\boldsymbol{x}}|u_{n\boldsymbol{k}}\rangle,$$

where $|\psi_{n\boldsymbol{k}}\rangle$ are the eigenfunctions for the wave vector $\boldsymbol{k}$ spanning the first Brillouin zone. The Bloch states $|u_{n\boldsymbol{k}}\rangle$ have the same periodicity as the lattice and the eigenfunctions of the Bloch Hamiltonian.

To find the condition for $\widehat{H}$ invariance under $\widehat{\Theta}$, we consider $\widehat{H}_\Theta(\boldsymbol{k}) = \widehat{\Theta}\widehat{H}(\boldsymbol{k})\widehat{\Theta}^{-1}$ in the continuous $\boldsymbol{k}$-representation,

$$\begin{aligned}\widehat{\Theta}\widehat{H}(\boldsymbol{k})\widehat{\Theta}^{-1} &= \widehat{\Theta}\int_{BZ} d\boldsymbol{k}|\boldsymbol{k}\rangle\widehat{H}(\boldsymbol{k})\langle\boldsymbol{k}|\widehat{\Theta}^{-1} \\ &= \int_{BZ} d(-\boldsymbol{k})|-\boldsymbol{k}\rangle\widehat{\Theta}\widehat{H}(\boldsymbol{k})\widehat{\Theta}^{-1}\langle-\boldsymbol{k}| = \int_{BZ} d\boldsymbol{k}|\boldsymbol{k}\rangle\widehat{\Theta}\widehat{H}(-\boldsymbol{k})\widehat{\Theta}^{-1}\langle\boldsymbol{k}|.\end{aligned}$$

Therefore, the Bloch Hamiltonian is invariant under time reversal, when

$$\widehat{H}(\boldsymbol{k}) = \widehat{\Theta}\widehat{H}(-\boldsymbol{k})\widehat{\Theta}^{-1} = \widehat{U}\widehat{H}^*(-\boldsymbol{k})\widehat{U}^{-1}.$$

To give an explicit representation of the time-reversal operator in Bloch state representation, we recall that the Kramer's partners are $|u_{n\uparrow,\boldsymbol{k}}\rangle$ and $|u_{n\downarrow,-\boldsymbol{k}}\rangle$, for which

$$\Theta|u_{n\uparrow,\boldsymbol{k}}\rangle = e^{i\xi_n(\boldsymbol{k})}|u_{n\downarrow,-\boldsymbol{k}}\rangle \text{ and } \widehat{\Theta}|u_{n\downarrow,\boldsymbol{k}}\rangle = -e^{i\xi_n(-\boldsymbol{k})}|u_{n\uparrow,-\boldsymbol{k}}\rangle.$$

The phase $\xi_n(\boldsymbol{k})$ appears since there is no strict one-to-one correspondence between the Kramer's partners. Therefore, we have

$$\begin{aligned}\widehat{\Theta}|u_{n,s,\boldsymbol{k}}\rangle &= -sign(s)\; e^{i\xi_n(-sign(s)\boldsymbol{k})}|u_{n,-s,-\boldsymbol{k}}\rangle, \\ \langle u_{n,s,\boldsymbol{k}}|\widehat{\Theta}|u_{n',s',\boldsymbol{k}'}\rangle &= \langle u_{n,s,\boldsymbol{k}}| - sign(s')\; e^{i\xi_{n'}(-sign(s')\boldsymbol{k}')}|u_{n',-s',-\boldsymbol{k}'}\rangle \\ &= -sign(s')\; e^{i\xi_{n'}(-sign(s')\boldsymbol{k}')}\langle u_{n,s,\boldsymbol{k}}|u_{n',-s',-\boldsymbol{k}'}\rangle \\ &= -sign(s')\; e^{i\xi_{n'}(-sign(s')\boldsymbol{k}')}\delta_{n,n'}\delta_{s,-s'}\delta(\boldsymbol{k}+\boldsymbol{k}').\end{aligned}$$

The Bloch state representation for the time-reversal operator can be obtained simply by constructing the following matrix:

$$\widehat{\Theta} = \begin{pmatrix} \langle u_{1s,\boldsymbol{k}}|\Theta|u_{1s',\boldsymbol{k}'}\rangle & \langle u_{1s,\boldsymbol{k}}|\Theta|u_{2s',\boldsymbol{k}'}\rangle & \langle u_{1s,\boldsymbol{k}}|\Theta|u_{3s',\boldsymbol{k}'}\rangle & \cdots \\ \langle u_{2s,\boldsymbol{k}}|\Theta|u_{1s',\boldsymbol{k}'}\rangle & \langle u_{2s,\boldsymbol{k}}|\Theta|u_{2s',\boldsymbol{k}'}\rangle & \langle u_{2s,\boldsymbol{k}}|\Theta|u_{3s',\boldsymbol{k}'}\rangle & \cdots \\ \vdots & \vdots & \vdots & \ddots \end{pmatrix}$$

$$= \begin{pmatrix} 0 & e^{i\xi_1(\boldsymbol{k}')}\delta(\boldsymbol{k}+\boldsymbol{k}') & 0 & 0 & \cdots \\ -e^{i\xi_1(-\boldsymbol{k}')}\delta(\boldsymbol{k}+\boldsymbol{k}') & 0 & 0 & 0 & \cdots \\ 0 & 0 & 0 & e^{i\xi_2(\boldsymbol{k}')}\delta(\boldsymbol{k}+\boldsymbol{k}') & \cdots \\ 0 & 0 & -e^{i\xi_2(-\boldsymbol{k}')}\delta(\boldsymbol{k}+\boldsymbol{k}') & 0 & \cdots \\ \vdots & \vdots & \vdots & \vdots & \ddots \end{pmatrix}.$$

Problem 2.35

Consider the Bloch Hamiltonian and its eigenstates $\widehat{H}(\boldsymbol{k})|\psi_{n\boldsymbol{k}}\rangle = E_n(\boldsymbol{k})|\psi_{n\boldsymbol{k}}\rangle$; $|\psi_{n\boldsymbol{k}}\rangle = e^{i\boldsymbol{k}\cdot\boldsymbol{x}}|u_{n\boldsymbol{k}}\rangle$ for each band index n, where $|u_{n\boldsymbol{k}}\rangle$ are the periodic Bloch states. If the

Hamiltonian is invariant under time reversal, then find the eigenenergies and eigenstates of $\widehat{H}(-\boldsymbol{k})$.

As found in previous problems, for a time-reversal Hamiltonian, $\widehat{H}(\boldsymbol{k}) = \widehat{H}_\Theta(\boldsymbol{k}) = \widehat{\Theta}\widehat{H}(-\boldsymbol{k})\widehat{\Theta}^{-1}$. Then we have two equivalent expressions:

$$\widehat{H}(\boldsymbol{k})\,|\psi_{n\boldsymbol{k}}\rangle = \widehat{\Theta}\widehat{H}(-\boldsymbol{k})\widehat{\Theta}^{-1}\,|\psi_{n\boldsymbol{k}}\rangle \text{ and } \widehat{H}(\boldsymbol{k})\,|\psi_{n\boldsymbol{k}}\rangle = E_n(\boldsymbol{k})|\psi_{n\boldsymbol{k}}\rangle.$$

Thus,

$$\widehat{\Theta}\widehat{H}(-\boldsymbol{k})\widehat{\Theta}^{-1}\,|\psi_{n\boldsymbol{k}}\rangle = E_n(\boldsymbol{k})\,|\psi_{n\boldsymbol{k}}\rangle.$$

From here, we multiply both sides by $\widehat{\Theta}^{-1}$ and find that

$$\widehat{\Theta}^{-1}\widehat{\Theta}\widehat{H}(-\boldsymbol{k})\widehat{\Theta}^{-1}\,|\psi_{n\boldsymbol{k}}\rangle = \widehat{\Theta}^{-1}E_n(\boldsymbol{k})\,|\psi_{n\boldsymbol{k}}\rangle,$$
$$\widehat{H}(-\boldsymbol{k})\widehat{\Theta}^{-1}|\psi_{n\boldsymbol{k}}\rangle = E_n(\boldsymbol{k})\widehat{\Theta}^{-1}|\psi_{n\boldsymbol{k}}\rangle = E_n(\boldsymbol{k})\widehat{U}^{-1}|\psi_{n\boldsymbol{k}}\rangle^*.$$

Thus, the eigenenergy of $\widehat{H}(-\boldsymbol{k})$ is $E_n(\boldsymbol{k})$ and the corresponding eigenstate is $\widehat{U}^{-1}|\psi_{n\boldsymbol{k}}\rangle^*$.

Problem 2.36
Consider the case of electrons scattering from a potential $\widehat{V}$ that is invariant under the time-reversal operation. One example of such a potential is scattering from nonmagnetic impurities in a material. Can this potential elastically scatter the quantum mechanical state $|\alpha\rangle$ into its time-reversed state $\Theta|\alpha\rangle$?

The first thing to realize is what is really being asked here. We need to determine if $\alpha \xrightarrow{V} \widehat{\Theta}|\alpha\rangle$ is possible; in other words, we need to evaluate the matrix element

$$\langle\widehat{\Theta}\alpha|\widehat{V}|\alpha\rangle = ?.$$

Let us remember that for the (antiunitary) time-reversal operator, we have

$$\langle\widehat{\Theta}\psi|\widehat{\Theta}\phi\rangle = \langle\phi|\psi\rangle.$$

Thus, for $\langle\widehat{\Theta}\alpha|\widehat{V}|\alpha\rangle = \langle\widehat{\Theta}\alpha|\widehat{V}\alpha\rangle$, we find

$$\langle\widehat{\Theta}\alpha|\widehat{V}\alpha\rangle = \langle\widehat{\Theta}\widehat{V}\alpha|\widehat{\Theta}^2\alpha\rangle = -\langle\widehat{\Theta}\widehat{V}\alpha|\alpha\rangle,$$
$$\langle\widehat{\Theta}\alpha|(\widehat{V}\alpha)\rangle = \langle\widehat{\Theta}^2\alpha|(\widehat{\Theta}\widehat{V}\alpha)\rangle = \widehat{\Theta}^2\langle\alpha|(\widehat{\Theta}\widehat{V}\widehat{\Theta}^{-1}\widehat{\Theta}\alpha)\rangle$$
$$= -\left\langle\widehat{V}\widehat{\Theta}\alpha|\alpha\right\rangle = -\left\langle\widehat{\Theta}\alpha|\widehat{V}\alpha\right\rangle$$

where we have used that $\widehat{V}$ is a linear unitary operator invariant under time reversal $\widehat{\Theta}\widehat{V}\widehat{\Theta}^{-1} = \widehat{V}$, and that for electrons, $\widehat{\Theta}^2 = -\widehat{1}$. This means that

$$\langle\widehat{\Theta}\alpha|\widehat{V}|\alpha\rangle = 0.$$

Food for thought: The scattering of electrons from a given state to its time-reversal partner due to a potential that is invariant under time reversal cannot happen. This conclusion has important consequences for scattering involving topological materials, as recently considered by Xu and Moore (2006).

Problem 2.37

An electronic particle is subject to the following Hamiltonian:

$$\hat{H}(\boldsymbol{k}) = 2t_1\cos(ka)\sigma_3 + [t_2 - 2t_3\sin(ka)]\,\sigma_1,$$

where t_1, t_2, t_3 are real constants, $\boldsymbol{k}$ is a wave vector, a is a lattice constant, and σ_i are the Pauli matrices.

a) How does this Hamiltonian behave under time-reversal symmetry?

b) Compare the eigenvalues of the original Hamiltonian and its time-reversal counterpart.

---

a) We begin by remembering that the time-reversal operator for an electron is

$$\hat{\Theta} = \hat{U}\hat{K} = i\eta\sigma_2\hat{K},$$

where the unitary operation is $\hat{U} = i\sigma_2$ and $\hat{K}$ denotes complex conjugation. Therefore,

$$\begin{aligned}\hat{H}_\Theta(\boldsymbol{k}) &= \hat{\Theta}\hat{H}(\boldsymbol{k})\Theta^{-1} = 2t_1\cos(-ka)\hat{\Theta}\sigma_3\hat{\Theta}^{-1} + [t_2 - 2t_3\sin(-ka)]\,\hat{\Theta}\sigma_1\hat{\Theta}^{-1}\\ &= 2t_1\cos(ka)i\sigma_2\sigma_3 i\sigma_2^* + [t_2 + 2t_3\sin(ka)]\,i\sigma_2\sigma_1 i\sigma_2^*\\ &= -2t_1\cos(ka)\sigma_3 - [t_2 + 2t_3\sin(ka)]\,\sigma_1.\end{aligned}$$

The Hamiltonian is not invariant under time reversal.

b) Let's then compare the eigenvalues. For the original Hamiltonian, we have

$$\hat{H} = \begin{pmatrix} 2t_1\cos(ka) & t_2 - 2t_3\sin(ka) \\ t_2 - 2t_3\sin(ka) & -2t_1\cos(ka) \end{pmatrix}.$$

The eigenvalues are $E_{12}(k) = \pm\sqrt{(2t_1\cos(ka))^2 + (t_2 - 2t_3\sin(ka))^2}$.

The transformed Hamiltonian is

$$\hat{H}_\Theta = \begin{pmatrix} -2t_1\cos(ka) & -t_2 - 2t_3\sin(ka) \\ -t_2 - 2t_3\sin(ka) & 2t_1\cos(ka) \end{pmatrix}$$

and the eigenvalues are $E_{12}^{\Theta}(k) = \pm\sqrt{(2t_1\cos(ka))^2 + (t_2 + 2t_3\sin(ka))^2}$.

Problem 2.38

In physics, problems related to particle-hole symmetry occur often. For this purpose, one defines a particle-hole operator, which is antiunitary (similar to the time-reversal operation). More specifically, the particle-hole operation can be given as the tensor product between

$$\hat{C} = \sigma_2 \otimes \sigma_1\hat{K},$$

where the unitary part of the operation is a tensor product of Pauli matrices

$$\widehat{U}_C = \sigma_2 \otimes \sigma_1 = \begin{pmatrix} 0 & -i \\ i & 0 \end{pmatrix} \otimes \begin{pmatrix} 0 & 1 \\ 1 & 0 \end{pmatrix} = \begin{pmatrix} 0\begin{pmatrix} 0 & 1 \\ 1 & 0 \end{pmatrix} & -i\begin{pmatrix} 0 & 1 \\ 1 & 0 \end{pmatrix} \\ i\begin{pmatrix} 0 & 1 \\ 1 & 0 \end{pmatrix} & 0\begin{pmatrix} 0 & 1 \\ 1 & 0 \end{pmatrix} \end{pmatrix}$$

$$= \begin{pmatrix} 0 & 0 & 0 & -i \\ 0 & 0 & -i & 0 \\ 0 & i & 0 & 0 \\ i & 0 & 0 & 0 \end{pmatrix},$$

and $\widehat{K}$ is a complex conjugation operation. Verify that $\widehat{C}\widehat{C} = -1$.

Write an equivalent expression for a transformed Hamiltonian in reciprocal space defined by the wave vector $\boldsymbol{k}$, $\widehat{H}_C(\boldsymbol{k}) = \widehat{C}\widehat{H}(\boldsymbol{k})\widehat{C}^{-1}$, and discuss the implications for particle-hole symmetry.

We start with

$$\begin{aligned}\widehat{C}\widehat{C} &= \widehat{U}_C\widehat{K}\widehat{U}_C\widehat{K} = \sigma_2\otimes\sigma_1\widehat{K}\sigma_2\otimes\sigma_1\widehat{K} = (\sigma_2\otimes\sigma_1)(\sigma_2^*\otimes\sigma_1)\widehat{K}^2 \\ &= (\sigma_2\otimes\sigma_1)(\sigma_2^*\otimes\sigma_1) = ([\sigma_2\cdot\sigma_2^*]\otimes[\sigma_1\cdot\sigma_1]) = ([-\sigma_0]\otimes\sigma_0) \\ &= -(\sigma_0\otimes\sigma_0) = -1.\end{aligned}$$

Let us then consider the second question,

$$\widehat{H}_C(\boldsymbol{k}) = \widehat{C}\widehat{H}(\boldsymbol{k})\widehat{C}^{-1} = \widehat{U}_C\widehat{K}\widehat{H}(\boldsymbol{k})\widehat{K}^{-1}\widehat{U}_C^{-1} = \widehat{U}_C\widehat{H}^*(-\boldsymbol{k})\widehat{U}_C^{-1}.$$

Thus, the Hamiltonian that has particle-hole symmetry must satisfy $H(\boldsymbol{k}) = U_C H^*(-\boldsymbol{k})U_C^{-1}$.

Problem 2.39

In physics we often have to deal with problems related to chiral symmetry. For this purpose, one defines a chiral operator, which is another antiunitary operation. It is defined as a product between the time-reversal and particle-hole operators,

$$\widehat{S} = \widehat{T}\widehat{C} = \widehat{U}_T\widehat{K}\widehat{U}_C\widehat{K} = \widehat{U}_T\widehat{U}_C^*,$$

where $\widehat{U}_T = i\sigma_0\otimes\sigma_2$ and $\widehat{U}_C = \sigma_2\otimes\sigma_1$.

Show that $\widehat{S}\widehat{S} = \widehat{1}$. Write an equivalent expression for a transformed Hamiltonian in reciprocal space defined by the wave vector $\boldsymbol{k}$, $\widehat{H}_S(\boldsymbol{k}) = \widehat{S}\widehat{H}(\boldsymbol{k})\widehat{S}^{-1}$, and discuss the implications for chiral symmetry.

We start with

$$\widehat{S} = \widehat{U}_T\widehat{U}_C^* = i(\sigma_0\otimes\sigma_2)\cdot(\sigma_2^*\otimes\sigma_1) = i([\sigma_0\cdot\sigma_2^*]\otimes[\sigma_2\cdot\sigma_1]) = i(\sigma_2^*\otimes[-i\sigma_3]) = \sigma_2^*\otimes\sigma_3,$$

$$\widehat{S}\widehat{S} = \widehat{U}_T\widehat{U}_C^*\widehat{U}_T\widehat{U}_C^* = (\sigma_2^*\otimes\sigma_3)\cdot(\sigma_2^*\otimes\sigma_3) = [\sigma_2^*\cdot\sigma_2^*]\otimes[\sigma_3\cdot\sigma_3] = \sigma_0\otimes\sigma_0 = \widehat{1}.$$

Additionally,

$$\begin{aligned}\widehat{H}_S(\boldsymbol{k}) &= \widehat{S}\widehat{H}(\boldsymbol{k})\widehat{S}^{-1} = (\widehat{U}_T\widehat{U}_C^*)\widehat{H}(\boldsymbol{k})(\widehat{U}_T\widehat{U}_C^*)^{-1} \\ &= (\sigma_2^*\otimes\sigma_3)\,\widehat{H}(\boldsymbol{k})\,(\sigma_2^*\otimes\sigma_3)^{-1} = (\sigma_2^*\otimes\sigma_3)\,\widehat{H}(\boldsymbol{k})\,(\sigma_2^*\otimes\sigma_3) \\ &= (\sigma_2^*\otimes\sigma_3)\cdot(\sigma_2^*\otimes\sigma_3)\,\widehat{H}(\boldsymbol{k}) = \widehat{H}(\boldsymbol{k}).\end{aligned}$$

Therefore, there is chiral symmetry if $\widehat{H}_S(\boldsymbol{k}) = \widehat{S}\widehat{H}(\boldsymbol{k})\widehat{S}^{-1} = +\widehat{H}(\boldsymbol{k})$.

Problem 2.40

Is the following Hamiltonian invariant under time-reversal, particle-hole, or chiral symmetries?

$$\widehat{H}(\boldsymbol{k}) = \begin{pmatrix} t_1\cos(ka) & i\lambda\sin(ka) & t_2 - t_3\sin(ka) & -i\lambda \\ -i\lambda\sin(ka) & t_1\cos(ka) & -i\lambda & t_2 + t_3\sin(ka) \\ t_2 - t_3\sin(ka) & i\lambda & -t_1\cos(ka) & i\lambda\sin(ka) \\ i\lambda & t_2 + t_3\sin(ka) & -i\lambda\sin(ka) & -t_1\cos(ka) \end{pmatrix},$$

where $\boldsymbol{k}$ is a 1D wave vector and t_1, t_2, t_3, λ are real parameters.

The conditions for a Hamiltonian in reciprocal space to be invariant under time-reversal $\widehat{\Theta}$ chiral $\widehat{C}$ and particle-hole $\widehat{S}$ symmetries are

$$\begin{aligned}\widehat{H}_\Theta(\boldsymbol{k}) &= \widehat{\Theta}\widehat{H}(\boldsymbol{k})\widehat{\Theta}^{-1} = \widehat{U}_T(\widehat{K}\widehat{H}(\boldsymbol{k})\widehat{K}^{-1})\widehat{U}_T^{-1} = \widehat{U}_T\widehat{H}^*(-\boldsymbol{k})\widehat{U}_T^{-1}, \\ \widehat{H}_C(\boldsymbol{k}) &= \widehat{C}\widehat{H}(\boldsymbol{k})\widehat{C}^{-1} = \widehat{U}_C(\widehat{K}\widehat{H}(\boldsymbol{k})\widehat{K}^{-1})\widehat{U}_C^{-1} = \widehat{U}_C\widehat{H}^*(-\boldsymbol{k})\widehat{U}_C^{-1}, \\ \widehat{H}_S(\boldsymbol{k}) &= \widehat{S}\widehat{H}(\boldsymbol{k})\widehat{S}^{-1} = \widehat{U}_T\widehat{U}_C^*\widehat{H}(\boldsymbol{k})\widehat{U}_C^{*-1}\widehat{U}_C^{-1}.\end{aligned}$$

Thus, we find

$$\begin{aligned}\widehat{H}^*(-\boldsymbol{k}) &= \begin{pmatrix} t_1\cos(-ka) & -i\lambda\sin(-ka) & t_2 - t_3\sin(-ka) & i\lambda \\ i\lambda\sin(-ka) & t_1\cos(-ka) & i\lambda & t_2 + t_3\sin(-ka) \\ t_2 - t_3\sin(-ka) & -i\lambda & -t_1\cos(-ka) & -i\lambda\sin(-ka) \\ -i\lambda & t_2 + t_3\sin(-ka) & i\lambda\sin(-ka) & -t_1\cos(-ka) \end{pmatrix} \\ &= \begin{pmatrix} t_1\cos(ka) & i\lambda\sin(ka) & t_2 + t_3\sin(ka) & i\lambda \\ -i\lambda\sin(ka) & t_1\cos(ka) & i\lambda & t_2 - t_3\sin(ka) \\ t_2 + t_3\sin(ka) & -i\lambda & -t_1\cos(ka) & i\lambda\sin(ka) \\ -i\lambda & t_2 - t_3\sin(ka) & -i\lambda\sin(ka) & -t_1\cos(ka) \end{pmatrix}.\end{aligned}$$

We also use the following:

$$\widehat{U}_T = i\sigma_0\otimes\sigma_2 = \begin{pmatrix} 0 & 1 & 0 & 0 \\ -1 & 0 & 0 & 0 \\ 0 & 0 & 0 & 1 \\ 0 & 0 & -1 & 0 \end{pmatrix} = -\widehat{U}_T^{-1},$$

$$\widehat{U}_C = \sigma_2 \otimes \sigma_1 = \begin{pmatrix} 0 & 0 & 0 & -i \\ 0 & 0 & -i & 0 \\ 0 & i & 0 & 0 \\ i & 0 & 0 & 0 \end{pmatrix} = -\widehat{U}_C^{-1},$$

$$\widehat{S} = \sigma_2^* \otimes \sigma_3 = \begin{pmatrix} 0 & 0 & i & 0 \\ 0 & 0 & 0 & -i \\ -i & 0 & 0 & 0 \\ 0 & i & 0 & 0 \end{pmatrix} = \widehat{S}^{-1}.$$

Then, we obtain

$$\begin{aligned} \widehat{H}_\Theta(\boldsymbol{k}) &= \widehat{U}_T \widehat{H}^*(-\boldsymbol{k}) \widehat{U}_T^{-1} = \widehat{H}(\boldsymbol{k}), \\ \widehat{H}_C(\boldsymbol{k}) &= \widehat{U}_C \widehat{H}^*(-\boldsymbol{k}) \widehat{U}_C^{-1} = -\widehat{H}(\boldsymbol{k}), \\ \widehat{H}_S(\boldsymbol{k}) &= \widehat{S} \widehat{H}(\boldsymbol{k}) \widehat{S}^{-1} = -\widehat{H}(\boldsymbol{k}). \end{aligned}$$

We conclude that $\widehat{H}(\boldsymbol{k})$ is invariant under time-reversal symmetry. However, $\widehat{H}(\boldsymbol{k})$ is not invariant under chiral and particle-hole symmetries.

3 Geometrical Phases

The appearance of a geometrical phase due to the evolution of a given system under a time-dependent potential can be understood by realizing that in many problems the time-dependent Hamiltonian can be parametrized as $\widehat{H}(t) = \widehat{H}(\xi(t))$, where $\xi(t) = (\xi_1(t), \xi_2(t), \ldots, \xi_l(t))$ is a time-dependent n-dimensional vector. Connecting with the adiabatic approximation, the parameter $\xi(t)$ is a slow variable, while the fast variables are transitions associated with the electronic states. Interestingly, the parametrization in terms of $(\xi_1(t), \xi_2(t), \ldots, \xi_l(t))$ may not be unique.

The properties regarding the geometrical phases are associated with the eigenvalue problem $\widehat{H}(\xi(t))\,|\psi_n(\xi(t))\rangle = E_n(\xi(t))\,|\psi_n(\xi(t))\rangle$, where the eigenstates $|\psi_n(\xi(t))\rangle$ and eigenenergies $E_n(\xi(t))$ are also functions of $\xi(t)$ in general. In this approximation, the state of a given quantum mechanical system eigenstate can be represented as $|\Psi(\xi(t))\rangle = \sum_n a_n(\xi(t))\,|\psi_n(\xi(t))\rangle$, where $a_n(\xi(t)) = a_n(t=0)e^{-\frac{i}{\hbar}\int_0^t d\tau E_n(\tau)}e^{-i\Phi_n(\xi(t))}$.

The appearance of the phase factor $\Phi_n(\xi(t))$ is related to the geometry in the system and it is called the *Berry phase* (Berry, 1984). There are actually several properties related to $\Phi_n(\xi(t))$, which are used throughout the scientific literature. Additionally, these properties can be given in real and reciprocal spaces with different versions of the formulas. In the following section we summarize those expressions for the benefit of the reader.

3.1 Real Space Expressions

Berry phase for the nth eigenstate $|\psi_n(\xi(t))\rangle$ of Hamiltonian $\widehat{H}(\xi(t))$:

$$\Phi_n(t) = \int_0^t \langle\psi_n(\xi(t))|i\frac{d}{dt}\psi_n(\xi(t))\rangle dt = \int_{\xi(0)}^{\xi(t)} \langle\psi_n(\xi(t))|i\nabla_\xi\psi_n(\xi(t))\rangle\cdot d\xi(t)$$

Berry connection for the nth eigenstate $|\psi_n(\xi(t))\rangle$ of Hamiltonian $H(\xi(t))$:

$$A_n(\xi(t)) = \langle\psi_n(\xi(t))|i\nabla_\xi\psi_n(\xi(t))\rangle,$$
$$A_\mu^n(\xi(t)) = \langle\psi_n(\xi(t))|i\nabla_\mu\psi_n(\xi(t))\rangle.$$

In the second component expression one uses $\nabla_\mu = \frac{\partial}{\partial\xi^\mu}$. The Berry connection is a real, nonconservative vector field. It is also termed the "geometrical" vector potential. It is gauge-dependent, and sometimes it is also called "gauge potential."

Berry phase for the nth eigenstate in terms of the *Berry connection* for a *general* $\boldsymbol{\xi}(t)$:

$$\Phi_n(t) = \int_{\xi(0)}^{\xi(t)} \langle \psi_n(\xi(t)) | i \nabla_\xi \psi_n(\xi(t)) \rangle \cdot d\boldsymbol{\xi}(t) = \int_C \boldsymbol{A}_n(\boldsymbol{\xi}(t)) \cdot d\boldsymbol{\xi}(t),$$

$$\Phi_n(t) = \int_{\xi(0)}^{\xi(t)} \langle \psi_n(\xi(t)) | i \nabla_\mu \psi_n(\xi(t)) \rangle d\xi^\mu(t) = \int_C A_\mu^n(\xi(t)) d\xi^\mu(t).$$

In the preceding expressions, C is the path of $\boldsymbol{\xi}(t)$ along the followed trajectory for the particle in $\boldsymbol{\xi}$-space.

Berry phase for the nth eigenstate in terms of the *Berry connection* for a *closed path* $\boldsymbol{\xi}(t)$:

$$\Phi_n(t) = \oint_{\partial S} \boldsymbol{A}_n(\boldsymbol{\xi}(t)) \cdot d\boldsymbol{\xi}(t) = \int_S \left[\nabla_\xi \times \boldsymbol{A}_n(\boldsymbol{\xi})\right] \cdot d^2\boldsymbol{\xi},$$

$$\Phi_n(t) = \oint_{\partial S} A_\mu^n(\xi(t)) d\xi^\mu(t) = \int_S \varepsilon^{\mu\nu\rho} \nabla_\mu A_\nu^n(\xi) d^2\xi_\rho.$$

In the preceding $C = \partial S$ defines the closed path traced by $\boldsymbol{\xi}(t)$. In this case, $\Phi_n(t)$ cannot be eliminated by a local "gauge" transformation.
Berry curvature tensor for the nth eigenstate $|\psi_n(\xi(t))\rangle$ of Hamiltonian $\widehat{H}(\xi(t))$:

$$\Omega_{\mu\nu}^n(\xi(t)) = \nabla_\mu A_\nu^n - \nabla_\nu A_\mu^n.$$

This is a second-rank antisymmetric tensor.
Berry curvature vector for the nth eigenstate $|\psi_n(\xi(t))\rangle$ of Hamiltonian $\widehat{H}(\xi(t))$:

$$\boldsymbol{\Omega}_n(\boldsymbol{\xi}(t)) = \nabla_\xi \times \boldsymbol{A}_n(\boldsymbol{\xi}(t)) \rightarrow \Omega_n^\rho(\xi(t)) = i\left(\left\langle \nabla\psi_n(\xi(t)) \right| \times \left| \nabla\psi_n(\xi(t)) \right\rangle\right)^\rho.$$

This vector is defined only when $\boldsymbol{\xi}(t)$ is a 3D vector. Note the relation: $\Omega_n^\rho(\xi(t)) = \varepsilon^{\mu\nu\rho}\Omega_{\mu\nu}^n(\xi(t))$.
Berry phase for the nth eigenstate in terms of the *Berry curvature* for a closed path $\boldsymbol{\xi}(t)$:

$$\Phi_n(t) = \oint_{\partial S} \boldsymbol{A}_n(\boldsymbol{\xi}(t)) \cdot d\boldsymbol{\xi}(t) = \int_S \left[\nabla_\xi \times \boldsymbol{A}_n(\boldsymbol{\xi})\right] \cdot d^2\boldsymbol{\xi} = \int_S \boldsymbol{\Omega}_n(\boldsymbol{\xi}) \cdot d^2\boldsymbol{\xi},$$

$$\Phi_n(t) = \oint_{\partial S} A_\mu^n(\xi(t)) d\xi^\mu(t) = \int_S \varepsilon^{\mu\nu\rho} \nabla_\mu A_\nu^n(\xi) d^2\xi_\rho = \int_S \Omega_n^\rho(\xi) d^2\xi_\rho.$$

The Berry curvature is gauge-invariant, thus it is potentially observable. It is the geometrical analog of a magnetic field, and it is sometimes called "gauge field." The Berry curvature is singular at degenerate points.
Chern number for the nth eigenstate $|\psi_n(\xi(t))\rangle$ of Hamiltonian $\widehat{H}(\xi(t))$:

$$C_n = \frac{1}{2\pi} \oint_{\partial S} \boldsymbol{A}_n(\boldsymbol{\xi}) \cdot d\boldsymbol{\xi} = \frac{1}{2\pi} \int_S \boldsymbol{\Omega}_n(\boldsymbol{\xi}) \cdot d^2\boldsymbol{\xi}.$$

Here ∂S is the (simply connected) closed trajectory of $\boldsymbol{\xi}(t)$, which defines the boundary of a 2D surface S, also in $\boldsymbol{\xi}$-space. Note that $C_n \in \mathbb{Z}$ is a quantized topological number, as known from Chern's theorem.

3.2 Reciprocal Space Expressions

The same properties can be expressed for periodic systems in reciprocal space for a closed path traced by the wave vector $\boldsymbol{k}(t) = \boldsymbol{k}(\boldsymbol{\xi}(t))$. In this case, the eigenfunctions are given as $|\psi_{n\boldsymbol{k}}(\boldsymbol{\xi}(t))\rangle = e^{i\boldsymbol{k}(t)\cdot\boldsymbol{x}}\,|u^0_{n\boldsymbol{k}}(\boldsymbol{\xi}(t))\rangle \equiv e^{i\boldsymbol{k}(t)\cdot\boldsymbol{x}}\,|u^0_n(\boldsymbol{k}(t))\rangle$ with $|u^0_n(\boldsymbol{k}(t))\rangle$ being the nth Bloch state. The definitions just given can be adapted into the following:

Berry phase for the nth Bloch state $|u^0_n(\boldsymbol{k}(t))\rangle$:

$$\Phi_n(t) = \oint_{\partial S} \boldsymbol{A}_n(\boldsymbol{k}(t))\cdot d\boldsymbol{k}(t) = \oint_{\partial S} A^n_\mu(\boldsymbol{k}(t))dk^\mu(t).$$

Berry connection for the nth Bloch state $|u^0_n(\boldsymbol{k}(t))\rangle$:

$$A_{n\alpha}(\boldsymbol{k}(t)) = \langle u^0_n(\boldsymbol{k}(t))|i\frac{\partial}{\partial k_\alpha(t)}u^0_n(\boldsymbol{k}(t))\rangle = \langle u^0_n(\boldsymbol{k}(t))|i\nabla_\alpha u^0_n(\boldsymbol{k}(t))\rangle.$$

Berry curvature tensor for nth Bloch state $|u^0_n(\boldsymbol{k}(t))\rangle$:

$$\Omega^n_{\mu\nu}(\boldsymbol{k}(t)) = \nabla_\mu A^n_\nu - \nabla_\nu A^n_\mu.$$

Berry curvature vector for the nth Bloch state $|u^0_n(\boldsymbol{k}(t))\rangle$:

$$\boldsymbol{\Omega}_n(\boldsymbol{k}(t)) = \boldsymbol{\nabla}_{\boldsymbol{k}}\times\boldsymbol{A}_n(\boldsymbol{k}(t)) \to \Omega^\rho_n(\boldsymbol{k}(t)) = \left(i\langle\boldsymbol{\nabla}u^0_n(\boldsymbol{k}(t))|\times|\boldsymbol{\nabla}u^0_n(\boldsymbol{k}(t))\rangle\right)^\rho,$$

$$\Omega^\alpha_n(\boldsymbol{k}(t)) = i\varepsilon^{\alpha\beta\gamma}\left\langle\frac{\partial u^0_n(\boldsymbol{k}(t))}{\partial k_\beta}\middle|\frac{\partial u^0_n(\boldsymbol{k}(t))}{\partial k_\gamma}\right\rangle = \varepsilon^{\alpha\beta\gamma}\frac{\partial A_{n\gamma}}{\partial k_\beta}.$$

Chern number for the nth Bloch state $|u^0_n(\boldsymbol{k}(t))\rangle$:

$$C_n = \frac{1}{2\pi}\oint_{\partial S}\boldsymbol{A}_n(\boldsymbol{k})\cdot d\boldsymbol{k} = \frac{1}{2\pi}\int_S \boldsymbol{\Omega}_n(\boldsymbol{k})\cdot d^2\boldsymbol{k} \in \mathbb{Z}.$$

In the preceding formulas in reciprocal space, $S = BZ$ is the Brillouin zone and $\partial S = \partial BZ$ is its boundary.

Problem 3.1
Show that the Berry phase is a real quantity.

To demonstrate that $\Phi(t) = i\oint\langle\psi(\boldsymbol{\xi})|\boldsymbol{\nabla}_{\boldsymbol{\xi}}\psi(\boldsymbol{\xi})\rangle\cdot d\boldsymbol{\xi}$ is real means that we have to show that the imaginary part is zero or equivalently $\Phi = \Phi^*$.

This can be shown by starting with

$$\langle\psi(\boldsymbol{\xi}(t))|\psi(\boldsymbol{\xi}(t))\rangle = 1.$$

After differentiating with respect to $\boldsymbol{\xi}$ on both sides, one obtains

$$(\boldsymbol{\nabla}_{\boldsymbol{\xi}}\langle\psi(\boldsymbol{\xi})|)|\psi(\boldsymbol{\xi})\rangle + \langle\psi(\boldsymbol{\xi})|(|\boldsymbol{\nabla}_{\boldsymbol{\xi}}\psi(\boldsymbol{\xi})\rangle) = 0.$$

Since $(\nabla_\xi\langle\psi(\xi)|)|\psi(\xi)\rangle=\langle\psi(\xi)|(\nabla_\xi\psi(\xi))\rangle^*$, then $\langle\psi(\xi)|(|\nabla_\xi\psi(\xi)\rangle)=-\langle\psi(\xi)|(\nabla_\xi\psi(\xi))\rangle^*$, meaning that this quantity always imaginary. Therefore, $\Phi(t)=i\oint\langle\psi(\xi)|\nabla_\xi\psi(\xi)\rangle\cdot d\xi$ is always real.

Problem 3.2

Show that the Berry phase is gauge-invariant under the transformation $|\psi(\xi)\rangle\rightarrow|\bar{\psi}(\xi)\rangle=e^{i\gamma(\xi)}\,|\psi(\xi)\rangle$.

We remember that a given quantity is gauge-invariant under a specified transformation, which means that this quantity stays the same under the transformation. Thus, we have to show that $\Phi(t)$ corresponding to $|\psi(\xi)\rangle$ and $\bar{\Phi}(t)$ corresponding to $|\bar{\psi}(\xi)\rangle$ are the same.

The best way to show this is to use the definition for the Berry phase together with the Stokes theorem and the Berry connection.

Start with

$$\begin{aligned}\bar{\Phi}(t)&=\oint_{\partial S}\langle\bar{\psi}(\xi)|i\nabla_\xi\bar{\psi}(\xi)\rangle\cdot d\xi=\oint_{\partial S}\bar{A}(\xi)\cdot d\xi=\int_S\nabla_\xi\times\bar{A}(\xi)\cdot d^2s\\&=\int_S\nabla_\xi\times\left(\langle\bar{\psi}(\xi)|i\nabla_\xi\bar{\psi}(\xi)\rangle\right)\cdot d^2s.\end{aligned}$$

However,

$$\begin{aligned}\bar{A}(\xi)&=\langle\bar{\psi}(\xi)|i\nabla_\xi\bar{\psi}(\xi)\rangle=e^{-i\gamma(\xi)}\langle\psi(\xi)|i\nabla_\xi(e^{i\gamma(\xi)}|\psi(\xi)\rangle)\\&=e^{-i\gamma(\xi)}\left\langle\psi(\xi)\middle|\left(\left(i\nabla_\xi e^{i\gamma(\xi)}\right)\middle|\psi(\xi)\right\rangle+e^{i\gamma(\xi)}\middle|i\nabla_\xi\psi(\xi)\right\rangle\Big)\\&=e^{-i\gamma(\xi)}\left\langle\psi(\xi)\middle|\left(i^2e^{i\gamma(\xi)}\left(\nabla_\xi\gamma(\xi)\right)\middle|\psi(\xi)\right\rangle+e^{i\gamma(\xi)}\middle|i\nabla_\xi\psi(\xi)\right\rangle\Big)\\&=-\left(\nabla_\xi\gamma(\xi)\right)\langle\psi(\xi)|\psi(\xi)\rangle+\langle\psi(\xi)|i\nabla_\xi\psi(\xi)\rangle=-\nabla_\xi\gamma(\xi)+A(\xi).\end{aligned}$$

Therefore,

$$\bar{\Phi}(t)=\int_S\nabla_\xi\times\bar{A}(\xi)\cdot d^2s=\int_S\nabla_\xi\times\left(-\nabla_\xi\gamma(\xi)+A(\xi)\right)\cdot d^2s=\int_S\nabla_\xi\times A(\xi)\cdot d^2s=\Phi(t).$$

Here we have used that $\nabla_\xi\times\nabla_\xi\gamma(\xi)=\mathbf{0}$, as known from vector algebra. Thus, the Berry phase is gauge invariant; thus it can be understood as a gauge potential.

Problem 3.3

Show that the Berry curvature can be written as $\boldsymbol{\Omega}_n(\xi(t))=Im\sum\limits_{m\neq n}\frac{\langle\psi_n|\nabla\hat{H}|\psi_m\rangle\times\langle\psi_m|\nabla\hat{H}|\psi_n\rangle}{(E_n-E_m)^2}$, where the notation follows the one given at the beginning of the chapter.

The preceding property appears in many textbooks already. Here we repeat the solution for the benefit of the reader. We start with the components of the *Berry curvature vector*,

$$\begin{aligned}\Omega_n^\rho(\xi(t))&=\varepsilon^{\mu\nu\rho}\Omega_{\mu\nu}^n(\xi(t))=\varepsilon^{\mu\nu\rho}\nabla_\mu A_\nu^n=\left(\nabla_\mu A_\nu^n-\nabla_\nu A_\mu^n\right)^\rho\\&=\varepsilon^{\mu\nu\rho}\nabla_\mu\langle\psi_n(\xi(t))|i\nabla_\nu\psi_n(\xi(t))\rangle\\&=i\varepsilon^{\mu\nu\rho}\langle\nabla_\mu\psi_n(\xi(t))|\nabla_\nu\psi_n(\xi(t))\rangle=\left(i\langle\nabla\psi_n(\xi(t))\,|\times|\,\nabla\psi_n(\xi(t))\rangle\right)^\rho.\end{aligned}$$

Using that $\widehat{\mathbb{I}} = \sum_m |\psi_m(\xi(t))\rangle\langle\psi_m(\xi(t))|$, we obtain

$$\Omega_n^\rho(\xi(t)) = i\varepsilon^{\mu\nu\rho}\sum_m \langle\nabla_\mu\psi_n(\xi(t))|\psi_m(\xi(t))\rangle\langle\psi_m(\xi(t))|\nabla_\nu\psi_n(\xi(t))\rangle.$$

Applying the operator ∇_μ to the **eigenvalue problem** $\widehat{H}(\xi(t))\ |\psi_n(\xi(t))\rangle = E_n\ |\psi_n(\xi(t))\rangle$, we find

$$\left(\nabla_\mu\widehat{H}\right)|\psi_n\rangle + \widehat{H}\,\nabla_\mu\,|\psi_n\rangle = \left(\nabla_\mu E_n\right)|\psi_n\rangle + E_n\,\nabla_\mu\,|\psi_n\rangle.$$

Multiplying both sides of the preceding equation by the bra $\langle\psi_m(\xi(t))|$ for $m \neq n$, we obtain

$$\begin{aligned}
\left\langle\psi_m\middle|\nabla_\mu\widehat{H}\middle|\psi_n\right\rangle + \left\langle\psi_m\middle|\widehat{H}\middle|\nabla_\mu\psi_n\right\rangle &= \langle\psi_m|(\nabla_\mu E_n)|\psi_n\rangle + \langle\psi_m|E_n|\nabla_\mu\psi_n\rangle,\\
\left\langle\psi_m\middle|\nabla_\mu\widehat{H}\middle|\psi_n\right\rangle + \langle\psi_m|E_m|\nabla_\mu\psi_n\rangle &= (\nabla_\mu E_n)\langle\psi_m|\psi_n\rangle + E_n\langle\psi_m|\nabla_\mu\psi_n\rangle,\\
\left\langle\psi_m\middle|\nabla_\mu\widehat{H}\middle|\psi_n\right\rangle &= (\nabla_\mu E_n)\,\delta_{mn} + (E_n - E_m)\langle\psi_m|\nabla_\mu\psi_n\rangle,\\
\langle\psi_m|\nabla_\mu\psi_n\rangle &= \frac{\left\langle\psi_m\middle|\nabla_\mu\widehat{H}\middle|\psi_n\right\rangle}{(E_n - E_m)} - \frac{(\nabla_\mu E_n)\,\delta_{mn}}{(E_n - E_m)}.
\end{aligned}$$

Multiplying by the bra $\langle\psi_n(\xi(t))|$, we find

$$\left\langle\psi_n\middle|\nabla_\mu\widehat{H}\middle|\psi_n\right\rangle = (\nabla_\mu E_n)\,\delta_{nn} + (E_n - E_n)\langle\psi_n|\nabla_\mu\psi_n\rangle = (\nabla_\mu E_n).$$

Applying the operator ∇_μ to the normalization condition $\langle\psi_m|\psi_n\rangle = \delta_{mn}$, we obtain

$$\begin{aligned}
\nabla_\mu\left(\langle\psi_m|\psi_n\rangle\right) &= \nabla_\mu\delta_{mn},\\
\langle\nabla_\mu\psi_m|\psi_n\rangle + \langle\psi_m|\nabla_\mu\psi_n\rangle &= 0,\\
\langle\nabla_\mu\psi_m|\psi_n\rangle &= -\langle\psi_m|\nabla_\mu\psi_n\rangle.
\end{aligned}$$

Therefore,

$$\langle\nabla_\mu\psi_n|\psi_m\rangle = -\frac{\left\langle\psi_n\middle|\nabla_\mu\widehat{H}\middle|\psi_m\right\rangle}{(E_m - E_n)} + \frac{(\nabla_\mu E_m)\,\delta_{mn}}{(E_m - E_n)}.$$

Now we continue with the expression for $\Omega_n^\rho(\xi(t))$,

$$\Omega_n^\rho(\xi(t)) = i\varepsilon^{\mu\nu\rho}\sum_m\left[\frac{\left\langle\psi_n\middle|\nabla_\mu\widehat{H}\middle|\psi_m\right\rangle}{(E_n - E_m)} - \frac{(\nabla_\mu E_m)\,\delta_{mn}}{(E_n - E_m)}\right]\left[\frac{\left\langle\psi_m\middle|\nabla_\nu\widehat{H}\middle|\psi_n\right\rangle}{(E_n - E_m)} - \frac{(\nabla_\nu E_n)\,\delta_{mn}}{(E_n - E_m)}\right],$$

$$\begin{aligned}
\Omega_n^\rho(\xi(t)) = i\varepsilon^{\mu\nu\rho}\sum_m\Bigg[&\frac{\left\langle\psi_n\middle|\nabla_\mu\widehat{H}\middle|\psi_m\right\rangle}{(E_n - E_m)}\frac{\left\langle\psi_m\middle|\nabla_\nu\widehat{H}\middle|\psi_n\right\rangle}{(E_n - E_m)} - \frac{(\nabla_\mu E_m)\,\delta_{mn}}{(E_n - E_m)}\frac{\left\langle\psi_m\middle|\nabla_\nu\widehat{H}\middle|\psi_n\right\rangle}{(E_n - E_m)}\\
&-\frac{\left\langle\psi_n\middle|\nabla_\mu\widehat{H}\middle|\psi_m\right\rangle}{(E_n - E_m)}\frac{(\nabla_\nu E_n)\,\delta_{mn}}{(E_n - E_m)} + \frac{(\nabla_\mu E_m)\,\delta_{mn}}{(E_n - E_m)}\frac{(\Delta_\nu E_n)\,\delta_{mn}}{(E_n - E_m)}\Bigg],
\end{aligned}$$

$$\Omega_n^{\rho}(\boldsymbol{\xi}(t)) = i\varepsilon^{\mu\nu\rho}\sum_m \left[\frac{\left\langle \psi_n \middle| \nabla_\mu \widehat{H} \middle| \psi_m \right\rangle \left\langle \psi_m \middle| \nabla_\nu \widehat{H} \middle| \psi_n \right\rangle}{(E_n - E_m)^2} - \frac{(\nabla_\mu E_m)(\nabla_\nu E_n)\delta_{mn}}{(E_n - E_m)^2}\right].$$

Using $\varepsilon^{\mu\nu\rho}F_\mu F_\nu = 0$, we obtain $\varepsilon^{\mu\nu\rho}(\nabla_\mu E_m)(\nabla_\nu E_n)\delta_{mn} = 0$ and, for $m = n$,

$$\varepsilon^{\mu\nu\rho}\left\langle \psi_n \middle| \nabla_\mu \widehat{H} \middle| \psi_m \right\rangle \left\langle \psi_m \middle| \nabla_\nu \widehat{H} \middle| \psi_n \right\rangle \delta_{mn} = \varepsilon^{\mu\nu\rho}\left\langle \psi_n \middle| \nabla_\mu \widehat{H} \middle| \psi_n \right\rangle \left\langle \psi_n \middle| \nabla_\nu \widehat{H} \middle| \psi_n \right\rangle = 0.$$

From here, we find

$$\Omega_n^{\rho}(\boldsymbol{\xi}(t)) = i\varepsilon^{\mu\nu\rho}\sum_{m\neq n} \frac{\left\langle \psi_n \middle| \nabla_\mu \widehat{H} \middle| \psi_m \right\rangle \left\langle \psi_m \middle| \nabla_\nu \widehat{H} \middle| \psi_n \right\rangle}{(E_n - E_m)^2}$$

$$= Im\left(\varepsilon^{\mu\nu\rho}\sum_{m\neq n} \frac{\left\langle \psi_n \middle| \nabla_\mu \widehat{H} \middle| \psi_m \right\rangle \left\langle \psi_m \middle| \nabla_\nu \widehat{H} \middle| \psi_n \right\rangle}{(E_n - E_m)^2}\right),$$

$$\boldsymbol{\Omega}_n(\boldsymbol{\xi}(t)) = i\sum_{m\neq n} \frac{\left\langle \psi_n \middle| \nabla \widehat{H} \middle| \psi_m \right\rangle \times \left\langle \psi_m \middle| \nabla \widehat{H} \middle| \psi_n \right\rangle}{(E_n - E_m)^2}$$

$$= Im\left(\sum_{m\neq n} \frac{\left\langle \psi_n \middle| \nabla \widehat{H} \middle| \psi_m \right\rangle \times \left\langle \psi_m \middle| \nabla \widehat{H} \middle| \psi_n \right\rangle}{(E_n - E_m)^2}\right),$$

where we have made use of the fact that $\langle \nabla_\mu \psi_m | \psi_n \rangle = \langle \psi_m | \nabla_\mu \psi_n \rangle^*$.

Problem 3.4

Consider a particle of spin 1/2 under a time-dependent magnetic field along an arbitrary direction $\widehat{\boldsymbol{n}}$, as given in the sketch in Figure 3.1.

a) Write the Hamiltonian by taking into account only the spin properties of the particle and find the corresponding eigenvalues and eigenfunctions.
b) Find the Berry curvature and Berry connection. Is there a region where these are not defined?
c) Express the eigenfunctions in a different gauge and calculate the Berry curvature and Berry connection again.
d) What can you say about the gauge dependence of the Berry curvature and the Berry connection?

---

a) The Hamiltonian for the electron under a magnetic field $\boldsymbol{B}(t)$ is

$$\widehat{H} = -\gamma \boldsymbol{B}\cdot\widehat{\boldsymbol{S}} = -\frac{\gamma\hbar}{2}\boldsymbol{B}\cdot\boldsymbol{\sigma}.$$

Here γ is the gyromagnetic ratio, $\widehat{\boldsymbol{S}}$ is the spin operator, and $\boldsymbol{\sigma} = (\sigma_1, \sigma_2, \sigma_3)$ are the Pauli matrices.

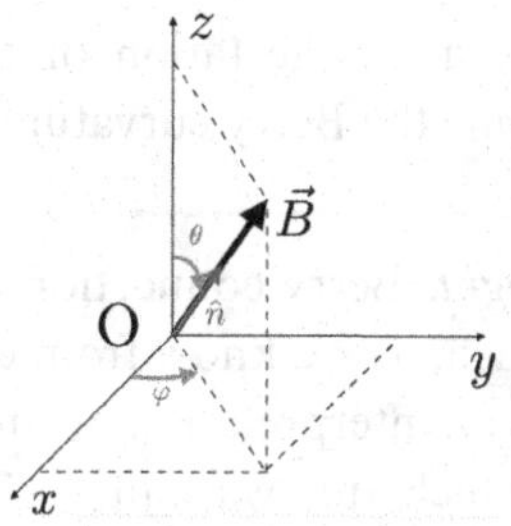

Figure 3.1 Schematic of a spin 1/2 particle under an external magnetic field directed along an arbitrary $\widehat{\boldsymbol{n}}$ direction.

The best way to solve this problem is in spherical coordinates with $\widehat{\boldsymbol{n}} = (\sin(\theta)\cos(\phi),\ \sin(\theta)\sin(\phi),\ \cos(\theta))$ (see Figure 3.1). Note that these angles change in time. The Hamiltonian and its eigenvalues and eigenstates are

$$\widehat{H} = -\frac{\gamma\hbar B}{2}\begin{pmatrix} \cos(\theta) & \sin(\theta)e^{-i\phi} \\ \sin(\theta)e^{i\phi} & -\cos(\theta) \end{pmatrix},$$

$$E_- = -\frac{\gamma\hbar B}{2};\quad \chi_- = \begin{pmatrix} \cos(\theta/2) \\ e^{i\phi}\sin(\theta/2) \end{pmatrix};\quad E_+ = \frac{\gamma\hbar B}{2};\quad \chi_+ = \begin{pmatrix} -\sin(\theta/2) \\ e^{i\phi}\cos(\theta/2) \end{pmatrix}.$$

b) To find the Berry curvature and the Berry connection, one needs to calculate $\boldsymbol{\nabla}\chi_\pm$, and since we are working in spherical coordinates, we need to use $\boldsymbol{\nabla} = \widehat{\boldsymbol{r}}\frac{\partial}{\partial r} + \frac{\widehat{\boldsymbol{\theta}}}{r}\frac{\partial}{\partial\theta} + \frac{\widehat{\boldsymbol{\phi}}}{r\sin(\theta)}\frac{\partial}{\partial\phi}$. Thus,

$$\boldsymbol{\nabla}\chi_- = \frac{\widehat{\boldsymbol{\theta}}}{2r}\begin{pmatrix} -\sin(\theta/2) \\ e^{i\phi}\cos(\theta/2) \end{pmatrix} + \frac{i\widehat{\boldsymbol{\phi}}}{r\sin(\theta)}\begin{pmatrix} 0 \\ e^{i\phi}\sin(\theta/2) \end{pmatrix},$$

$$\boldsymbol{\nabla}\chi_+ = -\frac{\widehat{\boldsymbol{\theta}}}{2r}\begin{pmatrix} \cos(\theta/2) \\ e^{i\phi}\sin(\theta/2) \end{pmatrix} + \frac{i\widehat{\boldsymbol{\phi}}}{r\sin(\theta)}\begin{pmatrix} 0 \\ e^{i\phi}\cos(\theta/2) \end{pmatrix}.$$

To find the Berry connection, we can use $\boldsymbol{A}_\pm(\boldsymbol{\xi}) = i\langle\chi_\pm|\boldsymbol{\nabla}\chi_\pm\rangle$, and to find the Berry curvature, we can use $\boldsymbol{\Omega}_\pm = i(\boldsymbol{\nabla}\langle\chi_\pm|) \times (\boldsymbol{\nabla}|\chi_\pm\rangle)$. The results are

$$\boldsymbol{A}_- = -\frac{\widehat{\boldsymbol{\phi}}}{2r}\tan\left(\frac{\theta}{2}\right);\quad \boldsymbol{A}_+ = -\frac{\widehat{\boldsymbol{\phi}}}{2r}\frac{1}{\tan\left(\frac{\theta}{2}\right)};\quad \boldsymbol{\Omega}_- = -\frac{\widehat{\boldsymbol{r}}}{2r^2};\quad \boldsymbol{\Omega}_+ = \frac{\widehat{\boldsymbol{r}}}{2r^2}.$$

All quantities diverge at $r \to 0$.

c) Let us choose a different gauge, which means we have to use a different representation for the eigenvectors. In addition to the results in (a), we also have another valid solution as

$$E_- = -\frac{\gamma\hbar B}{2};\quad \chi_- = \begin{pmatrix} e^{-i\phi}\cos(\theta/2) \\ \sin(\theta/2) \end{pmatrix};\quad E_+ = \frac{\gamma\hbar B}{2};\quad \chi_+ = \begin{pmatrix} -e^{-i\phi}\sin(\theta/2) \\ \cos(\theta/2) \end{pmatrix}.$$

Thus, using the definitions for the Berry connection and curvature, as in part (b), we find

$$\boldsymbol{A}_- = \frac{\widehat{\boldsymbol{\phi}}}{2r}\frac{1}{\tan\left(\frac{\theta}{2}\right)};\quad \boldsymbol{A}_+ = \frac{\widehat{\boldsymbol{\phi}}}{2r}\tan\left(\frac{\theta}{2}\right);\quad \boldsymbol{\Omega}_- = -\frac{\widehat{\boldsymbol{r}}}{2r^2};\quad \boldsymbol{\Omega}_+ = \frac{\widehat{\boldsymbol{r}}}{2r^2}.$$

d) This is simply a manifestation of the fact that the Berry connection is gauge dependent, while the Berry curvature is gauge independent.

> *Food for thought:* Berry connection has the same interpretation as a "vector" potential, which, as we know from electrodynamics, is not unique. The Berry curvature has an interpretation of a magnetic field of "monopole" with (+) and (−) charges, which are invariant.

Problem 3.5

Consider the case of a spin $S = 1$ particle placed under a time-dependent magnetic field $\boldsymbol{B} = B_0(\sin(\alpha)\cos(\omega t)\widehat{\boldsymbol{x}} + \sin(\alpha)\sin(\omega t)\widehat{\boldsymbol{y}} + \cos(\alpha)\widehat{\boldsymbol{z}})$. Calculate the Berry connection and the Berry phase for this particle.

For the Berry connection and phase one needs the eigenstates of the Hamiltonian, which can be written as $\widehat{H} = -\gamma \boldsymbol{B}\cdot\widehat{\boldsymbol{S}}$, where γ is the gyromagnetic ratio and

$$S_x = \frac{\hbar}{\sqrt{2}}\begin{pmatrix} 0 & 1 & 0 \\ 1 & 0 & 1 \\ 0 & 1 & 0 \end{pmatrix}, S_y = \frac{\hbar}{\sqrt{2}}\begin{pmatrix} 0 & -i & 0 \\ i & 0 & -i \\ 0 & i & 0 \end{pmatrix}, S_z = \hbar\begin{pmatrix} 1 & 0 & 0 \\ 0 & 0 & 0 \\ 0 & 0 & -1 \end{pmatrix}.$$

The matrix form of the Hamiltonian then becomes

$$\widehat{H} = -\frac{\hbar\gamma B_0}{\sqrt{2}}\begin{pmatrix} \sqrt{2}\cos(\alpha) & \sin(\alpha)e^{-i\omega t} & 0 \\ \sin(\alpha)e^{i\omega t} & 0 & \sin(\alpha)e^{-i\omega t} \\ 0 & \sin(\alpha)e^{i\omega t} & -\sqrt{2}\cos(\alpha) \end{pmatrix},$$

whose eigenvalues and eigenstates are

$$E_0 = 0, \quad \chi_0 = \frac{1}{\sqrt{2}}\begin{pmatrix} -e^{-i\omega t}\sin(\alpha) \\ \sqrt{2}\cos(\alpha) \\ e^{i\omega t}\sin(\alpha) \end{pmatrix}; \quad E_- = -\hbar\gamma B_0, \quad \chi_- = \begin{pmatrix} e^{-i\omega t}\cos^2\left(\frac{\alpha}{2}\right) \\ \frac{1}{\sqrt{2}}\sin(\alpha) \\ e^{i\omega t}\sin^2\left(\frac{\alpha}{2}\right) \end{pmatrix};$$

$$E_+ = \hbar\gamma B_0, \quad \chi_+ = \begin{pmatrix} e^{-i\omega t}\sin^2\left(\frac{\alpha}{2}\right) \\ \frac{-1}{\sqrt{2}}\sin(\alpha) \\ e^{i\omega t}\cos^2\left(\frac{\alpha}{2}\right) \end{pmatrix}.$$

Then using the definition for Berry connections (in vector form) and phases, we find

$$\boldsymbol{A}_0 = \boldsymbol{0}; \quad \boldsymbol{A}_- = \frac{1}{B_0\tan(\alpha)}\widehat{\boldsymbol{\phi}}; \quad \boldsymbol{A}_+ = -\frac{1}{B_0\tan(\alpha)}\widehat{\boldsymbol{\phi}},$$

$$\Phi_0 = 0; \quad \Phi_- = +\cos(\alpha)\omega t; \quad \Phi_+ = -\cos(\alpha)\omega t.$$

Problem 3.6

Consider a material whose periodic lattice has time reversal and inversion symmetries at the same time. Show that for such a system the Berry curvature vanishes.

This problem concerns the Berry curvature expressed in Bloch states, since the question is about a material with a periodic lattice. Thus, we work in the first Brillouin zone (BZ) and the following definition: $\Omega_{n\alpha}(\boldsymbol{k}) = i\varepsilon_{\alpha\beta\gamma}\left\langle \frac{\partial u_n^0(\boldsymbol{k})}{\partial k_\beta} \middle| \frac{\partial u_n^0(\boldsymbol{k})}{\partial k_\gamma} \right\rangle$, where $u_n^0(\boldsymbol{k})$ are the Bloch states for the nth orbital and $\varepsilon_{\alpha\beta\gamma}$ is the Levi–Civita symbol $(\alpha,\beta,\gamma = x,y,z)$.

From the properties of time-reversal operation, we see that

$$
\begin{aligned}
\widehat{\Theta}\Omega_{n\alpha}(\boldsymbol{k})\widehat{\Theta}^{-1} &= \widehat{\Theta} i\varepsilon_{\alpha\beta\gamma}\left\langle \frac{\partial u_n^0(\boldsymbol{k})}{\partial k_\beta} \middle| \frac{\partial u_n^0(\boldsymbol{k})}{\partial k_\gamma} \right\rangle \widehat{\Theta}^{-1} = \widehat{\Theta} i\varepsilon_{\alpha\beta\gamma}\int_{BZ} \frac{\partial u_n^{0*}(\boldsymbol{k})}{\partial k_\beta}\frac{\partial u_n^0(\boldsymbol{k})}{\partial k_\gamma} d\boldsymbol{k}\widehat{\Theta}^{-1} \\
&= -i\varepsilon_{\alpha\beta\gamma}\int_{BZ} \frac{\partial u_n^0(-\boldsymbol{k})}{\partial(-k_\beta)}\frac{\partial u_n^{0*}(-\boldsymbol{k})}{\partial(-k_\gamma)} d(-\boldsymbol{k})\widehat{\Theta}\widehat{\Theta}^{-1} \\
&= i\varepsilon_{\alpha\beta\gamma}\int_{BZ} \frac{\partial u_n^0(-\boldsymbol{k})}{\partial k_\beta}\frac{\partial u_n^{0*}(-\boldsymbol{k})}{\partial k_\gamma} d\boldsymbol{k} = i\varepsilon_{\alpha\beta\gamma}\left\langle \frac{\partial u_n^0(-\boldsymbol{k})}{\partial k_\gamma} \middle| \frac{\partial u_n^0(-\boldsymbol{k})}{\partial k_\beta} \right\rangle \\
&= -i\varepsilon_{\alpha\beta\gamma}\left\langle \frac{\partial u_n^0(-\boldsymbol{k})}{\partial k_\beta} \middle| \frac{\partial u_n^0(-\boldsymbol{k})}{\partial k_\gamma} \right\rangle = -\Omega_{n\alpha}(-\boldsymbol{k}),
\end{aligned}
$$

$$
\Omega_{n\alpha}(\boldsymbol{k}) \xrightarrow{\widehat{\Theta}} -\Omega_{n\alpha}(-\boldsymbol{k}).
$$

From the properties of inversion operation, we have

$$
\begin{aligned}
\widehat{\pi}\Omega_{n\alpha}(\boldsymbol{k})\widehat{\pi}^{-1} &= \widehat{\pi} i\varepsilon_{\alpha\beta\gamma}\left\langle \frac{\partial u_n^0(\boldsymbol{k})}{\partial k_\beta} \middle| \frac{\partial u_n^0(\boldsymbol{k})}{\partial k_\gamma} \right\rangle \widehat{\pi}^{-1} = \widehat{\pi} i\varepsilon_{\alpha\beta\gamma}\int_{BZ} \frac{\partial u_n^{0*}(\boldsymbol{k})}{\partial k_\beta}\frac{\partial u_n^0(\boldsymbol{k})}{\partial k_\gamma} d\boldsymbol{k}\widehat{\pi}^{-1} \\
&= i\varepsilon_{\alpha\beta\gamma}\int_{BZ} \frac{\partial u_n^0(-\boldsymbol{k})}{\partial(-k_\beta)}\frac{\partial u_n^{0*}(-\boldsymbol{k})}{\partial(-k_\gamma)} d(-\boldsymbol{k})\widehat{\pi}\widehat{\pi}^{-1} \\
&= -i\varepsilon_{\alpha\beta\gamma}\int_{BZ} \frac{\partial u_n^0(-\boldsymbol{k})}{\partial k_\beta}\frac{\partial u_n^{0*}(-\boldsymbol{k})}{\partial k_\gamma} d\boldsymbol{k} \\
&= -i\varepsilon_{\alpha\beta\gamma}\left\langle \frac{\partial u_n^0(-\boldsymbol{k})}{\partial k_\gamma} \middle| \frac{\partial u_n^0(-\boldsymbol{k})}{\partial k_\beta} \right\rangle = i\varepsilon_{\alpha\beta\gamma}\left\langle \frac{\partial u_n^0(-\boldsymbol{k})}{\partial k_\beta} \middle| \frac{\partial u_n^0(-\boldsymbol{k})}{\partial k_\gamma} \right\rangle \\
&= \Omega_{n\alpha}(-\boldsymbol{k}),
\end{aligned}
$$

$$
\Omega_{n\alpha}(\boldsymbol{k}) \xrightarrow{\widehat{\pi}} \Omega_{n\alpha}(-\boldsymbol{k}).
$$

Since the Berry curvature is invariant under both operations, we have

$$
\Omega_{n\alpha}(\boldsymbol{k}) = \widehat{\Theta}\widehat{\pi}\Omega_{n\alpha}(\boldsymbol{k})\widehat{\pi}^{-1}\widehat{\Theta}^{-1} = \widehat{\Theta}\Omega_{n\alpha}(-\boldsymbol{k})\widehat{\Theta}^{-1} = -\Omega_{n\alpha}(\boldsymbol{k}).
$$

Therefore, $\Omega_{n\alpha}(\boldsymbol{k}) = 0$.

> *Food for thought:* It is important to observe that the Berry curvature does not vanish if the periodic system has inversion symmetry, but it does not have time-reversal symmetry. The same is true if the periodic system has inversion, but not time-reversal symmetry.

Problem 3.7

The anomalous Hall conductivity for a given material with d-dimension is given as $\sigma_{xy} = \frac{e^2}{\hbar(2\pi)^d}\sum_n \int_{BZ} d\boldsymbol{k}\Omega_{n,xy}(\boldsymbol{k}) n_F(E_n(\boldsymbol{k}))$, where $n_F(E_n(\boldsymbol{k}))$ is the Fermi distribution function for the nth Bloch band $u_n^0(\boldsymbol{k})$ with energy $E_n(\boldsymbol{k})$ and $\Omega_{n,xy}(\boldsymbol{k})$ is the xy component of the Berry curvature tensor. The summation is over all Bloch states and the integration is over the first Brillouin zone.

Show that for a lattice that has time-reversal symmetry, the anomalous Hall conductivity vanishes.

Here we notice that the expression for the anomalous Hall conductivity is given using *the tensor form* of the Berry curvature. Nevertheless, if the lattice is invariant under time reversal, then

$$\widehat{\Theta}\widehat{H}(\boldsymbol{k})\widehat{\Theta}^{-1} = \widehat{H}(-\boldsymbol{k}); \quad E_n(\boldsymbol{k}) = E_n(-\boldsymbol{k}).$$

Therefore, $n_F(E_n(\boldsymbol{k})) = n_F(E_n(-\boldsymbol{k}))$.

From the last exercise, we also have $\widehat{\Theta}\Omega_{n,xy}(\boldsymbol{k})\widehat{\Theta}^{-1} = -\Omega_{n,xy}(-\boldsymbol{k})$.

Under time reversal, then

$$\begin{aligned}
\widehat{\Theta}\sigma_{xy}\widehat{\Theta}^{-1} &= \frac{e^2}{\hbar(2\pi)^d}\sum_n \int_{BZ} \widehat{\Theta} d\boldsymbol{k}\Omega_{n,xy}(\boldsymbol{k}) n_F(E_n(\boldsymbol{k}))\widehat{\Theta}^{-1} \\
&= \frac{e^2}{\hbar(2\pi)^d}\sum_n \int_{BZ} d(-\boldsymbol{k})\left(-\Omega_{n,xy}(-\boldsymbol{k})\right) n_F(E_n(-\boldsymbol{k})) \\
&= -\frac{e^2}{\hbar(2\pi)^d}\sum_n \int_{BZ} d\boldsymbol{k}_0\Omega_{n,xy}(\boldsymbol{k}_0) n_F(E_n(\boldsymbol{k}_0)) = -\sigma_{xy},
\end{aligned}$$

where we have made the $\boldsymbol{k} = -\boldsymbol{k}_0$ change of variable in the last step. Thus, comparing σ_{xy} and $\widehat{\Theta}\sigma_{xy}\widehat{\Theta}^{-1}$, we realize that σ_{xy} vanishes.

Problem 3.8

The Hamiltonian for an electron in a periodic environment with wave vector $\boldsymbol{k}$ is given in the following general form,

$$\widehat{H}(\boldsymbol{k}) = a_0(\boldsymbol{k})\sigma_0 + \boldsymbol{a}(\boldsymbol{k})\cdot\boldsymbol{\sigma},$$

where σ_0 is the identity matrix in 2D and $\boldsymbol{\sigma} = (\sigma_1, \sigma_2, \sigma_3)$ are the Pauli matrices. Also, $\boldsymbol{a}(\boldsymbol{k}) = (a_1(\boldsymbol{k}), a_2(\boldsymbol{k}), a_3(\boldsymbol{k}))$ is $\boldsymbol{k}$-dependent and real, and the real scalar function $a_0(\boldsymbol{k})$ is also $\boldsymbol{k}$-dependent. Find the eigenvalues and eigenstates of this Hamiltonian, and then obtain the corresponding Berry connections and Berry curvatures.

Working directly with the Hamiltonian in the xyz-coordinate system, we find that

$$\widehat{H} = \begin{pmatrix} a_0 + a_3 & a_1 - ia_2 \\ a_1 + ia_2 & a_0 - a_3 \end{pmatrix},$$

where $a_\mu(\boldsymbol{k}) = a_\mu$ ($\mu = 0,1,2,3$) for clarity. The two eigenvalues and their corresponding eigenfunctions are found as

$$E_\pm = a_0 \pm a; \quad |\chi_\pm\rangle = \frac{1}{N_\pm}\begin{pmatrix} a_3 \pm a \\ a_1 + ia_2 \end{pmatrix},$$

where $N_\pm = \sqrt{2a(a \pm a_3)}$ and $a = \sqrt{a_1^2 + a_2^2 + a_3^2}$. For the Berry curvature calculation, we need

$$\nabla|\chi_\pm\rangle = -\frac{\nabla N_\pm}{N_\pm^2}\begin{pmatrix} a_3 \pm a \\ a_1 + ia_2 \end{pmatrix} + \frac{1}{N_\pm}\begin{pmatrix} \nabla(a_3 \pm a) \\ \nabla(a_1 + ia_2) \end{pmatrix},$$

$$\langle\chi_\pm| = \frac{1}{N_\pm}(a_3 \pm a,\ a_1 - ia_2).$$

The Berry connection for each state is

$$\boldsymbol{A}_\pm = \frac{\mathrm{i}}{N_\pm^2}(a_1 - ia_2)\nabla(a_1 + ia_2) + (a \pm a_3)\nabla(a \pm a_3) - \frac{i}{N_\pm}\nabla N_\pm,$$

$$\boldsymbol{A}_\pm = \frac{a_2\nabla a_1 - a_1\nabla a_2}{N_\pm^2},$$

and the Berry curvature vector for each state is then found, using $\boldsymbol{\Omega}_\pm(\boldsymbol{k}) = \nabla_{\boldsymbol{k}} \times \boldsymbol{A}_\pm(\boldsymbol{k})$, as

$$\boldsymbol{\Omega}_\pm = \frac{2}{N_\pm^3}(N_\pm\nabla a_1 \times \nabla a_2 - a_1\nabla N_\pm \times \nabla a_2 + a_2\nabla N_\pm \times \nabla a_1).$$

Note: This result can further be given in an alternative form. In fact, we can show that

$$\Omega_\pm^\rho = \varepsilon^{\mu\nu\rho}\Omega_{\mu\nu;\pm}^\rho = \pm\varepsilon^{\mu\nu\rho}\frac{\boldsymbol{a}\cdot(\nabla_\mu\boldsymbol{a} \times \nabla_\nu\boldsymbol{a})}{2a^3} = \pm\varepsilon^{\mu\nu\rho}\varepsilon^{\alpha\beta\gamma}\frac{a_\alpha\nabla_\mu a_\beta\nabla_\nu a_\gamma}{2a^3}$$

$$= \pm\frac{1}{2}\varepsilon^{\mu\nu\rho}\varepsilon^{\alpha\beta\gamma}\frac{a_\alpha}{a}\nabla_\mu\frac{a_\beta}{a}\nabla_\nu\frac{a_\gamma}{a}.$$

The proof is left for an independent exercise by the skillful reader.

> *Food for thought:* This is an important problem, especially in condensed matter physics. It captures many cases in which topologically nontrivial features, such as Berry phase properties, need to be calculated. Here we consider the solution in Cartesian coordinates.

Problem 3.9

The Hamiltonian for an electron in a periodic environment with wave vector $\boldsymbol{k}$ is given in the following general form,

$$\hat{H}(\boldsymbol{k}) = a_0(\boldsymbol{k})\sigma_0 + \boldsymbol{a}(\boldsymbol{k})\cdot\boldsymbol{\sigma},$$

where σ_0 is the identity matrix in 2D and $\boldsymbol{\sigma} = (\sigma_1, \sigma_2, \sigma_3)$ are the Pauli matrices. The vector $\boldsymbol{a}(\boldsymbol{k})$ is of the form $\boldsymbol{a}(\boldsymbol{k}) = (a_1(\boldsymbol{k}), a_2(\boldsymbol{k}), a_3(\boldsymbol{k})) = a(\boldsymbol{k})(\sin(\theta)\cos(\phi), \sin(\theta)\sin(\phi), \cos(\theta))$, where $a(\boldsymbol{k}) = \sqrt{a_1^2(\boldsymbol{k}) + a_2^2(\boldsymbol{k}) + a_3^2(\boldsymbol{k})}$. Find the eigenvalues and eigenstates of this Hamiltonian, and then obtain the corresponding Berry connections and Berry curvatures.

Clearly this problem is very similar to the previous one. The Hamiltonian is actually simpler due to the condition given for $a(\mathbf{k})$ implying that the magnitude of this vector is constant. Given this, the problem can be worked out not only in Cartesian coordinates by simply substituting in the solution of the previous problem, but also in spherical coordinates.

Let's solve this problem in *spherical coordinates*. The eigenenergies and eigenvectors can be written as

$$E_{\pm} = a_0(\boldsymbol{k}) \pm a(\boldsymbol{k}); \quad \zeta_+ = \begin{pmatrix} \cos\left(\frac{\theta}{2}\right) e^{-i\phi} \\ \sin\left(\frac{\theta}{2}\right) \end{pmatrix}; \quad \zeta_- = \begin{pmatrix} \sin\left(\frac{\theta}{2}\right) e^{-i\phi} \\ -\cos\left(\frac{\theta}{2}\right) \end{pmatrix}.$$

Using the definition of Berry connection given earlier, we arrive at

$$\boldsymbol{A}_+ = \frac{\widehat{\boldsymbol{\phi}}}{2k\tan\left(\frac{\theta}{2}\right)}; \qquad \boldsymbol{A}_- = \frac{\tan\left(\frac{\theta}{2}\right)\widehat{\boldsymbol{\phi}}}{2k},$$

Using the definition for the Berry connection, we find

$$\boldsymbol{\Omega}_- = -\frac{\widehat{\boldsymbol{k}}}{2k^2}; \qquad \boldsymbol{\Omega}_+ = \frac{\widehat{\boldsymbol{k}}}{2k^2}.$$

Problem 3.10

We often have to deal with Hamiltonians of the kind $\widehat{H} = v\boldsymbol{\sigma}\cdot\boldsymbol{k}$, as is the case for certain topological materials, called Dirac semimetals. Find the Berry connection and Berry curvature for the eigenstates corresponding to $\widehat{H}$.

Comparing with the previous two problems, we see that this is essentially the same exercise, but for a specific and simpler $a(\boldsymbol{k}) = \boldsymbol{k}$ whose magnitude is always $\boldsymbol{k}$.

From Problem 3.9, the eigenenergies and eigenvectors can be written using *spherical coordinates*,

$$E_{\pm} = a_0(\boldsymbol{k}) \pm a(\boldsymbol{k}) = \pm vk; \quad \zeta_+ = \begin{pmatrix} \cos\left(\frac{\theta}{2}\right) e^{-i\phi} \\ \sin\left(\frac{\theta}{2}\right) \end{pmatrix}; \quad \zeta_- = \begin{pmatrix} \sin\left(\frac{\theta}{2}\right) e^{-i\phi} \\ -\cos\left(\frac{\theta}{2}\right) \end{pmatrix}.$$

Based on the definition of Berry connection in spherical coordinates given earlier, we arrive at

$$\mathbf{A}_- = \sin\left(\frac{\theta}{2}\right)^2 \nabla\phi = \frac{\tan\left(\frac{\theta}{2}\right)\widehat{\boldsymbol{\phi}}}{2k}; \quad \mathbf{A}_+ = \cos\left(\frac{\theta}{2}\right)^2 \nabla\phi = \frac{\widehat{\boldsymbol{\phi}}}{2k\tan\left(\frac{\theta}{2}\right)}.$$

Using the definition for the Berry connection, we find

$$\boldsymbol{\Omega}_{\pm} = \pm\frac{\widehat{\boldsymbol{k}}}{2k^2}.$$

The same problem can be considered in *Cartesian coordinates* also,

$$\mathrm{A}_{\pm,\mu} = \frac{a_1\nabla_\mu a_2 - a_2\nabla_\mu a_1}{N_\pm^2} = \frac{v^2 k_1\delta_{\mu 2} - v^2 k_2\delta_{\mu 1}}{N_\pm^2} \rightarrow \mathbf{A}_\pm = \frac{v^2}{N_\pm^2}\left(k_1\widehat{\boldsymbol{k}}_2 - k_2\widehat{\boldsymbol{k}}_1\right).$$

Again, from the previous problem the Berry curvature is

$$\Omega_\pm^\rho = \pm\varepsilon^{\mu\nu\rho}\varepsilon^{\alpha\beta\gamma}\frac{a_\alpha\nabla_\mu a_\beta\nabla_\nu a_\gamma}{2a^3} = \pm\varepsilon^{\mu\nu\rho}\varepsilon^{\alpha\beta\gamma}\frac{v^3 k_\alpha\delta_{\mu\beta}\delta_{\nu\gamma}}{2v^3k^3} = \pm\varepsilon_{\beta\gamma\sigma}\varepsilon^{\alpha\beta\gamma}\delta^{\rho\sigma}\frac{k_a}{2k^3}.$$

Using $\varepsilon_{\beta\gamma\sigma}\varepsilon^{\alpha\beta\gamma} = \delta_\sigma^\alpha$, we obtain

$$\Omega_\pm^\rho = \pm\delta_\sigma^\alpha\delta^{\rho\sigma}\frac{k_\alpha}{2k^3} = \pm\frac{k^\rho}{2k^3} \rightarrow \boldsymbol{\Omega}_\pm = \pm\frac{\boldsymbol{k}}{2k^3} = \pm\frac{\widehat{\boldsymbol{k}}}{2k^2}.$$

Food for thought: This Hamiltonian corresponds to graphene, indicating a Dirac-like physics for low-energy carriers. We also note that this problem is very similar to Problem 3.4, where we are asked to solve for similar things but in real space. Actually, the radial variable $\boldsymbol{r}$ in the Berry curvature and connection in real space (Problem 3.4) is now substituted by the wave vector $\boldsymbol{k}$ in reciprocal space.

Problem 3.11

Let us consider an electron moving in a periodic environment under the following effective Hamiltonian in Bloch space $\widehat{H} = \boldsymbol{B_k}\cdot\boldsymbol{\sigma}$, where $\boldsymbol{B_k} = -2t\cos(ka)\widehat{\boldsymbol{x}} - 2t\sin(ka)\widehat{\boldsymbol{y}} + \Delta\widehat{\boldsymbol{z}}$, $\boldsymbol{\sigma} = (\sigma_1,\sigma_2,\sigma_3)$ are the Pauli matrices, $\boldsymbol{k}$ is the wave vector, and t,Δ are positive constants. Find the Berry connections and phases for this case.

This problem is another variation of Problem 3.8. We will provide solutions in Cartesian and spherical coordinates for the benefit of the reader.

Cartesian coordinates: In the Bloch state basis, the given Hamiltonian can be written as

$$\widehat{H} = \begin{pmatrix} \Delta & -2te^{-ika} \\ -2te^{ika} & -\Delta \end{pmatrix}.$$

The solutions of the eigenproblem $\widehat{H}\,|\chi_\pm(\boldsymbol{k})\rangle = E_\pm\,|\chi_\pm(\boldsymbol{k})\rangle$ are

$$E_\pm = \pm\sqrt{\Delta^2+4t^2}; \qquad |\chi_\pm(\boldsymbol{k})\rangle = \frac{1}{N_\pm}\begin{pmatrix} \mp 2te^{-ika} \\ \sqrt{\Delta^2+4t^2}\mp\Delta \end{pmatrix},$$

with $N_\pm = \sqrt{2}\sqrt{\Delta^2+4t^2\pm\Delta\sqrt{\Delta^2+4t^2}}$. Taking into account that t and Δ are independent of the wave vector, we obtain

$$A_\pm = \langle\chi_\pm(\boldsymbol{k})\,|i\partial_k|\,\chi_\pm(\boldsymbol{k})\rangle = \frac{4t^2a}{N_\pm^2}\partial_k k = \frac{4t^2a}{N_\pm^2}.$$

Similarly, the Berry phases can also be found:

$$\Phi_\pm = \int_{-\frac{\pi}{2a}}^{\frac{\pi}{2a}} \boldsymbol{A}_\pm \cdot d\boldsymbol{k} = \int_{-\frac{\pi}{2a}}^{\frac{\pi}{2a}} \frac{4t^2 a}{N_\pm^2} \widehat{\boldsymbol{u}}_{\boldsymbol{k}} \cdot d\boldsymbol{k} = \frac{4t^2 a}{N_\pm^2} \int_{-\frac{\pi}{2a}}^{\frac{\pi}{2a}} dk = \frac{4t^2 \pi}{N_\pm^2}.$$

Using $\frac{4t^2}{N_\pm^2} = \frac{1}{2}\left(1 \mp \frac{\Delta}{\sqrt{\Delta^2+4t^2}}\right)$, we obtain

$$\Phi_\pm = \frac{\pi}{2}\left(1 \mp \frac{\Delta}{\sqrt{\Delta^2+4t^2}}\right).$$

Note that the Berry curvature is zero in this case:

$$\boldsymbol{\Omega}_\pm = \boldsymbol{\nabla} \times \boldsymbol{A}_\pm = 0.$$

Spherical coordinates: The eigenvalues and eigenstates can alternatively be obtained as

$$E_+ = \sqrt{\Delta^2+4t^2}; \qquad |\chi_+(\boldsymbol{k})\rangle = \begin{pmatrix} \cos\left(\frac{\theta}{2}\right) e^{-ika} \\ -\sin\left(\frac{\theta}{2}\right) \end{pmatrix},$$

$$E_- = -\sqrt{\Delta^2+4t^2}; \qquad |\chi_-(\boldsymbol{k})\rangle = \begin{pmatrix} \sin\left(\frac{\theta}{2}\right) e^{-ika} \\ \cos\left(\frac{\theta}{2}\right) \end{pmatrix},$$

where $\tan(\theta) = 2t/\Delta$. From here, the Berry connections are

$$\boldsymbol{A}_\pm = i\langle\chi_\pm(\boldsymbol{k})|\boldsymbol{\nabla}_{\boldsymbol{k}}\chi_\pm(\boldsymbol{k})\rangle = \widehat{\boldsymbol{k}} a \cos^2(\theta/2) = \frac{\widehat{\boldsymbol{k}} a}{2}\left(1 \pm \frac{\Delta}{\sqrt{\Delta^2+4t^2}}\right).$$

The phases are

$$\Phi_\pm = \int_{-\pi/2}^{\pi/2} \frac{1}{2}\left(1 \pm \frac{\Delta}{\sqrt{\Delta^2+4t^2}}\right) d(ka) = \frac{\pi}{2}\left(1 \pm \frac{\Delta}{\sqrt{\Delta^2+4t^2}}\right),$$

giving zero Berry curvatures $\boldsymbol{\Omega}_\pm = \boldsymbol{\nabla} \times \boldsymbol{A}_\pm = 0$.

> *Food for thought:* Notice that the geometrical phase, often called the "Zac" phase (Zak, 1989), of the electron moving in this periodic environment can take any value in the $[0, 2\pi]$ interval. It is not quantized, and in this sense it is not topological. It is interesting to note that the Zac phase actually depends on the origin of space.

Problem 3.12

Here is a model Hamiltonian for a topological semimetal phase given as $\widehat{H} = v\left(k_x\sigma_x + k_y\sigma_y\right) + M\left(k_w^2 - k^2\right)\sigma_z$. Here $\boldsymbol{k} = (k_x, k_y, k_z)$, $k = \sqrt{k_x^2+k_y^2+k_z^2}$, and k_w is a constant in momentum space. Also, A, M are positive constants. Calculate the Berry curvature.

Again, this is another variation of Problem 3.8. This model Hamiltonian was considered in Lu and Shen (2017) as an example of a two-node Dirac semimetal.

The eigenenergies and corresponding eigenvectors are

$$E_{\pm} = \pm\sqrt{v^2\left(k_x^2+k_y^2\right)+\Delta^2(\boldsymbol{k})}, \qquad \chi_{\pm} = \frac{1}{N_{\pm}(\boldsymbol{k})}\begin{pmatrix} \pm v(k_x - ik_y) \\ E_{\pm} \mp \Delta(\boldsymbol{k}) \end{pmatrix},$$

where $N_{\pm}(\boldsymbol{k}) = \sqrt{2E_{+}(E_{+} \mp \Delta(\boldsymbol{k}))}$ and $\Delta(\boldsymbol{k}) = M\left(k_w^2 - k^2\right)$. Then, using the result $\boldsymbol{\Omega}_{\pm} = \frac{1}{2a^3}\left(a_1 \boldsymbol{\nabla} a_2 \times \boldsymbol{\nabla} a_3 - a_2 \boldsymbol{\nabla} a_1 \times \boldsymbol{\nabla} a_3 + a_3 \boldsymbol{\nabla} a_1 \times \boldsymbol{\nabla} a_2\right)$ found in Problem 3.8, we obtain

$$\boldsymbol{\Omega}_{\pm} = \pm\frac{2M\,v^2}{E_{+}^3}\begin{pmatrix} -k_x k_z \\ -k_y k_z \\ -k_z^2 + \frac{k_w^2+k^2}{2} \end{pmatrix} = \frac{2M\,v^2}{E_{\pm}^3}\left(\frac{k_w^2+k^2}{2}\widehat{\boldsymbol{u}}_{k_z} - k_z\boldsymbol{k}\right).$$

Problem 3.13

Consider the Hamiltonian $\widehat{H}(\boldsymbol{k}) = A_{\mu\nu}k^{\nu}\sigma^{\mu} + \Delta\sigma^3 = d_{\mu}\sigma^{\mu}$, for a system with a 2D wave vector $\boldsymbol{k} = (k^x, k^y)$ and where σ^{μ} are the Pauli matrices. Note that in this particular Hamiltonian A is a 2×2 constant matrix. Using $\Omega_{\mu\nu} = \frac{\boldsymbol{a}(\boldsymbol{k})\cdot(\nabla_{\mu}\boldsymbol{a}(\boldsymbol{k})\times\nabla_{\nu}\boldsymbol{a}(\boldsymbol{k}))}{2a(\boldsymbol{k})^3}$, show that the Chern number C can be written as $C = sign(\Delta)sign(|A|)$, where $|A|$ is the determinant of the matrix A.

The given Hamiltonian specifies an energy dispersion of a band crossing with a finite gap Δ. The Chern number depends on Ω^{ρ}, the ρ-component of the Berry curvature vector; thus we find

$$\Omega^{\rho} = \varepsilon^{\mu\nu\rho}\Omega_{\mu\nu} = \varepsilon^{\mu\nu\rho}\frac{\left[\partial_{k_\mu}\boldsymbol{d}(\boldsymbol{k})\times\partial_{k_\nu}\boldsymbol{d}(\boldsymbol{k})\right]\cdot\boldsymbol{d}(\boldsymbol{k})}{2|\boldsymbol{d}(\boldsymbol{k})|^3} = \frac{\varepsilon^{\mu\nu\rho}\varepsilon^{\alpha\beta\gamma}d_{\alpha}\partial_{\mu}d_{\beta}\partial_{\nu}d_{\gamma}}{2|d|^3}$$

$$= \frac{\varepsilon^{\mu\nu\rho}\varepsilon^{\alpha\beta\gamma}\Delta\,\delta_{\alpha 3}A_{\beta\mu}A_{\gamma\nu}}{2\left|\Delta^2+(A\cdot\boldsymbol{k})^2\right|^3} = \frac{\Delta|A|\delta^{\rho 3}}{|\Delta^2+(A\cdot\boldsymbol{k})^2|^3},$$

where we have used the property $\varepsilon^{\mu\nu\rho}\varepsilon^{\alpha\beta\gamma}A_{\beta\mu}A_{\gamma\nu} = 2|A|\delta^{\rho\alpha}$. Now we proceed to carry out the integral over the Brillion zone to obtain

$$C = \frac{1}{2\pi}\int_{BZ} d^2k_{\rho}\Omega^{\rho} = \frac{1}{2\pi}\int_{BZ} d^2\boldsymbol{k}\frac{\Delta|A|}{|\Delta^2+(A\cdot\boldsymbol{k})^2|^3}.$$

After applying the change of variables $\boldsymbol{q} = A\cdot\boldsymbol{k}$, we find that

$$C = \frac{1}{2\pi}\int_{BZ}\frac{d^2\boldsymbol{q}}{||A||}\frac{\Delta|A|}{|\Delta^2+\boldsymbol{q}^2|^3},$$

where $||A||$ is the absolute value of the determinant $|A|$. After a second change of variables $\boldsymbol{v} = \Delta\boldsymbol{q}$, one obtains

$$C = \frac{1}{2\pi}\int_{BZ}\frac{d^2\boldsymbol{v}}{|\Delta|\,||A||}\frac{\Delta\,|A|}{|1+\boldsymbol{v}^2|^3} = \frac{1}{2\pi}\frac{\Delta|A|}{|\Delta|\,||A||}\int_{BZ}\frac{d^2\boldsymbol{v}}{|1+\boldsymbol{v}^2|^3}.$$

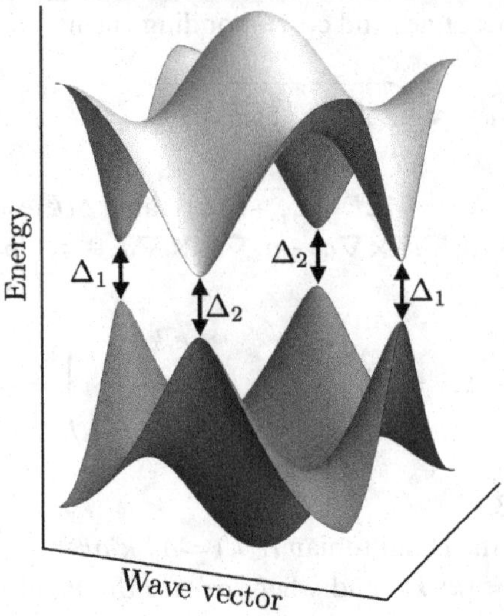

Figure 3.2 Schematic of energy eigenvalues dispersion for the considered Hamiltonian with four crossing points in reciprocal space with finite gaps.

This last 2D integral can be easily carried out, since

$$\int_{BZ} \frac{d^2 \nu}{|1+\nu^2|^3} = \int_0^{2\pi} d\varphi \int_0^{\infty} d\nu \frac{\nu}{(1+\nu^2)^{\frac{3}{2}}} = 2\pi,$$

which yields the desired result:

$$C = sign(\Delta) sign(|A|).$$

Food for thought: The preceding result can be generalized for Hamiltonians with crossing points in the band structure $\widehat{H}(\boldsymbol{k}) = \boldsymbol{d}(\boldsymbol{k}) \cdot \boldsymbol{\sigma} \approx \sum_i A_{\mu\nu}(\boldsymbol{k}_{D_i})\left(k^\nu - k^\nu_{D_i}\right)\sigma^\mu + \Delta(\boldsymbol{k}_{D_i})\sigma^3$, where $A_{\mu\nu}(\boldsymbol{k}_{D_i}) = \left.\frac{\partial d_\mu(\boldsymbol{k})}{\partial k^\nu}\right|_{\boldsymbol{k}\to\boldsymbol{k}_{D_i}}$. The energy dispersion then corresponds to several crossing points in reciprocal space $\boldsymbol{k}_{D_i}$ characterized by finite gaps $\Delta(\boldsymbol{k}_{D_i})$. In Figure 3.2 a sketch for four such crossing points is shown for illustration. In this case,

$$C = \sum_{\boldsymbol{k}_{D_i}} sign\left(\Delta(\boldsymbol{k}_{D_i})\right) sign\left(\left|A(\boldsymbol{k}_{D_i})\right|\right).$$

Problem 3.14

Consider the Hamiltonian $\widehat{H}(\boldsymbol{k}) = v\boldsymbol{\sigma}\cdot\boldsymbol{k}$, where $\boldsymbol{v}$ is a constant, $\boldsymbol{\sigma} = \sigma_x\widehat{\boldsymbol{x}} + \sigma_y\widehat{\boldsymbol{y}}$, and $\boldsymbol{k}$ is a 2D wave vector. Calculate the Chern number.

The solution to this problem is a direct consequence of Problem 3.13. Since here $\Delta = 0$, then the Chern number is

$$C = 0.$$

> *Food for thought:* As an exercise, calculate the result directly. This indicates that the Chern number for graphene is zero, since this Hamiltonian corresponds to the case of a graphene system.

Problem 3.15

Consider the Hamiltonian $\widehat{H}(\boldsymbol{k}) = t\sin(k_x a)\,\sigma_x + t\sin(k_y a)\,\sigma_y + t\left(\cos(k_x a) + \cos(k_y a)\right)\sigma_z$, where t is a positive constant, $\boldsymbol{k} = (k_x, k_y)$ is the 2D wave vector, and a is a lattice constant. Calculate the Chern numbers.

For the solution of this problem, we make a connection with Problem 3.13 and look for crossing points in the energy dispersion. With this in mind, let's consider the Hamiltonian in a linear Taylor expansion with respect to the wave vector to find these crossing points at the minima placed at $\boldsymbol{k_D} = \boldsymbol{0}$ for the first Brillouin zone,

$$\widehat{H}(\boldsymbol{k_D}) \cong A_{\mu\nu}(\boldsymbol{k_D} = \boldsymbol{0})\, k^\nu \sigma^\mu + \Delta(\boldsymbol{k_D} = \boldsymbol{0})\,\sigma^3,$$

where $A_{\mu\nu}(\boldsymbol{k_D}) = \left.\frac{\partial d_\mu(\boldsymbol{k})}{\partial k^\nu}\right|_{\boldsymbol{k}\to\boldsymbol{k_D}=\boldsymbol{0}}$. Thus, the Hamiltonian is expressed as

$$\widehat{H}(\boldsymbol{k_D}) \cong ta\,k^1\sigma^1 + ta\,k^2\sigma^2 + 2t\,\sigma^3 = \begin{pmatrix} k^1 & k^2 \end{pmatrix} \cdot \begin{pmatrix} ta & 0 \\ 0 & ta \end{pmatrix} \cdot \begin{pmatrix} \sigma^1 \\ \sigma^2 \end{pmatrix} + 2t\,\sigma^3.$$

This problem now becomes a direct application of the formula $C = sign(\Delta) sign(|A|)$, where $\Delta = 2t$ and $A = \begin{pmatrix} ta & 0 \\ 0 & ta \end{pmatrix}$. Therefore:

$$C = sign(\Delta) sign(|A|) = sign(2t) sign\left(\begin{vmatrix} ta & 0 \\ 0 & ta \end{vmatrix}\right) = sign(t) sign(t^2 a^2) = sign(t).$$

Problem 3.16

Consider the Hamiltonian $\widehat{H} = A\left(k_x\sigma_x + k_y\sigma_y\right) + Mk^2\sigma_z$. Here $\boldsymbol{k} = (k_x, k_y)$ is a 2D wave vector, $k = \sqrt{k_x^2 + k_y^2}$, and A, M are positive constants. Calculate the Chern number.

This problem is a direct application of the formula

$$C = \sum_{k_D \in D_i} sign\left(\Delta(\boldsymbol{k}_{D_i})\right) sign\left(\left|A(\boldsymbol{k}_{D_i})\right|\right),$$

with the Dirac point at $\boldsymbol{k_D} = (0,0)$ and $\Delta(\boldsymbol{k_D}) = 0$. Therefore, $C = 0$.

Problem 3.17

The eigenvalue equation for a periodic crystal with nondegenerate states is $\widehat{H}_\lambda|\,\Phi_n(\boldsymbol{k},\lambda)\rangle = E_n(\boldsymbol{k})\,|\Phi_n(\boldsymbol{k},\lambda)\rangle$, where the Hamiltonian is $\widehat{H}_\lambda = \frac{\widehat{p}^2}{2m} + V_\lambda(\widehat{r})$ with λ being an adiabatically changing parameter, $|\Phi_n(\boldsymbol{k},\lambda)\rangle$ are the eigenstates with a wave vector $\boldsymbol{k}$ for the nth band, and $E_n(\boldsymbol{k})$ are the corresponding eigenvalues. Show that by using the difference of the electric polarization $\boldsymbol{P}$ of a material with a periodic lattice defined as

$$\Delta \boldsymbol{P} = \int_0^1 d\lambda \frac{d\boldsymbol{P}}{d\lambda}, \text{ where } \boldsymbol{P}(\lambda) = \frac{q}{V}\sum_i \langle \Phi_i(\lambda)|\widehat{\boldsymbol{r}}|\Phi_i(\lambda)\rangle,$$

one can obtain an equivalent expression in terms of the Berry phase for an nth Bloch band,

$$\Delta \boldsymbol{P} = \frac{\hbar q}{V}\sum_n \int_{BZ} d\boldsymbol{k} \int_0^1 d\lambda \Omega^n_{\lambda,k_\mu}(\boldsymbol{k},\lambda).$$

In the preceding expressions, λ is an adiabatically changing parameter, V is the volume, q is the charge, and Ω^n_{λ,k_μ} is the Berry phase tensor component for the nth band. Here we note that the polarization $\boldsymbol{P}$ of an infinite periodic crystal depends on the choice of the unit cell, thus it is not a well-defined quantity. However, $\Delta\boldsymbol{P}$ is well defined, since it is connected to the Berry phase of Bloch electronic states as implied by this problem.

Here we provide two different solutions to this problem, which may be beneficial to the reader.

Version 1: The solution in this case can be given as a multistep process, which we start by executing the derivative $\frac{d\boldsymbol{P}}{d\lambda}$ and rewriting (note that $\frac{\partial \mathrm{r}}{\partial \lambda} = 0$):

$$\begin{aligned}\frac{d\boldsymbol{P}}{d\lambda} &= \frac{d}{d\lambda}\frac{q}{V}\sum_{n,k}\langle \Phi_n(\boldsymbol{k},\lambda)|\widehat{\boldsymbol{r}}|\Phi_n(\boldsymbol{k},\lambda)\rangle \\ &= \frac{q}{V}\sum_{n,k}\left[\left\langle \frac{\partial \Phi_n(\boldsymbol{k},\lambda)}{\partial\lambda}|\widehat{\boldsymbol{r}}|\Phi_n(\boldsymbol{k},\lambda)\right\rangle + \left\langle \Phi_n(\boldsymbol{k},\lambda)|\widehat{\boldsymbol{r}}|\frac{\partial \Phi_n(\boldsymbol{k},\lambda)}{\partial\lambda}\right\rangle\right].\end{aligned}$$

Using the identity, $\sum_m \left|\Phi_m(\boldsymbol{k},\lambda)\right\rangle \left\langle \Phi_m(\boldsymbol{k},\lambda)\right|) = 1$, we further write

$$\begin{aligned}\frac{d\boldsymbol{P}}{d\lambda} = \frac{q}{V}\sum_{n,k}\sum_m \Bigg[&\left\langle \frac{\partial \Phi_n(\boldsymbol{k},\lambda)}{\partial\lambda}|\Phi_m(\boldsymbol{k},\lambda)\right\rangle\langle\Phi_m(\boldsymbol{k},\lambda)|\widehat{\boldsymbol{r}}|\Phi_n(\boldsymbol{k},\lambda)\rangle \\ &+ \langle\Phi_n(\boldsymbol{k},\lambda)|\widehat{\boldsymbol{r}}|\Phi_m(\boldsymbol{k},\lambda)\rangle\left\langle\Phi_m(\boldsymbol{k},\lambda)|\frac{\partial \Phi_n(\boldsymbol{k},\lambda)}{\partial\lambda}\right\rangle\Bigg].\end{aligned}$$

To proceed further, we recognize the following:

Property 1: $\langle\Phi_m|\Phi_n\rangle = \delta_{mn}$,

$$\begin{aligned}0 = \boldsymbol{\nabla}\langle\Phi_m|\Phi_n\rangle &= \langle\boldsymbol{\nabla}\Phi_m|\Phi_n\rangle + \langle\Phi_m|\boldsymbol{\nabla}\Phi_n\rangle. \\ \langle\boldsymbol{\nabla}\Phi_m|\Phi_n\rangle &= -\langle\Phi_m|\boldsymbol{\nabla}\Phi_n\rangle.\end{aligned}$$

Property 2: $\widehat{r}_\mu = i\hbar\frac{\partial}{\partial k^\mu} = i\hbar\nabla_\mu$,

$$\langle\Phi_n(\boldsymbol{k},\lambda)|\widehat{r}_\mu|\Phi_m(\boldsymbol{k},\lambda)\rangle = \langle\Phi_n(\boldsymbol{k},\lambda)|i\hbar\nabla_\mu|\Phi_m(\boldsymbol{k},\lambda)\rangle = i\hbar\langle\Phi_n(\boldsymbol{k},\lambda)|\nabla_\mu\Phi_m(k,\lambda)\rangle.$$

Rewriting the derivative of the polarization again, we find

$$\frac{dP_\mu}{d\lambda} = \frac{i\hbar q}{V}\sum_{n,k}\sum_{m}\Bigg[\Big\langle\frac{\partial\Phi_n(\boldsymbol{k},\lambda)}{\partial\lambda}|\Phi_m(\boldsymbol{k},\lambda)\rangle\langle\Phi_m(\boldsymbol{k},\lambda)|\nabla_\mu\Phi_n(\boldsymbol{k},\lambda)\rangle$$
$$- \langle\nabla_\mu\Phi_n(\boldsymbol{k},\lambda)|\Phi_m(\boldsymbol{k},\lambda)\rangle\langle\Phi_m(\boldsymbol{k},\lambda)|\frac{\partial\Phi_n(\boldsymbol{k},\lambda)}{\partial\lambda}\Big\rangle\Bigg]$$
$$= \frac{i\hbar q}{V}\sum_{n,k}\Bigg[\Big\langle\frac{\partial\Phi_n(\boldsymbol{k},\lambda)}{\partial\lambda}\Big|\frac{\partial\Phi_n(\boldsymbol{k},\lambda)}{\partial k_\mu}\Big\rangle - \Big\langle\frac{\partial\Phi_n(\boldsymbol{k},\lambda)}{\partial k_\mu}\Big|\frac{\partial\Phi_n(\boldsymbol{k},\lambda)}{\partial\lambda}\Big\rangle\Bigg].$$

Under the sum, we simply give the Berry curvature tensor, obtaining

$$\frac{dP_\mu}{d\lambda} = \frac{i\hbar q}{V}\sum_{n,k}\Bigg[\Big\langle\frac{\partial\Phi_n(\boldsymbol{k},\lambda)}{\partial\lambda}\Big|\frac{\partial\Phi_n(\boldsymbol{k},\lambda)}{\partial k_\mu}\Big\rangle - \Big\langle\frac{\partial\Phi_n(\boldsymbol{k},\lambda)}{\partial k_\mu}\Big|\frac{\partial\Phi_n(\boldsymbol{k},\lambda)}{\partial\lambda}\Big\rangle\Bigg],$$

where we recognize that $\frac{dP_\mu}{d\lambda} = \frac{\hbar q}{V}\sum_{n,k}\Omega^n_{\lambda,k_\mu}(\boldsymbol{k},\lambda)$. This can finally be expressed in an integral form:

$$\Delta\boldsymbol{P} = \int_0^1 d\lambda\frac{d\boldsymbol{P}}{d\lambda} = \frac{\hbar q}{V}\sum_n\int_{BZ}d\boldsymbol{k}\int_0^1 d\lambda\Omega^n_{\lambda,k_\mu}(\boldsymbol{k},\lambda).$$

Version 2: For this version of the solution process, we start by executing the derivative $\frac{d\boldsymbol{P}}{d\lambda}$ and rewriting (note that $\frac{\partial \mathbf{r}}{\partial\lambda} = 0$):

$$\frac{d\boldsymbol{P}}{d\lambda} = \frac{d}{d\lambda}\frac{q}{V}\sum_{n,k}\langle\Phi_n(\boldsymbol{k},\lambda)|\widehat{\boldsymbol{r}}|\Phi_n(\boldsymbol{k},\lambda)\rangle$$
$$= \frac{q}{V}\sum_{n,k}\Bigg[\Big\langle\frac{\partial\Phi_n(\boldsymbol{k},\lambda)}{\partial\lambda}|\widehat{\boldsymbol{r}}|\Phi_n(\boldsymbol{k},\lambda)\rangle + \langle\Phi_n(\boldsymbol{k},\lambda)|\widehat{\boldsymbol{r}}|\frac{\partial\Phi_n(\boldsymbol{k},\lambda)}{\partial\lambda}\Big\rangle\Bigg].$$

Using that $\widehat{r}^\mu = +i\hbar\frac{\partial}{\partial k_\mu} = i\hbar\nabla^\mu$ and applying this operator to the eigenfunction of each bra-ket, we get

$$\frac{dP^\mu}{d\lambda} = \frac{q}{V}\sum_{n,k}\Bigg[\Big\langle\frac{\partial\Phi_n(\boldsymbol{k},\lambda)}{\partial\lambda}|i\hbar\nabla^\mu|\Phi_n(\boldsymbol{k},\lambda)\rangle + \langle\Phi_n(\boldsymbol{k},\lambda)\,|i\hbar\nabla^\mu|\,\frac{\partial\Phi_n(\boldsymbol{k},\lambda)}{\partial\lambda}\Big\rangle\Bigg]$$
$$= \frac{q}{V}\sum_{n,k}\Bigg[\Big\langle\frac{\partial\Phi_n(\boldsymbol{k},\lambda)}{\partial\lambda}|i\hbar\nabla^\mu\Phi_n(\boldsymbol{k},\lambda)\rangle + \langle -i\hbar\nabla^\mu\Phi_n(\boldsymbol{k},\lambda)|\frac{\partial\Phi_n(\boldsymbol{k},\lambda)}{\partial\lambda}\Big\rangle\Bigg]$$
$$= \frac{i\hbar q}{V}\sum_{n,k}\Bigg[\Big\langle\frac{\partial\Phi_n(\boldsymbol{k},\lambda)}{\partial\lambda}\Big|\frac{\partial\Phi_n(\boldsymbol{k},\lambda)}{\partial k_\mu}\Big\rangle - \Big\langle\frac{\partial\Phi_n(\boldsymbol{k},\lambda)}{\partial k_\mu}\Big|\frac{\partial\Phi_n(\boldsymbol{k},\lambda)}{\partial\lambda}\Big\rangle\Bigg].$$

Here we recognize again $\frac{dP_\mu}{d\lambda} = \frac{\hbar q}{V}\sum_{n,k}\Omega^n_{\lambda,k_\mu}(\boldsymbol{k},\lambda)$, which can finally be expressed in an integral form, $\Delta\boldsymbol{P} = \int_0^1 d\lambda\frac{d\boldsymbol{P}}{d\lambda} = \frac{\hbar q}{V}\sum_n\int_{BZ}d\boldsymbol{k}\int_0^1 d\lambda\Omega^n_{\lambda,k_\mu}(\boldsymbol{k},\lambda)$.

Food for thought: This result for the polarization, now obtained in two ways, has far-reaching consequences in condensed matter physics, as first recognized in Vanderbilt (2018).

Problem 3.18

The momentum operator $\widehat{\boldsymbol{p}}$ in position space is represented as $\widehat{\boldsymbol{p}} = -i\hbar\boldsymbol{\nabla}_{\boldsymbol{r}}$. The position operator $\widehat{\boldsymbol{r}}$ in momentum space is represented as $\widehat{\boldsymbol{r}} = i\hbar\boldsymbol{\nabla}_{\boldsymbol{p}}$. These definitions are valid in free space. In a periodic environment (such as a lattice), these definitions need to be reexamined, however. Show that for a periodic lattice, the momentum and position operators are

$$\widehat{\boldsymbol{r}} = i\boldsymbol{\nabla}_{\boldsymbol{k}} + \boldsymbol{A}_n,$$
$$\widehat{\boldsymbol{p}} = -i\hbar\boldsymbol{\nabla}_{\boldsymbol{r}},$$

where $\boldsymbol{k}$ is the wave vector and $\boldsymbol{A}_{nm}(\boldsymbol{k},\boldsymbol{r}) = i\langle u_{n,\boldsymbol{k}}(\boldsymbol{r})|\boldsymbol{\nabla}_{\boldsymbol{k}} u_{m,\boldsymbol{k}}(\boldsymbol{r})\rangle$ is the interband Berry connection for the Bloch states $u_{n,\boldsymbol{k}}(\boldsymbol{r})$. In the preceding we have taken the single band approximation, meaning that the Berry connection is $\boldsymbol{A}_n(\boldsymbol{k}) = i\langle u_{n,\boldsymbol{k}}(\boldsymbol{r})|\boldsymbol{\nabla}_{\boldsymbol{k}} u_{n,\boldsymbol{k}}(\boldsymbol{r})\rangle$.

To show the preceding relation, one can consider what the action of $\widehat{\boldsymbol{r}}$ is on a general wave function $\Psi(\boldsymbol{r})$ representing the Bloch band basis for the Hamiltonian. For this purpose, we recall the eigenstates of a Hamiltonian in a periodic environment, $\widehat{H}\psi_{n,\boldsymbol{k}}(\boldsymbol{r}) = E_{n,\boldsymbol{k}}\psi_{n,\boldsymbol{k}}(\boldsymbol{r})$, where $\psi_{n,\boldsymbol{k}}(\boldsymbol{r}) = u_{n,\boldsymbol{k}}(\boldsymbol{r})e^{i\boldsymbol{k}\cdot\boldsymbol{r}}$. Thus, by expanding $\Psi(\boldsymbol{r})$ in the eigenstates of the Hamiltonian and using $\widehat{\boldsymbol{r}}|\psi_\alpha(\boldsymbol{x}_n)\rangle = \boldsymbol{x}_n|\psi_\alpha(\boldsymbol{x}_n)\rangle$, we write

$$\begin{aligned}
\widehat{\boldsymbol{r}}\Psi(\boldsymbol{x}) &= \widehat{\boldsymbol{r}}\left[\sum_n \int d\boldsymbol{k}\alpha_n(\boldsymbol{k})\psi_{n,\boldsymbol{k}}(\boldsymbol{x})\right] = \sum_n \int d\boldsymbol{k}\alpha_n(\boldsymbol{k})u_{n,\boldsymbol{k}}(\boldsymbol{x})\boldsymbol{x}e^{i\boldsymbol{k}\cdot\boldsymbol{x}} \\
&= \sum_n \int d\boldsymbol{k}\alpha_n(\boldsymbol{k})u_{n,\boldsymbol{k}}(\boldsymbol{x})\left(-i\boldsymbol{\nabla}_{\boldsymbol{k}}e^{i\boldsymbol{k}\cdot\boldsymbol{x}}\right) \\
&= \sum_n \int d\boldsymbol{k}\left(i\boldsymbol{\nabla}_{\boldsymbol{k}}\left[\alpha_n(\boldsymbol{k})u_{n,\boldsymbol{k}}(\boldsymbol{x})\right]\right)e^{i\boldsymbol{k}\cdot\boldsymbol{x}}.
\end{aligned}$$

Using the following property,

$$\begin{aligned}
\sum_n \int d\boldsymbol{k}\left(i\boldsymbol{\nabla}_{\boldsymbol{k}}\left[\alpha_n(\boldsymbol{k})u_{n,\boldsymbol{k}}(\boldsymbol{x})\right]\right)e^{i\boldsymbol{k}\cdot\boldsymbol{x}} &= \sum_n \int d\boldsymbol{k}\left(i\boldsymbol{\nabla}_{\boldsymbol{k}}\alpha_n(\boldsymbol{k})\right)u_{n,\boldsymbol{k}}(\boldsymbol{x})e^{i\boldsymbol{k}\cdot\boldsymbol{x}} \\
&\quad + \sum_n \int d\boldsymbol{k}\alpha_n(\boldsymbol{k})\left(i\boldsymbol{\nabla}_{\boldsymbol{k}}u_{n,\boldsymbol{k}}(\boldsymbol{x})\right)e^{i\boldsymbol{k}\cdot\boldsymbol{x}},
\end{aligned}$$

we rewrite:

$$\widehat{\boldsymbol{r}}\Psi(\boldsymbol{x}) = \sum_n \int d\boldsymbol{k}\left[\left(i\boldsymbol{\nabla}_{\boldsymbol{k}}\alpha_n(\boldsymbol{k})\right)u_{n,\boldsymbol{k}}(\boldsymbol{x}) + \alpha_n(\boldsymbol{k})\left(i\boldsymbol{\nabla}_{\boldsymbol{k}}u_{n,\boldsymbol{k}}(\boldsymbol{x})\right)\right]e^{i\boldsymbol{k}\cdot\boldsymbol{x}}.$$

Noting that

$$\sum_m u^*_{m,\boldsymbol{k}}(\boldsymbol{x}_0)u_{m,\boldsymbol{k}}(\boldsymbol{x}) = \delta(\boldsymbol{x}-\boldsymbol{x}_0),$$

we obtain

$$\begin{aligned}
\widehat{\boldsymbol{r}}\Psi(\boldsymbol{x}) &= \sum_n \int d\boldsymbol{k}\left[(i\boldsymbol{\nabla}_{\boldsymbol{k}}\alpha_n(\boldsymbol{k}))u_{n,\boldsymbol{k}}(\boldsymbol{x}) + \alpha_n(\boldsymbol{k})\int d\boldsymbol{x}_0\left(i\boldsymbol{\nabla}_{\boldsymbol{k}}u_{n,\boldsymbol{k}}(\boldsymbol{x}_0)\right)\delta(\boldsymbol{x}-\boldsymbol{x}_0)\right]e^{i\boldsymbol{k}\cdot\boldsymbol{x}} \\
&= \sum_n \int d\boldsymbol{k}\left[(i\boldsymbol{\nabla}_{\boldsymbol{k}}\alpha_n(\boldsymbol{k}))u_{n,\boldsymbol{k}}(\boldsymbol{x}) + \alpha_n(\boldsymbol{k})\int d\boldsymbol{x}_0\left[\left(i\boldsymbol{\nabla}_{\boldsymbol{k}}u_{n,\boldsymbol{k}}(\boldsymbol{x}_0)\right)\sum_m u^*_{m,\boldsymbol{k}}(\boldsymbol{x}_0)u_{m,\boldsymbol{k}}(\boldsymbol{x})\right]\right]e^{i\boldsymbol{k}\cdot\boldsymbol{x}}.
\end{aligned}$$

Rewriting again, using the definition $A_{nm}(\boldsymbol{k},\boldsymbol{r}) = \langle u_{n,\boldsymbol{k}}(\boldsymbol{r})|i\nabla_{\boldsymbol{k}} u_{m,\boldsymbol{k}}(\boldsymbol{r})\rangle$,

$$\begin{aligned}\widehat{\boldsymbol{r}}\Psi(\boldsymbol{r}) &= \sum_n \int d\boldsymbol{k}\left[(i\nabla_{\boldsymbol{k}}\alpha_n(\boldsymbol{k}))\,u_{n,\boldsymbol{k}}(\boldsymbol{x})e^{i\boldsymbol{k}\cdot\boldsymbol{x}} + \alpha_n(\boldsymbol{k})\sum_m e^{i\boldsymbol{k}\cdot\boldsymbol{x}}u_{m,\boldsymbol{k}}(\boldsymbol{x})\int d\boldsymbol{x}_0\left[u^*_{m,\boldsymbol{k}}(\boldsymbol{x}_0)\left(i\nabla_{\boldsymbol{k}}u_{n,\boldsymbol{k}}(\boldsymbol{x}_0)\right)\right]\right]\\ &= \sum_n \int d\boldsymbol{k}\left[(i\nabla_{\boldsymbol{k}}\alpha_n(\boldsymbol{k}))\,u_{n,\boldsymbol{k}}(\boldsymbol{x})e^{i\boldsymbol{k}\cdot\boldsymbol{x}} + \alpha_n(\boldsymbol{k})\sum_m e^{i\boldsymbol{k}\cdot\boldsymbol{x}}u_{m,\boldsymbol{k}}(\boldsymbol{x})\int d\boldsymbol{x}_0 A_{mn}(\boldsymbol{k},\boldsymbol{x}_0)\right].\end{aligned}$$

Now using $\int d\boldsymbol{x}_0 A_{mn}(\boldsymbol{k},\boldsymbol{x}_0) = A_{mn}(\boldsymbol{k})$, we obtain

$$\begin{aligned}\widehat{\boldsymbol{r}}\Psi(\boldsymbol{x}) &= \sum_n \int d\boldsymbol{k}\left[(i\nabla_{\boldsymbol{k}}\alpha_n(\boldsymbol{k}))\,u_{n,\boldsymbol{k}}(\boldsymbol{x})e^{i\boldsymbol{k}\cdot\boldsymbol{x}} + \alpha_n(\boldsymbol{k})\sum_m e^{i\boldsymbol{k}\cdot\boldsymbol{x}}u_{m,\boldsymbol{k}}(\boldsymbol{x})A_{mn}(\boldsymbol{k})\right]\\ &= \sum_{n,m}\int d\boldsymbol{k}\left[(i\nabla_{\boldsymbol{k}}\delta_{mn} + A_{mn}(\boldsymbol{k}))\,\alpha_n(\boldsymbol{k})\right]u_{m,\boldsymbol{k}}(\boldsymbol{x})e^{i\boldsymbol{k}\cdot\boldsymbol{x}}.\end{aligned}$$

Finally, applying the single band approximation $A_{mn}(\boldsymbol{k}) = \delta_{nm}A_n(\boldsymbol{k})$, we obtain

$$\begin{aligned}\widehat{\boldsymbol{r}}\Psi(\boldsymbol{x}) &= \sum_{n,m}\int d\boldsymbol{k}\left[\delta_{mn}\left(i\nabla_{\boldsymbol{k}} + A_n(\boldsymbol{k})\right)\alpha_n(\boldsymbol{k})\right]u_{m,\boldsymbol{k}}(\boldsymbol{x})e^{i\boldsymbol{k}\cdot\boldsymbol{x}}\\ &= \sum_n \int d\boldsymbol{k}\left[(i\nabla_{\boldsymbol{k}} + A_n(\boldsymbol{k}))\,\alpha_n(\boldsymbol{k})\right]u_{n,\boldsymbol{k}}(\boldsymbol{x})e^{i\boldsymbol{k}\cdot\boldsymbol{x}}.\end{aligned}$$

Therefore, we find that $\widehat{\boldsymbol{r}} = i\nabla_{\boldsymbol{k}} + A_n(\boldsymbol{k})$ in the single band approximation, and $\widehat{\boldsymbol{r}} = i\delta_{mn}\nabla_{\boldsymbol{k}} + A_{mn}(\boldsymbol{k})$ in the general case.

Consider now the momentum operator by keeping in mind that $\widehat{\boldsymbol{p}} = \hbar\boldsymbol{k}$,

$$\begin{aligned}\widehat{\boldsymbol{p}}\tilde{\Psi}(\boldsymbol{k}) &= \widehat{\boldsymbol{p}}\int d\boldsymbol{x}\Psi(\boldsymbol{x})e^{-i\boldsymbol{k}\cdot\boldsymbol{x}} = \int d\boldsymbol{x}\Psi(\boldsymbol{x})\hbar\boldsymbol{k}e^{-i\boldsymbol{k}\cdot\boldsymbol{x}} = \int d\boldsymbol{x}\Psi(\boldsymbol{x})\left(i\hbar\nabla_{\boldsymbol{x}}e^{-i\boldsymbol{k}\cdot\boldsymbol{x}}\right)\\ &= \int d\boldsymbol{x}\left(-i\hbar\nabla_{\boldsymbol{x}}\Psi(\boldsymbol{x})\right)e^{-i\boldsymbol{k}\cdot\boldsymbol{x}},\end{aligned}$$

thus $\widehat{\boldsymbol{p}} = -i\hbar\nabla_{\boldsymbol{x}}$.

Problem 3.19

a) Let a charged quantum mechanical particle move in the presence of external electric and magnetic fields. Derive the equations of motion for the position and momentum operators of this particle.

b) Let the charged quantum mechanical particle move in a periodic environment in the presence of external electric and magnetic fields. Obtain the equations of motion for $\widehat{\boldsymbol{r}}$ and $\widehat{\boldsymbol{p}}$ in this case by considering a Hamiltonian expanded about a given point $\boldsymbol{r}_0$ of the trajectory and retaining only the first two terms in this series.

---

a) To derive the equations of motion for the position and momentum operators, we recall that the equation of motion for any operator $\widehat{A}$ is $\frac{d\widehat{A}}{dt} = \frac{1}{i\hbar}\left[\widehat{A},\widehat{H}\right] + \frac{\partial\widehat{A}}{\partial t}$, where $\widehat{H}$ is the Hamiltonian of the system and $\frac{\partial\widehat{A}}{\partial t}$ is the partial time derivative of a time-dependent operator $\widehat{A}$.

The Hamiltonian for a charged particle in external fields is

$$\widehat{H}=\frac{(\widehat{\boldsymbol{p}}-q\boldsymbol{A}(\widehat{\boldsymbol{r}}))^2}{2m}-\frac{q\hbar}{2m}\boldsymbol{\sigma}\cdot\boldsymbol{B}(\widehat{\boldsymbol{r}})+V(\widehat{\boldsymbol{r}})-q\Phi(\widehat{\boldsymbol{r}}),$$

where q is the charge of the particle, $\boldsymbol{A}(\boldsymbol{r}),\Phi(\boldsymbol{r})$ are the vector and scalar potentials associated with the magnetic $\boldsymbol{B}(\boldsymbol{r})$ and electric $\boldsymbol{E}(\boldsymbol{r})$ fields respectively. Therefore,

$$\begin{aligned}\frac{d\widehat{r}_\mu}{dt}=\frac{1}{i\hbar}\left[\widehat{r}_\mu,\widehat{H}\right]+\frac{\partial\widehat{r}_\mu}{\partial t}&=\frac{1}{i\hbar}\frac{1}{2m}\left[\widehat{r}_\mu,(\widehat{\boldsymbol{p}}-q\boldsymbol{A}(\widehat{\boldsymbol{r}}))^2\right]\\&+\left[\widehat{r}_\mu,-\frac{q\hbar}{2m}\boldsymbol{\sigma}\cdot\boldsymbol{B}(\widehat{\boldsymbol{r}})+V(\widehat{\boldsymbol{r}})-q\Phi(\widehat{\boldsymbol{r}})\right]+0.\end{aligned}$$

Using $[\widehat{\boldsymbol{r}},f(\widehat{\boldsymbol{r}})]=0$ and $\left[\widehat{r}_\mu,\widehat{p}_\nu\right]=i\hbar\delta_{\mu\nu}$, we obtain

$$\begin{aligned}\frac{d\widehat{r}_\mu}{dt}=\frac{1}{i\hbar}\frac{1}{2m}\delta^{\nu\rho}\Big(&\left[\widehat{r}_\mu,\widehat{p}_\nu\widehat{p}_\rho\right]+q^2\left[\widehat{r}_\mu,A_\nu(\widehat{\boldsymbol{r}})A_\rho(\widehat{\boldsymbol{r}})\right]-q\left[\widehat{r}_\mu,\widehat{p}_\nu A_\rho(\widehat{\boldsymbol{r}})\right]\\&-q\left[\widehat{r}_\mu,A_\nu(\widehat{\boldsymbol{r}})\widehat{p}_\rho\right]\Big),\end{aligned}$$

$$\begin{aligned}\left[\widehat{r}_\mu,\widehat{p}_\nu\widehat{p}_\rho\right]\delta^{\nu\rho}&=\delta^{\nu\rho}\left[\widehat{r}_\mu,\widehat{p}_\nu\right]\widehat{p}_\rho+\delta^{\nu\rho}\widehat{p}_\nu\left[\widehat{r}_\mu,\widehat{p}_\rho\right]\\&=\delta^{\nu\rho}\left(i\hbar\delta_{\mu\nu}\widehat{p}_\rho+\widehat{p}_\nu i\hbar\delta_{\mu\rho}\right)=2i\hbar\widehat{p}_\mu,\\\left[\widehat{r}_\mu,\widehat{p}_\nu A_\rho(\widehat{\boldsymbol{r}})\right]\delta^{\nu\rho}&=\delta^{\nu\rho}\left[\widehat{r}_\mu,\widehat{p}_\nu\right]A_\rho(\widehat{\boldsymbol{r}})+\delta^{\nu\rho}\widehat{p}_\nu\left[\widehat{r}_\mu,A_\rho(\widehat{\boldsymbol{r}})\right]\\&=\delta^{\nu\rho}i\hbar\delta_{\mu\nu}A_\rho(\widehat{\boldsymbol{r}})+0=i\hbar A_\mu(\widehat{\boldsymbol{r}}),\\\left[\widehat{r}_\mu,A_\nu(\widehat{\boldsymbol{r}})\widehat{p}_\rho\right]\delta^{\nu\rho}&=\delta^{\nu\rho}\left[\widehat{r}_\mu,A_\nu(\widehat{\boldsymbol{r}})\right]\widehat{p}_\rho+\delta^{\nu\rho}A_\nu(\widehat{\boldsymbol{r}})\left[\widehat{r}_\mu,\widehat{p}_\rho\right]\\&=0+\delta^{\nu\rho}A_\nu(\widehat{\boldsymbol{r}})i\hbar\delta_{\mu\rho}=i\hbar A_\mu(\widehat{\boldsymbol{r}}).\end{aligned}$$

Then,

$$\begin{aligned}\frac{d\widehat{r}_\mu}{dt}=\frac{1}{i\hbar}\frac{1}{2m}\left(2i\hbar\widehat{p}_\mu+q^2 0-qi\hbar A_\mu(\widehat{\boldsymbol{r}})-qi\hbar A_\mu(\widehat{\boldsymbol{r}})\right)&=\frac{1}{m}\left(\widehat{p}_\mu-qA_\mu(\widehat{\boldsymbol{r}})\right),\\\widehat{\dot{\boldsymbol{r}}}=\frac{1}{m}\left(\widehat{\boldsymbol{p}}-q\boldsymbol{A}(\widehat{\boldsymbol{r}})\right).\end{aligned}$$

To obtain the dynamic equation for the momentum operator, we are going to study the dynamical evolution of the conjugate momentum,

$$\widehat{\boldsymbol{\pi}}=m\widehat{\dot{\boldsymbol{r}}}=\widehat{\boldsymbol{p}}-q\boldsymbol{A}(\widehat{\boldsymbol{r}}),$$

then write the Hamiltonian as

$$\widehat{H}=\frac{\widehat{\boldsymbol{\pi}}^2}{2m}-\frac{q\hbar}{2m}\boldsymbol{\sigma}\cdot\boldsymbol{B}(\widehat{\boldsymbol{r}})+V(\widehat{\boldsymbol{r}})-q\Phi(\widehat{\boldsymbol{r}}).$$

The equation of motion for components of $\widehat{\pi}$ becomes

$$\frac{d\widehat{\pi}_\mu}{dt}=\frac{1}{i\hbar}\left[\widehat{\pi}_\mu,\widehat{H}\right]+\frac{\partial\widehat{\pi}_\mu}{\partial t}=\frac{1}{i\hbar}\left[\widehat{\pi}_\mu,\widehat{H}\right],$$

$$\left[\widehat{\pi}_\mu,\widehat{H}\right]=\left[\widehat{\pi}_\mu,\frac{\widehat{\boldsymbol{\pi}}^2}{2m}\right]+\left[\widehat{\pi}_\mu,\left(-\frac{q\hbar}{2m}\boldsymbol{\sigma}\cdot\boldsymbol{B}(\widehat{\boldsymbol{r}})+V(\widehat{\boldsymbol{r}})-q\Phi(\widehat{\boldsymbol{r}})\right)\right],$$

$$\left[\widehat{\pi}_\mu,\widehat{H}\right] = \frac{\delta^{\nu\rho}}{2m}\left[\widehat{\pi}_\mu,\widehat{\pi}_\nu\widehat{\pi}_\rho\right] + \left[(\widehat{p}_\mu - qA_\mu(\widehat{\boldsymbol{r}})),\left(-\frac{q\hbar}{2m}\boldsymbol{\sigma}\cdot\boldsymbol{B}(\widehat{\boldsymbol{r}}) + V(\widehat{\boldsymbol{r}}) - q\Phi(\widehat{\boldsymbol{r}})\right)\right].$$

After evaluating the commutators,

$$\begin{aligned}
\delta^{\nu\rho}\left[\widehat{\pi}_\mu,\widehat{\pi}_\nu\widehat{\pi}_\rho\right] &= \delta^{\nu\rho}\widehat{\pi}_\nu\left[\widehat{\pi}_\mu,\widehat{\pi}_\rho\right] + \delta^{\nu\rho}\left[\widehat{\pi}_\mu,\widehat{\pi}_\nu\right]\widehat{\pi}_\rho,\\
\left[\widehat{\pi}_\mu,\widehat{\pi}_\nu\right] &= \left[\widehat{p}_\mu - qA_\mu(\widehat{\boldsymbol{r}}),\widehat{p}_\nu - qA_\nu(\widehat{\boldsymbol{r}})\right]\\
&= \left[\widehat{p}_\mu,\widehat{p}_\nu\right] - q\left[\widehat{p}_\mu,A_\nu(\widehat{\boldsymbol{r}})\right] - q\left[A_\mu(\widehat{\boldsymbol{r}}),\widehat{p}_\nu\right] + q^2\left[A_\mu(\widehat{\boldsymbol{r}}),A_\nu(\widehat{\boldsymbol{r}})\right]\\
&= 0 - q\left[-i\hbar\nabla_\mu,A_\nu(\widehat{\boldsymbol{r}})\right] - q\left[A_\mu(\widehat{\boldsymbol{r}}),-i\hbar\nabla_\nu\right] + q^2 0\\
&= i\hbar q\left(\nabla_\mu A_\nu(\widehat{\boldsymbol{r}}) - \nabla_\nu A_\mu(\widehat{\boldsymbol{r}})\right),
\end{aligned}$$

we can write

$$\left[\widehat{\pi}_\mu,\widehat{\pi}_\nu\right] = i\hbar q\varepsilon_{\mu\nu\alpha}\varepsilon^{\alpha\beta\gamma}\nabla_\beta A_\gamma(\widehat{\boldsymbol{r}}).$$

Because $\varepsilon_{\mu\nu\alpha}\varepsilon^{\alpha\beta\gamma} = \delta^\beta_\mu\delta^\gamma_\nu - \delta^\gamma_\mu\delta^\beta_\nu$, we obtain

$$\begin{aligned}
\delta^{\nu\rho}\left[\widehat{\pi}_\mu,\widehat{\pi}_\nu\widehat{\pi}_\rho\right] &= \delta^{\nu\rho}\widehat{\pi}_\nu i\hbar q\varepsilon_{\mu\rho\alpha}\varepsilon^{\alpha\beta\gamma}\nabla_\beta A_\gamma(\widehat{\boldsymbol{r}}) + \delta^{\nu\rho}i\hbar q\varepsilon_{\mu\nu\alpha}\varepsilon^{\alpha\beta\gamma}\nabla_\beta A_\gamma(\widehat{\boldsymbol{r}})\widehat{\pi}_\rho\\
&= 2i\hbar q\widehat{\pi}^\nu\varepsilon_{\mu\nu\alpha}\varepsilon^{\alpha\beta\gamma}\nabla_\beta A_\gamma(\widehat{\boldsymbol{r}}).
\end{aligned}$$

This commutator can be written in a vector form as

$$\delta^{\nu\rho}\left[\widehat{\pi}_\mu,\widehat{\pi}_\nu\widehat{\pi}_\rho\right] = 2i\hbar q\varepsilon_{\mu\nu\alpha}\widehat{\pi}^\nu\left[\boldsymbol{\nabla}_{\boldsymbol{r}}\times\boldsymbol{A}(\widehat{\boldsymbol{r}})\right]^\alpha = 2i\hbar q\left[\widehat{\boldsymbol{\pi}}\times\left[\boldsymbol{\nabla}_{\boldsymbol{r}}\times\boldsymbol{A}(\widehat{\boldsymbol{r}})\right]\right]_\mu.$$

The other term of the commutator is

$$\left[\widehat{\pi}_\mu,\left(-\frac{q\hbar}{2m}\boldsymbol{\sigma}\cdot\boldsymbol{B}(\widehat{\boldsymbol{r}}) + V(\widehat{\boldsymbol{r}}) - q\Phi(\widehat{\boldsymbol{r}})\right)\right] = -i\hbar\nabla_\mu\left(-\frac{q\hbar}{2m}\boldsymbol{\sigma}\cdot\boldsymbol{B}(\widehat{\boldsymbol{r}}) + V(\widehat{\boldsymbol{r}}) - q\Phi(\widehat{\boldsymbol{r}})\right).$$

Finally, we group all the commutators:

$$\begin{aligned}
\frac{d\widehat{\pi}_\mu}{dt} &= \frac{1}{i\hbar}\left[\widehat{\pi}_\mu,\widehat{H}\right] = \frac{1}{i\hbar}\left(\frac{i\hbar q}{m}\left[\widehat{\boldsymbol{\pi}}\times(\boldsymbol{\nabla}_{\boldsymbol{r}}\times\boldsymbol{A}(\widehat{\boldsymbol{r}}))\right]_\mu - i\hbar\nabla_\mu\left(-\frac{q\hbar}{2m}\boldsymbol{\sigma}\cdot\boldsymbol{B}(\widehat{\boldsymbol{r}}) + V(\widehat{\boldsymbol{r}}) - q\Phi(\widehat{\boldsymbol{r}})\right)\right)\\
&= \frac{q}{m}\left[\widehat{\boldsymbol{\pi}}\times(\boldsymbol{\nabla}_{\boldsymbol{r}}\times\boldsymbol{A}(\widehat{\boldsymbol{r}}))\right]_\mu - \nabla_\mu\left(-\frac{q\hbar}{2m}\boldsymbol{\sigma}\cdot\boldsymbol{B}(\widehat{\boldsymbol{r}}) + V(\widehat{\boldsymbol{r}}) - q\Phi(\widehat{\boldsymbol{r}})\right),\\
\frac{d\widehat{\boldsymbol{\pi}}}{dt} &= -q\boldsymbol{\nabla}_{\boldsymbol{r}}\Phi(\widehat{\boldsymbol{r}}) + \frac{q}{m}\widehat{\boldsymbol{\pi}}\times(\boldsymbol{\nabla}_{\boldsymbol{r}}\times\boldsymbol{A}(\widehat{\boldsymbol{r}})) - \boldsymbol{\nabla}_{\boldsymbol{r}}V(\widehat{\boldsymbol{r}}) + \frac{q\hbar}{2m}\boldsymbol{\nabla}_{\boldsymbol{r}}(\boldsymbol{\sigma}\cdot\boldsymbol{B}(\widehat{\boldsymbol{r}})).
\end{aligned}$$

Using that the electric field is $E_\mu(\widehat{\boldsymbol{r}}) = -\nabla_\mu\Phi(\widehat{\boldsymbol{r}})$ and $\boldsymbol{B}(\widehat{\boldsymbol{r}}) = \boldsymbol{\nabla}_{\boldsymbol{r}}\times\boldsymbol{A}(\widehat{\boldsymbol{r}})$, we find

$$\frac{d\widehat{\boldsymbol{p}}}{dt} = q\boldsymbol{E}(\widehat{\boldsymbol{r}}) + q\dot{\widehat{\boldsymbol{r}}}\times\boldsymbol{B}(\widehat{\boldsymbol{r}}) - \boldsymbol{\nabla}_{\boldsymbol{r}}V(\widehat{\boldsymbol{r}}) + \frac{q\hbar}{2m}\boldsymbol{\nabla}_{\boldsymbol{r}}(\boldsymbol{\sigma}\cdot\boldsymbol{B}(\widehat{\boldsymbol{r}})).$$

We note that the first two terms correspond to the Lorentz force, the third one is the effect of the external potential, and the fourth one is the Zeeman effect with nonclassical analogue. Of course, if the external fields are zero ($\boldsymbol{A}(\widehat{\boldsymbol{r}}) = \boldsymbol{0}$, $\Phi(\widehat{\boldsymbol{r}}) = 0$), then we simply have the equations of motion for a free particle.

b) For a charged particle in a periodic environment, the natural basis for the Hamiltonian is the Bloch functions $\widehat{H}\psi_{n,\boldsymbol{k}}(\boldsymbol{x}) = E_{n,\boldsymbol{k}}\psi_{n,\boldsymbol{k}}(\boldsymbol{x})$, where $\psi_{n,\boldsymbol{k}}(\boldsymbol{x}) = u_{n,\boldsymbol{k}}(\boldsymbol{x})e^{i\boldsymbol{k}\cdot\boldsymbol{x}}$. In this case, the displacement operator becomes $\widehat{\boldsymbol{r}} = i\hbar\boldsymbol{\nabla}_{\boldsymbol{p}} + \boldsymbol{A}_n(\widehat{\boldsymbol{p}})$, where $\boldsymbol{A}_n(\widehat{\boldsymbol{p}})$ is the Berry connection for the nth Bloch state, as we showed in the previous problem. The presence of a periodic environment introduces some difficulties in obtaining the equations of motion for $\widehat{\boldsymbol{r}}$ and $\widehat{\boldsymbol{p}}$ operators. The reason is that now the position $\widehat{\boldsymbol{r}} = i\hbar\boldsymbol{\nabla}_{\boldsymbol{p}} + \boldsymbol{A}_n(\widehat{\boldsymbol{p}})$, analogously to the canonical momentum $\widehat{\boldsymbol{\pi}} = m\widehat{\boldsymbol{p}} - q\boldsymbol{A}(\widehat{\boldsymbol{r}})$ before, is a noncommutative operator; let's check this:

$$
\begin{aligned}
\left[\widehat{r}_\mu,\widehat{r}_\nu\right] &= \left[i\hbar\nabla_\mu + A_\mu(\widehat{\boldsymbol{p}}), i\hbar\nabla_\nu + A_\nu(\widehat{\boldsymbol{p}})\right] \\
&= \left[i\hbar\nabla_\mu, i\hbar\nabla_\nu\right] + \left[i\hbar\nabla_\mu, A_\nu(\widehat{\boldsymbol{p}})\right] + \left[A_\mu(\widehat{\boldsymbol{p}}), i\hbar\nabla_\nu\right] + \left[A_\mu(\widehat{\boldsymbol{p}}), A_\nu(\widehat{\boldsymbol{p}})\right] \\
&= 0 + i\hbar\left[\nabla_\mu, A_\nu(\widehat{\boldsymbol{p}})\right] + i\hbar\left[A_\mu(\widehat{\boldsymbol{p}}), \nabla_\nu\right] + 0 = i\hbar\left(\nabla_\mu A_\nu(\widehat{\boldsymbol{p}}) - \nabla_\nu A_\mu(\widehat{\boldsymbol{p}})\right) \\
&= i\hbar\varepsilon_{\mu\nu\alpha}\varepsilon^{\alpha\beta\gamma}\nabla_\beta A_\gamma(\widehat{\boldsymbol{p}}).
\end{aligned}
$$

This is reminiscent of the canonical momentum $\widehat{\boldsymbol{\pi}} = m\widehat{\boldsymbol{p}} - q\boldsymbol{A}(\widehat{\boldsymbol{r}})$, whose components also do not commute, as shown in part (a) of the problem. The equation of motion of the position is

$$
\frac{d\widehat{r}_\mu}{dt} = \frac{i}{\hbar}\left[\widehat{H},\widehat{r}_\mu\right] + \frac{\partial\widehat{r}_\mu}{\partial t}.
$$

We carry out the calculation of this commutator by using the property

$$
\left[\widehat{H}(\widehat{\boldsymbol{p}},\widehat{\boldsymbol{r}}),\widehat{r}_\mu\right] = \frac{\delta\widehat{H}}{\delta\widehat{p}_\nu}\left[\widehat{p}_\nu,\widehat{r}_\mu\right] + \frac{\delta\widehat{H}}{\delta\widehat{r}_\nu}\left[\widehat{r}_\nu,\widehat{r}_\mu\right],
$$

where $\frac{\delta\widehat{H}}{\delta\widehat{p}_\nu}$ are functional derivatives

$$
\frac{\delta\widehat{H}}{\delta\widehat{p}_\nu} = \frac{\partial\widehat{H}}{\partial\widehat{p}_\nu} - \frac{\partial}{\partial\widehat{x}^\mu}\left(\frac{\partial\widehat{H}}{\partial\left(\nabla_\mu\widehat{p}_\nu\right)}\right) + \ldots.
$$

Then, it is clear that $\frac{\delta\widehat{H}}{\delta\widehat{p}_\nu} = \frac{\partial\widehat{H}}{\partial\widehat{p}_\nu}$ and, using the Hamilton's equations $\dot{\widehat{p}}^\nu = -\frac{\delta\widehat{H}}{\delta\widehat{r}_\nu}$, we obtain

$$
\left[\widehat{H}(\widehat{\boldsymbol{p}},\widehat{\boldsymbol{r}}),\widehat{r}_\mu\right] = \frac{\partial\widehat{H}}{\partial\widehat{p}_\nu}\left[\widehat{p}_\nu,\widehat{r}_\mu\right] - \dot{\widehat{p}}^\nu\left[\widehat{r}_\nu,\widehat{r}_\mu\right].
$$

Further, using the commutation relation $\left[\widehat{r}_\mu,\widehat{p}_\nu\right] = i\hbar\delta_{\mu\nu}$ and $\left[\widehat{r}_\mu,\widehat{r}_\nu\right] = i\hbar\varepsilon_{\mu\nu\alpha}\varepsilon^{\alpha\beta\gamma}\nabla_\beta A_\gamma(\widehat{p})$,

$$
\begin{aligned}
\left[H(\widehat{\boldsymbol{p}},\widehat{\boldsymbol{r}}),\widehat{r}_\mu\right] &= -i\hbar\delta_{\mu\nu}\frac{\partial\widehat{H}}{\partial\widehat{p}_\nu} + \dot{\widehat{p}}^\nu i\hbar\varepsilon_{\mu\nu\alpha}\varepsilon^{\alpha\beta\gamma}\nabla_\beta A_\gamma(\widehat{p}) \\
&= -i\hbar\left(\delta_{\mu\nu}\frac{\partial\widehat{H}}{\partial\widehat{p}_\nu} - \dot{\widehat{p}}^\nu\varepsilon_{\mu\nu\alpha}\left(\boldsymbol{\nabla}_{\boldsymbol{p}}\times\boldsymbol{A}(\widehat{\boldsymbol{p}})\right)^\alpha\right) \\
&= -i\hbar\left(\frac{\partial\widehat{H}}{\partial\widehat{p}_\mu} - \left[\dot{\widehat{\boldsymbol{p}}}\times\left(\boldsymbol{\nabla}_{\boldsymbol{p}}\times\boldsymbol{A}(\widehat{\boldsymbol{p}})\right)\right]_\mu\right).
\end{aligned}
$$

Finally, the equation of motion of the position is

$$\frac{d\hat{r}_\mu}{dt} = \frac{i}{\hbar}\left[\hat{H},\hat{r}_\mu\right] + \frac{\partial \hat{r}_\mu}{\partial t} = \frac{i}{\hbar}\left[-i\hbar\left(\frac{\partial \hat{H}}{\partial \hat{p}_\mu} - \left[\hat{\boldsymbol{p}}\times\left(\boldsymbol{\nabla}_p\times\boldsymbol{A}(\hat{\boldsymbol{p}})\right)\right]_\mu\right)\right] + 0$$
$$= \frac{\partial \hat{H}}{\partial \hat{p}_\mu} - \left[\hat{\boldsymbol{p}}\times\left(\boldsymbol{\nabla}_p\times\boldsymbol{A}(\hat{\boldsymbol{p}})\right)\right]_\mu .$$

In vector form, and noting that the Berry curvature vector $\boldsymbol{\Omega}_n(\hat{\boldsymbol{p}}) = \boldsymbol{\nabla}_p\times\boldsymbol{A}_n(\hat{\boldsymbol{p}})$ plays the role of a magnetic field in the momentum space, we obtain

$$\frac{d\hat{\boldsymbol{r}}}{dt} = \boldsymbol{\nabla}_p\hat{H} - \dot{\hat{\boldsymbol{p}}}\times\left(\boldsymbol{\nabla}_p\times\boldsymbol{A}_n(\hat{\boldsymbol{p}})\right) = \frac{(\hat{\boldsymbol{p}} - q\boldsymbol{A}(\hat{\boldsymbol{r}}))}{m} - \dot{\hat{\boldsymbol{p}}}\times\left(\boldsymbol{\nabla}_p\times\boldsymbol{A}_n(\hat{\boldsymbol{p}})\right),$$
$$\frac{d\hat{\boldsymbol{p}}}{dt} = -\boldsymbol{\nabla}_r\hat{H} + q\dot{\hat{\boldsymbol{r}}}\times\left(\boldsymbol{\nabla}_r\times\boldsymbol{A}(\hat{\boldsymbol{r}})\right)$$
$$= -\boldsymbol{\nabla}_r\left(V(\hat{\boldsymbol{r}}) - \frac{q\hbar}{2m}\boldsymbol{\sigma}\cdot\boldsymbol{B}(\hat{\boldsymbol{r}}) - q\Phi(\hat{\boldsymbol{r}})\right) + q\dot{\hat{\boldsymbol{r}}}\times\left(\boldsymbol{\nabla}_r\times\boldsymbol{A}(\hat{\boldsymbol{r}})\right).$$

If we compare the dynamical equations with their classical analogs, we observe that the double cross product in the dynamical equations appears when the position and the canonical momentum do not commute with each other in the different directions of space.

Food for thought: Examples and specific consequences for the equations of motion in periodic environments in the presence of electromagnetic fields can be found in Xiao et al. (2010).

4 Density Operator

The concept of a *density operator* was first introduced by von Neumann in the context of quantum mechanical ensembles. The density operator is quite useful when we deal with randomness and statistics. With the emergence of quantum communication, quantum information, and other areas of quantum science, we need to be able to describe the information contained in a quantum mechanical system. It turns out that the foundation of entanglement, entropy, teleportation, and other relevant properties are understood using density operators.

Let's review some basic definitions and properties of a density operator $\widehat{\rho}$ and build our foundation with various problems. The following summary is not intended to provide a thorough representation of quantum information; rather, the properties given here are directly related to the density operator, giving an opportunity to practice knowledge from quantum mechanics. Specifically, there are many types of entropies defined in quantum information. Here, however, we introduce only the von Neumann and entanglement entropies to illustrate basic problems involving the density operator. We hope that emphasizing different aspects will be helpful for building better understanding of this unique concept.

4.1 Density Operator for a Single Particle

Density Operator $\widehat{\rho}$:

$$\widehat{\rho}=\sum_{n}\sum_{m}\rho_{n,m}|\phi_n\rangle\langle\phi_m|$$

This is a positive semidefinite Hermitian operator, such that $\widehat{\rho}=\widehat{\rho}^T$ with $\rho_{m,n}=(\rho_{n,m})^*$. Its definition relies on the eigenstates $|\phi_n\rangle$ of a given Hamiltonian. The $\widehat{\rho}$ operator has unity trace, $Tr(\widehat{\rho})=\sum_{n}\rho_{n,n}=1$.

Density Operator for a *Pure State*:

$$\widehat{\rho}=|\psi\rangle\langle\psi|$$

The quantum mechanical state $|\psi\rangle$ can be any superposition of eigenstates of a given Hamiltonian. For a pure state, $\widehat{\rho}^2=\widehat{\rho}$ with a purity parameter $\mathscr{P}(\widehat{\rho})=Tr(\widehat{\rho}^2)=1$.

Density Operator for a *Mixed State*:

$$\widehat{\rho}=\sum_{n}p_n|\psi_n\rangle\langle\psi_n|=\sum_{n}p_n\widehat{\rho}_n$$

Here $|\psi_i\rangle$ are individual pure states and p_i are their distribution probabilities, such that $\sum_n p_n = 1$. For a mixed state, $\widehat{\rho}^2 \neq \widehat{\rho}$ and its purity $\mathcal{P}(\widehat{\rho}) = Tr(\widehat{\rho}^2) < 1$.

Density Operator for a *Maximally Mixed State*:

$$\widehat{\rho} = \frac{1}{N}\sum_{i=1}^{N} |\psi_i\rangle\langle\psi_i|$$

The density operator is proportional to the identity matrix with diagonal elements $\rho_{nn} = \frac{1}{N}$ (N – space dimension).

Measurement of an Observable $\widehat{O}$: $\langle\widehat{O}\rangle = Tr[\widehat{\rho}\widehat{O}]$

Measuring an observable is defined as an ensemble average over the quantum mechanical distribution of states.

4.2 Density Operator for a Two-Particle System

For composite systems, quantum mechanical states contain two or more particles. As an example, we consider a two-particle system composed of A and B particles.

The Hamiltonian for a two-particle system:

$$\widehat{H}_{AB} = \widehat{H}_A \otimes \widehat{H}_B.$$

This is a tensor product of the Hamiltonians for the two subsystems, A and B.

Density operator for a two-particle system:

$$\widehat{\rho} \equiv \widehat{\rho}_{AB} = \sum_{n,m,j,k} \rho_{n,m}^{j,k} |a_n\rangle\langle a_m| \otimes |b_j\rangle\langle b_k| = \sum_{n,m,j,k} \rho_{n,m}^{j,k} |a_n b_j\rangle\langle a_m b_k|.$$

Here, $|a_n\rangle$ are the eigenstates corresponding to $\widehat{H}_A$ and $|b_j\rangle$ are the eigenstates corresponding to $\widehat{H}_B$. The composite eigenbasis $|a_n b_j\rangle$ corresponds to $\widehat{H}_{AB}$, and it can equivalently be represented as $|a_n b_j\rangle = |a_n\rangle \otimes |b_j\rangle = |a_n\rangle|b_j\rangle$.

Reduced density operator for subsystems A and B: $\widehat{\rho}_A = Tr_B(\widehat{\rho})$; $\widehat{\rho}_B = Tr_A(\widehat{\rho})$;

$$\begin{aligned}
\widehat{\rho}_B = Tr_A(\widehat{\rho}_{AB}) &= \sum_{\alpha=1} \langle a_\alpha|\widehat{\rho}_{AB}|a_\alpha\rangle \\
&= \sum_{\alpha=1} \langle a_\alpha| \sum_{n,m,j,k} \rho_{n,m}^{j,k} |a_n\rangle\langle a_m| \otimes |b_j\rangle\langle b_k| |a_\alpha\rangle \\
&= \sum_{\alpha=1} \sum_{n,m,j,k} \rho_{n,m}^{j,k} \langle a_\alpha|a_n\rangle\langle a_m|a_\alpha\rangle \otimes |b_j\rangle\langle b_k| \\
&= \sum_{\alpha=1} \sum_{n,m,j,k} \rho_{n,m}^{j,k} \delta_{\alpha n}\delta_{\alpha m} \otimes |b_j\rangle\langle b_k| \\
&= \sum_{j,k} \left(\sum_{\alpha=1} \rho_{\alpha,\alpha}^{j,k} \right) |b_j\rangle\langle b_k| = \sum_{j,k} \rho_B^{j,k} |b_j\rangle\langle b_k|.
\end{aligned}$$

The *partial trace* $\widehat{\rho}_B = Tr_A(\widehat{\rho}_{AB})$ is the density matrix that accounts for all the experimental observations done in subsystem B that do not involve subsystem A.

Density operator for a pure noncorrelated two-particle state: $|\psi_{AB}\rangle = |\psi_A\rangle \otimes |\psi_B\rangle$: $\widehat{\rho}_{AB} = \widehat{\rho}_A \otimes \widehat{\rho}_B$.

The noncorrelated state can be decomposed into two independent parts characterized by their own states $|\psi_A\rangle$, $|\psi_B\rangle$. As a result, the density operator is a tensor product of the individual density operators $\widehat{\rho}_A$ and $\widehat{\rho}_B$.

Density operator for a mixed classically correlated state: $\widehat{\rho}_{AB} = \sum_n p_n \widehat{\rho}^n_A \otimes \widehat{\rho}^n_B$.

Here $\widehat{\rho}^n_A = |\psi_{A,n}\rangle\langle\psi_{A,n}|$ and $\widehat{\rho}^n_B = |\psi_{B,n}\rangle\langle\psi_{B,n}|$, meaning that each term in the summation is a noncorrelated state. The populations $p_n \in [0,1]$, and $\sum_n p_n = 1$.

Density operator for an entangled, quantum correlated state: $\widehat{\rho}_{AB} = \sum_n p_n \widehat{\rho}^n_{AB}$.

Entangled states are the ones that cannot be written as a sum of product states. The populations $p_n \in [0,1]$, and $\sum_n p_n = 1$, and at least one $\widehat{\rho}^n_{AB}$ is not separable: $\widehat{\rho}^n_{Ab} \neq \widehat{\rho}^n_A \otimes \widehat{\rho}^n_B$.

Schmidt decomposition for $\widehat{H}_{AB} = \widehat{H}_A \otimes \widehat{H}_B$: $|\psi_{AB}\rangle = \sum_{n=1}^{r} g_n |\psi^n_A\rangle \otimes |\psi^n_B\rangle$.

The quantum mechanical state $|\psi_{AB}\rangle$ corresponding to $\widehat{H}_{AB}$ can always be given as in the preceding relation, where $|\psi^n_{A,B}\rangle$ are orthonormal and associated with $\widehat{H}_{AB}$. Also, g_n are nonnegative numbers with $\sum_{n=1}^{r} |g_n|^2 = 1$. The number r in the summation is termed as the *Schmidt number*. For a two-particle state that is not entangled, the Schmidt number $r = 1$, and $|\psi_{AB}\rangle = |\psi_A\rangle \otimes |\psi_B\rangle$. For an entangled two-particle state, $r \geq 2$, then $|\psi_{AB}\rangle \neq |\psi_A\rangle \otimes |\psi_B\rangle$.

4.3 Entropy

Von Neumann entropy:

$$S_N(\widehat{\rho}) = -Tr\left[\widehat{\rho}\ln(\widehat{\rho})\right] = -\sum_n \lambda_n \ln(\lambda_n).$$

The λ_n are the eigenvalues of the density matrix $\widehat{\rho}$. This quantity measures how much information the state has.

Entanglement entropy for a bipartite system $\widehat{H}_{AB} = \widehat{H}_A \otimes \widehat{H}_B$:

$$E = S(\widehat{\rho}_A) = -Tr_A\left[\widehat{\rho}_A \ln(\widehat{\rho}_A)\right] = -Tr_B\left[\widehat{\rho}_B \ln(\widehat{\rho}_B)\right] = S(\widehat{\rho}_B)$$
$$E \in [0, \ln(2)].$$

The $\widehat{\rho}_A$ and $\widehat{\rho}_B$ are the partial traces of $\widehat{\rho}_{AB}$. If $E = 0$, then $\widehat{\rho}_{AB}$ corresponds to a pure (nonentangled) state $\widehat{\rho}_{AB} = \widehat{\rho}_A \otimes \widehat{\rho}_B$. If $E = \ln(2)$, then $\widehat{\rho}_{AB}$ corresponds to a maximally entangled state $\widehat{\rho}_{AB} \neq \widehat{\rho}_A \otimes \widehat{\rho}_B$.

Bell states for two particles, each one with spin up $|\uparrow\rangle = \begin{pmatrix}1\\0\end{pmatrix}$ and down, $|\downarrow\rangle = \begin{pmatrix}0\\1\end{pmatrix}$:

$$|a,\pm\rangle = \frac{1}{\sqrt{2}}(|\uparrow\downarrow\rangle \pm |\downarrow\uparrow\rangle); \quad |c,\pm\rangle = \frac{1}{\sqrt{2}}(|\uparrow\uparrow\rangle \pm |\downarrow\downarrow\rangle).$$

Bell states correspond to maximal entanglement between two subsystems. Spins are *anticorrelated* for $|a,\pm\rangle$ and *correlated* for $|c,\pm\rangle$.

Problem 4.1

We have several examples of 2×2 matrices given in what follows. Which of these can be density matrices? From the possible density matrices, which one can correspond to a pure state and which one to a mixed state? Compare the von Neumann entropy for the possible density matrices:

$$\rho_1 = \begin{pmatrix} 1 & 1 \\ -1 & 1 \end{pmatrix}; \quad \rho_2 = \begin{pmatrix} 1 & \frac{i}{2} \\ -\frac{i}{2} & 0 \end{pmatrix}; \quad \rho_3 = \frac{1}{2}\begin{pmatrix} 3 & 1 \\ 1 & -1 \end{pmatrix};$$

$$\rho_4 = \frac{1}{2}\begin{pmatrix} 1 & 1 \\ 1 & 1 \end{pmatrix}; \quad \rho_5 = \frac{1}{2}\begin{pmatrix} 1 & 0 \\ 0 & 1 \end{pmatrix}.$$

This problem probes our knowledge of basic properties of density operators. Thus, we have to check the following for each matrix:

- Is it Hermitian?
- Is $Tr(\rho) = 1$?
- Is $\mathscr{P}(\widehat{\rho}) = Tr(\rho^2) = 1$ to indicate a pure state?
- Is $\mathscr{P}(\widehat{\rho}) = Tr(\rho^2) < 1$ to indicate a mixed state?

It is easy to see the following:

$$\text{All matrices are Hermitian}: \rho_i^+ = \rho_i \text{ for i} = 1,2,3,4,5.$$

$$Tr(\rho_1) = 2; \quad Tr(\rho_{2,3,4,5}) = 1.$$

$$Tr\left(\rho_2^2\right) = \frac{3}{2}; \quad Tr\left(\rho_3^2\right) = 3; \quad Tr\left(\rho_4^2\right) = 1; \quad Tr\left(\rho_5^2\right) = \frac{1}{2}.$$

Thus, only ρ_4 and ρ_5 can represent density matrices, such that ρ_4 corresponds to a pure state, while ρ_5 corresponds to a mixed state.

Since the von Neumann entropy is defined as $S_N(\widehat{\rho}) = -\sum_i \lambda_i \ln(\lambda_i)$, where λ_i are the eigenvalues of the density matrix, we find that for ρ_4, the eigenvalues are $\lambda_1 = 0; \lambda_2 = 1 \rightarrow S_N(\rho_4) = 0\ln(0) + 1\ln(1) = 0$. Note that here we have used the fact that $\lim_{x\to 0} x\ln(x) = 0$.

For ρ_5, the eigenvalues are $\lambda_{1,2} = \frac{1}{2} \rightarrow S_N(\rho_5) = -2\left(\frac{1}{2}\right)\ln\left(\frac{1}{2}\right) = \ln(2)$.

Problem 4.2

The density matrix for an ensemble of spin ½ particles in the S_z basis is given as

$$\rho = \begin{pmatrix} \frac{1}{3} & b \\ c & a \end{pmatrix}.$$

What values must a, b, c be so that ρ is a density matrix? Determine a, b, c such that the density matrix ρ would correspond to a pure state.

The objective of this problem is similar: basic properties of density operators. We remember:

$\widehat{\rho}^+ = \widehat{\rho}$ – must be Hermitian;

$Tr(\widehat{\rho}) = Tr(\widehat{\rho}^2) = 1$ for a pure state.

From the condition for $\widehat{\rho}$ being Hermitian, we find that $c = b^*$. Also,

$$1 = Tr(\widehat{\rho}) = \tfrac{1}{3} + a, \qquad a = \tfrac{2}{3},$$
$$1 = Tr(\widehat{\rho}^2) = \tfrac{5}{9} + 2|b|^2, \qquad b = \tfrac{\sqrt{2}}{3}e^{-i\phi}\ (\phi - \text{arbitrary phase factor}).$$

From here, we find

$$\rho = \begin{pmatrix} \frac{1}{3} & \frac{\sqrt{2}}{3}e^{i\phi} \\ \frac{\sqrt{2}}{3}e^{-i\phi} & \frac{2}{3} \end{pmatrix}.$$

The pure state can also be found simply by writing ρ in Dirac notation and using some straightforward rearrangements:

$$\begin{aligned}\widehat{\rho} &= \frac{1}{3}|\uparrow\rangle\langle\uparrow| + \frac{\sqrt{2}}{3}e^{i\phi}|\uparrow\rangle\langle\downarrow| + \frac{\sqrt{2}}{3}e^{-i\phi}|\downarrow\rangle\langle\uparrow| + \frac{2}{3}|\downarrow\rangle\langle\downarrow| \\ &= \left(\frac{1}{\sqrt{3}}|\uparrow\rangle + \sqrt{\frac{2}{3}}e^{i\phi}|\downarrow\rangle\right)\left(\frac{1}{\sqrt{3}}\langle\uparrow| + \sqrt{\frac{2}{3}}e^{-i\phi}\langle\uparrow|\right). \\ |\psi\rangle &= \frac{1}{\sqrt{3}}|\uparrow\rangle + \sqrt{\frac{2}{3}}e^{i\phi}|\downarrow\rangle.\end{aligned}$$

Problem 4.3

The expectation value of an arbitrary operator $\widehat{A}$ can be given with respect to a given quantum mechanical state $|\psi\rangle$. Show that this expectation value can be written as $\langle\widehat{A}\rangle = \langle\psi|\widehat{A}|\psi\rangle = Tr(\widehat{\rho}\widehat{A})$, where $\widehat{\rho}$ is the density operator.

This problem essentially reinforces the statistical representation of the ensemble average of an operator for a pure state.

This can easily be demonstrated by using the spectral representation of the operator $\widehat{A}$ in some eigenbasis $|\ell\rangle$. Take $\widehat{A}$ and multiply from left and right with $\widehat{\mathbb{1}} = \sum_{\ell}|\ell\rangle\langle\ell|$:

$$\widehat{A} = \widehat{\mathbb{1}}\,\widehat{A}\,\widehat{\mathbb{1}} = \sum_{\ell,m}|\ell\rangle\langle\ell|\widehat{A}|m\rangle\langle m| = \sum_{\ell,m}|\ell\rangle\widehat{A}_{\ell m}\langle m|.$$

This is nothing but the spectral representation of the operator $\widehat{A}$ in the eigenbasis $|\ell\rangle$. Furthermore,

$$\langle\widehat{A}\rangle = \langle\psi|\widehat{A}|\psi\rangle = \sum_{\ell,m}\langle\psi|\ell\rangle A_{\ell m}\langle m|\psi\rangle = \sum_{\ell,m}\langle m|\psi\rangle\langle\psi|\ell\rangle A_{\ell m}.$$

Since $\widehat{\rho} = |\psi\rangle\langle\psi|$, then

$$\langle\widehat{A}\rangle = \sum_{\ell,m}\rho_{m\ell}A_{\ell m} = \sum_{m}(\widehat{\rho}\widehat{A})_{mm} = Tr(\widehat{\rho}\widehat{A}).$$

Food for thought: Suppose you are asked a similar question to find the ensemble average of the operator $\widehat{A}$, but in the case of a mixed-state density operator. What changes in the proof? The answer, of course, is the same as above: $\langle\widehat{A}\rangle = Tr(\widehat{\rho}\widehat{A})$.

Problem 4.4
Consider a Hamiltonian $\widehat{H}$ for which $\widehat{H}|\psi_n\rangle = E_n|\psi_n\rangle$.

(a) What is the density operator that minimizes the von Neumann entropy as a function of the eigenenergies of $\widehat{H}$ if the state has an energy E?
(b) What is the purity in this case?

---

a) This problem is intended to make a connection between the quantum mechanical representation in terms of the canonical ensemble.

The problem can be solved with the help of Lagrange multipliers. We begin by constructing a functional $\mathscr{L}[\widehat{\rho}]$ that contains the von Neumann entropy (the quantity to be extremized).

We add two constraints with the help of Lagrange multipliers: λ corresponding to the measured energy $Tr[\widehat{H}\widehat{\rho}] = E$, and α corresponding to the $Tr[\widehat{\rho}] = 1$ requirement,

$$\begin{aligned}\mathscr{L}[\widehat{\rho}] &= -Tr[\ln(\widehat{\rho})\widehat{\rho}] + \lambda(Tr[\widehat{H}\widehat{\rho}] - E) + \alpha(Tr[\widehat{\rho}] - 1)\\ &= \sum_n \left(\lambda E_n\rho_{nn} - \rho_{nm}\left[\ln(\rho_{\alpha\beta})\right]_{mn} + \alpha\rho_{nn}\right) - \alpha - \lambda E\\ &= \sum_n\sum_m \left(\lambda E_n\rho_{nm}\delta_{nm} - \sum_p \left(\rho_{nm}[\ln(\rho)]_{mp}\delta_{np}\right) + \alpha\rho_{nm}\delta_{nm}\right) - \alpha - \lambda E.\end{aligned}$$

The condition of extremum requires that $\frac{\delta\mathscr{L}[\widehat{\rho}]}{\delta\rho_{ab}} = 0$, thus

$$\begin{aligned}\frac{\delta\mathscr{L}[\widehat{\rho}]}{\delta\rho_{ab}} &= \frac{\delta}{\delta\rho_{ab}}\sum_n\sum_m\left((\lambda E_n + \alpha)\rho_{nm}\delta_{nm} - \sum_p\left(\rho_{nm}[\ln(\rho)]_{mp}\delta_{np}\right)\right)\\ &= \sum_n\sum_m\left((\lambda E_n + \alpha)\frac{\delta\rho_{nm}}{\delta\rho_{ab}}\delta_{nm}\right)\\ &\quad - \sum_n\sum_m\sum_p\left(\frac{\delta\rho_{nm}}{\delta\rho_{ab}}[\ln(\rho)]_{mp}\delta_{np} + \rho_{nm}\frac{\delta[\ln(\rho)]_{mp}}{\delta\rho_{ab}}\delta_{np}\right)\\ &= \sum_n\sum_m\left((\lambda E_n + \alpha)\frac{\delta\rho_{nm}}{\delta\rho_{ab}}\delta_{nm}\right)\\ &\quad - \sum_n\sum_m\sum_p\left(\frac{\delta\rho_{nm}}{\delta\rho_{ab}}[\ln(\rho)]_{mp}\delta_{np} + \sum_q \rho_{nm}[\rho^{-1}]_{mq}\frac{\delta\rho_{qp}}{\delta\rho_{ab}}\delta_{np}\right) = 0.\end{aligned}$$

Now we exchange dummy indices m and q in the second summand to obtain

$$\sum_m\sum_q \rho_{nm}[\rho^{-1}]_{mq}\frac{\delta\rho_{qp}}{\delta\rho_{ab}}\delta_{np} = \sum_m\sum_q \rho_{nq}[\rho^{-1}]_{qm}\frac{\delta\rho_{mp}}{\delta\rho_{ab}}\delta_{np} = \sum_m \delta_{nm}\frac{\delta\rho_{mp}}{\delta\rho_{ab}}\delta_{np}.$$

After using the relation $\frac{\delta\rho_{nm}}{\delta\rho_{ab}} = \delta_{an}\delta_{bm}$, we obtain

$$\begin{aligned}\frac{\delta\mathscr{L}[\widehat{\rho}]}{\delta\rho_{ab}} &= \sum_n\sum_m\left((\lambda E_n + \alpha)\frac{\delta\rho_{nm}}{\delta\rho_{ab}}\delta_{nm}\right) - \sum_n\sum_m\sum_p\left(\frac{\delta\rho_{nm}}{\delta\rho_{ab}}[\ln(\rho)]_{mp}\delta_{np} + \delta_{nm}\frac{\delta\rho_{mp}}{\delta\rho_{ab}}\delta_{np}\right)\\ &= \sum_n\sum_m\left((\lambda E_n + \alpha)\delta_{an}\delta_{bm}\delta_{nm}\right)\\ &\quad - \sum_n\sum_m\sum_p\left(\delta_{an}\delta_{bm}[\ln(\rho)]_{mp}\delta_{np} + \delta_{nm}\delta_{am}\delta_{bp}\delta_{np}\right) = 0.\end{aligned}$$

Finally, using the properties of the Kronecker delta, we can carry out all the sums:

$$\begin{aligned}\frac{\delta\mathcal{L}[\widehat{\rho}]}{\delta\rho_{ab}} &= \sum_n\sum_m((\lambda E_n+\alpha)\delta_{an}\delta_{bm}\delta_{nm}-\delta_{an}\delta_{bm}[\ln(\rho)]_{mn}-\delta_{nm}\delta_{am}\delta_{bn})\\ &= \sum_n((\lambda E_n+\alpha)\delta_{an}\delta_{bn}-\delta_{an}[\ln(\rho)]_{bn}-\delta_{na}\delta_{bn})\\ &= (\lambda E_a+\alpha)\delta_{ba}-[\ln(\rho)]_{ba}-\delta_{ba}=0.\end{aligned}$$

Then, we obtain that $\widehat{\rho}$ maximizes the entropy (at a given energy) if

$$(\lambda E_a+\alpha)\delta_{ba}-[\ln(\rho)]_{ba}-\delta_{ba}=0\rightarrow[\ln(\rho)]_{ba}=(\lambda E_a+\alpha-1)\delta_{ba}.$$

For matrices, the inverse operation of the logarithm is the exponential, $\exp([\ln(\rho)]_{ba})=\rho_{ab}=\exp((\lambda E_a+\alpha-1)\delta_{ba})$, thus the exponential of a diagonal matrix is immediately $\rho_{ab}=e^{\lambda E_a-1+\alpha}\delta_{ab}$.

We define the trace of the density operator as

$$Z'=Tr[\rho]=e^{-1+\alpha}\sum_n e^{\lambda E_n}.$$

After identifying $\lambda=-\beta=-1/k_BT$, we obtain

$$\rho_n=\frac{e^{-1+\alpha}e^{-\beta E_n}}{Z'},$$

$$\rho_n=\frac{e^{-1+\alpha}e^{-\beta E_n}}{Z'}=\frac{e^{-1+\alpha}e^{-\beta E_n}}{e^{-1+\alpha}Tr\left[e^{-\beta E_n}\right]}=\frac{e^{-\beta E_n}}{Tr\left[e^{-\beta E_n}\right]}=\frac{e^{-\beta E_n}}{Z(\beta)},$$

with $Z(\beta)=Tr[e^{-\beta E_n}]=Tr[e^{-\beta\widehat{H}}]$.

b) Let's consider the purity of the state $\mathcal{P}(\widehat{\rho})=Tr(\widehat{\rho}^2)$, which in this particular case is

$$\mathcal{P}(\widehat{\rho})=Tr(\widehat{\rho}^2)=\sum_n\left(\frac{e^{-\beta E_n}}{Z(\beta)}\frac{e^{-\beta E_n}}{Z(\beta)}\right)=\frac{1}{Z(\beta)^2}\sum_n e^{-2\beta E_n}=\frac{Z(2\beta)}{Z(\beta)^2}.$$

The preceding result indicates that at $T=0$, $\mathcal{P}(\widehat{\rho})=1$, meaning that the state is always pure. On the other hand, as $T\rightarrow\infty, \mathcal{P}(\widehat{\rho})=0$, meaning that the state is maximally mixed.

> *Food for thought:* Making connections with information theory, we realize that minimizing the von Neumann entropy is the same as minimizing the mean quantity of information contained in the system. The preceding result indicates that a thermal state is the state with less information at a given energy E.

Problem 4.5

Consider the case when the density operator corresponds to the canonical ensemble, $\widehat{\rho}=\frac{e^{-\beta\widehat{H}}}{Z}$, where $\beta=1/k_BT$, $\widehat{H}$ is the Hamiltonian, and $Z=Tr(e^{-\beta\widehat{H}})$ is the partition function (k_B – Boltzmann constant; T – temperature). Calculate the von Neumann entropy and compare it with the thermodynamic entropy.

This problem is intended to make a connection between the quantum mechanical and thermodynamical nature of quantum mechanics as captured in entropy. It is closely related to the previous one.

The von Neumann entropy can be calculated simply by using its definition and the given $\widehat{\rho}$ and partition function,

$$S_N(\widehat{\rho}) = -Tr(\widehat{\rho}\ln(\widehat{\rho})) = -Tr\left(\frac{e^{-\beta\widehat{H}}}{Z}\ln\left(\frac{e^{-\beta\widehat{H}}}{Z}\right)\right) = -Tr\left(\frac{e^{-\beta\widehat{H}}}{Z}(\ln(e^{-\beta\widehat{H}}) - \ln(Z))\right)$$
$$= Tr\left(\beta\widehat{H}\frac{e^{-\beta\widehat{H}}}{Z}\right) + Tr\left(\ln(Z)\frac{e^{-\beta\widehat{H}}}{Z}\right) = \beta Tr(\widehat{\rho}\widehat{H}) + \ln(Z)Tr(\widehat{\rho}) = \beta\langle\widehat{H}\rangle + \ln(Z)$$
$$= \frac{\langle\widehat{H}\rangle - (-k_BT\ln(Z))}{k_BT},$$

where we have used the fact that $Tr(\widehat{\rho}) = 1$. From statistical mechanics, we recall that the Helmholtz free energy is defined as $F = -k_BT\ln(Z) = \langle H\rangle - TS_{th}$, where S_{th} is the thermodynamic entropy. We find that

$$S_N(\widehat{\rho}) = \frac{S_{th}}{k_B}.$$

In other words, the von Neumann entropy is actually the thermodynamic entropy divided by the Boltzmann constant.

Problem 4.6

Consider a system of N electrons per unit volume in thermal equilibrium that are placed in an external magnetic field $\boldsymbol{B} = B\widehat{\boldsymbol{u}}_z$.

a) What is the measured magnetization of this system in the high-temperature limit?

b) Obtain the measured magnetization for all temperatures.

The words "thermal equilibrium" are indicative that this problem is related to the previous one since the canonical ensemble $\widehat{\rho} = \frac{e^{-\beta\widehat{H}}}{Z}$, where $\beta = 1/k_BT$, $\widehat{H}$ is the Hamiltonian, and $Z = Tr\left(e^{-\beta\widehat{H}}\right)$ is the partition function (k_B – Boltzmann constant; T – temperature), corresponds to precisely this situation.

Let's begin with the definition for the magnetization for a Hamiltonian $\widehat{H} = -\widehat{\mu}\cdot\boldsymbol{B}$ and its observable,

$$\widehat{M}_\mu = \left(\frac{\partial\widehat{H}}{\partial B^\mu}\right)_{T,V} = \frac{\partial}{\partial B^\mu}\left(-\sum_{n=1}^{N}\frac{\gamma}{2}B^\mu\widehat{\sigma}_\mu\right) = -\sum_{n=1}^{N}\frac{\gamma}{2}\widehat{\sigma}_\mu = -\frac{N\gamma}{2}\widehat{\sigma}_\mu,$$
$$\langle\widehat{M}_\mu\rangle = -N\langle\widehat{\mu}_\mu\rangle,$$

where $\widehat{\mu}_\mu = \frac{1}{2}\gamma\widehat{\sigma}_\mu$, with γ being the gyromagnetic ratio and σ_μ the Pauli matrix in the μ-direction. For a system in thermal equilibrium, we use

$$\langle\widehat{M}_z\rangle = Tr(\widehat{\rho}\widehat{M}_z),$$
$$\widehat{\rho} = \frac{e^{-\beta\widehat{H}}}{Z}, \quad Z = Tr(e^{-\beta\widehat{H}}).$$

From here, we write

$$\langle \widehat{M}_z \rangle = -\frac{N\gamma}{2Z} Tr\left[e^{-\beta\widehat{H}}\widehat{\sigma}_z\right].$$

In the high-temperature limit, we have

$$e^{-\beta\widehat{H}} \approx \widehat{\mathbb{1}} - \beta\widehat{H} = \widehat{\mathbb{1}} + \frac{\gamma B\widehat{\sigma}_z}{2k_BT}, \quad Z \approx Tr\left(\widehat{\mathbb{1}} + \frac{\gamma B\widehat{\sigma}_z}{2k_BT}\right) \approx 2, \quad \widehat{\mathbb{1}} = \begin{pmatrix} 1 & 0 \\ 0 & 1 \end{pmatrix},$$

$$\langle \widehat{M}_z \rangle \approx \frac{N\gamma}{4} Tr\left[\widehat{\sigma}_z + \frac{\gamma B\widehat{\sigma}_z^2}{2k_BT}\right] = \frac{N\gamma^2 B}{4k_BT}.$$

This, of course, is the well-known *Curie law* for spin 1/2 particles, whose magnetization in thermal equilibrium has the $M_z \sim \frac{1}{T}$ behavior with respect to temperature.

b) Obtaining the magnetization for any T can be done by using the general results for $\langle M_z \rangle$ without making any assumptions:

$$\langle \widehat{M}_z \rangle = -\frac{N\gamma}{2Z} Tr\left[e^{-\beta\widehat{H}}\widehat{\sigma}_z\right].$$

We further find

$$Z = Tr\left[e^{-\beta\widehat{H}}\right] = Tr\left[e^{\beta\frac{\gamma}{2}B\widehat{\sigma}_z}\right] = Tr\left[e^{\beta\frac{\gamma}{2}B\left(\begin{smallmatrix} 1 & 0 \\ 0 & -1 \end{smallmatrix}\right)}\right]$$

$$= Tr\left[\begin{pmatrix} e^{\frac{\beta\gamma B}{2}} & 0 \\ 0 & e^{-\frac{\beta\gamma B}{2}} \end{pmatrix}\right] = 2\cosh\left(\frac{\beta\gamma B}{2}\right),$$

$$Tr\left[e^{-\beta\widehat{H}}\sigma_z\right] = Tr\left[e^{\beta\frac{\gamma}{2}B\widehat{\sigma}_z}\widehat{\sigma}_z\right] = Tr\left[e^{\beta\frac{\gamma}{2}B\left(\begin{smallmatrix} 1 & 0 \\ 0 & -1 \end{smallmatrix}\right)}\begin{pmatrix} 1 & 0 \\ 0 & -1 \end{pmatrix}\right]$$

$$= Tr\left[\begin{pmatrix} e^{\frac{\beta\gamma B}{2}} & 0 \\ 0 & e^{-\frac{\beta\gamma B}{2}} \end{pmatrix}\begin{pmatrix} 1 & 0 \\ 0 & -1 \end{pmatrix}\right]$$

$$= Tr\left[\begin{pmatrix} e^{\frac{\beta\gamma B}{2}} & 0 \\ 0 & -e^{-\frac{\beta\gamma B}{2}} \end{pmatrix}\right] = 2\sinh\left(\frac{\beta\gamma B}{2}\right).$$

Therefore:

$$\langle \widehat{M}_z \rangle = -\frac{N\gamma}{2}\frac{Tr\left[e^{-\beta\widehat{H}}\widehat{\sigma}_z\right]}{Z} = -\frac{N\gamma}{2}\frac{2\sinh\left(\frac{\beta\gamma B}{2}\right)}{2\cosh\left(\frac{\beta\gamma B}{2}\right)} = -\frac{N\gamma}{2}\tanh\left(\frac{\gamma B}{2k_BT}\right).$$

From this result, we can obtain the high- and low-temperature regimes easily:

$$\lim_{k_BT \gg \gamma B} \langle \widehat{M}_z \rangle = -\frac{N\gamma^2 B}{4k_BT},$$

$$\lim_{k_BT \ll \gamma B} \langle \widehat{M}_z \rangle = -\frac{N\gamma}{2}.$$

Problem 4.7

Often, it is necessary to consider the time dependence of the density operator. What is the time evolution of the density operator $\widehat{\rho}(t) = \sum_n \rho_n |\psi_n(t)\rangle\langle\psi_n(t)|$, where $|\psi_n(t)\rangle$ satisfies the Schrödinger equation?

As $|\psi_n(t)\rangle$ satisfies the Schrödinger equation, we are working in the Schrödinger picture; then, we start with $|\psi_n(t)\rangle = \widehat{U}(t)|\psi_n(0)\rangle$, where $\widehat{U}(t)$ is the time evolution operator. This is a consequence of the fact that $|\psi_n(t)\rangle$ satisfies the Schrödinger equation: $i\hbar\frac{\partial}{\partial t}|\psi_n(t)\rangle = \widehat{H}|\psi_n(t)\rangle$.

It is easy then to find

$$\begin{aligned}\widehat{\rho}(t) &= \sum\nolimits_n \rho_n |\psi_n(t)\rangle\langle\psi_n(t)| = \sum\nolimits_n \rho_n \widehat{U}(t)|\psi_n(0)\rangle\langle\psi_n(0)|\widehat{U}^+(t)\\ &= \widehat{U}(t)\sum\nolimits_n \rho_n|\psi_n(0)\rangle\langle\psi_n(0)|\widehat{U}^+(t)\\ &= \widehat{U}(t)\widehat{\rho}(0)\widehat{U}^+(t).\end{aligned}$$

Note that in the preceding relation ρ_n are taken to be time-independent, which is in line with the Schrödinger picture and its statistical interpretation for a closed system. We can now consider the derivative of the density operator with respect to time,

$$\begin{aligned}\frac{\partial}{\partial t}\widehat{\rho}(t) &= \frac{\partial}{\partial t}\sum\nolimits_n \rho_n|\psi_n(t)\rangle\langle\psi_n(t)| = \sum\nolimits_n \rho_n \frac{\partial}{\partial t}|\psi_n(t)\rangle\langle\psi_n(t)| + \sum\nolimits_n \rho_n|\psi_n(t)\rangle\frac{\partial}{\partial t}\langle\psi_n(t)|\\ &= \sum\nolimits_n \rho_n \frac{1}{i\hbar}\widehat{H}|\psi_n(t)\rangle\langle\psi_n(t)| + \sum\nolimits_n \rho_n|\psi_n(t)\rangle\langle\psi_n(t)|\frac{-1}{i\hbar}\widehat{H}\\ &= \frac{1}{i\hbar}\left(\widehat{H}\sum\nolimits_n \rho_n|\psi_n(t)\rangle\langle\psi_n(t)| - \sum\nolimits_n \rho_n|\psi_n(t)\rangle\langle\psi_n(t)|\widehat{H}\right)\\ &= \frac{1}{i\hbar}(\widehat{H}\widehat{\rho}(t) - \widehat{\rho}(t)\widehat{H}) = \frac{1}{i\hbar}[\widehat{H},\widehat{\rho}(t)].\end{aligned}$$

> *Food for thought:* Solve the same problem in the Heisenberg and Interaction (Dirac) pictures.

Problem 4.8

A spin $s = 1/2$ particle placed in a magnetic field $\boldsymbol{B}$ has the Hamiltonian $\widehat{H} = -\gamma \boldsymbol{B}\cdot\widehat{\boldsymbol{S}}$ (where γ is the gyromagnetic ratio and $\widehat{\boldsymbol{S}} = \hbar s\widehat{\boldsymbol{\sigma}}$ is the spin operator vector). The time-dependent polarization vector can be given in terms of the density operator as $\boldsymbol{P}(t) = \langle\widehat{\boldsymbol{\sigma}}(t)\rangle = Tr(\widehat{\boldsymbol{\sigma}}\widehat{\rho}(t))$, where $\boldsymbol{\sigma} = (\sigma_1, \sigma_2, \sigma_3)$ are the Pauli matrices and $\widehat{\rho}(t)$ is the time-dependent density operator. Find an equation for the time dynamics of $\boldsymbol{P}(t)$.

Here again we are dealing with an electronic system in a magnetic field given with the familiar Hamiltonian $\widehat{H} = -\gamma\boldsymbol{B}\cdot\widehat{\boldsymbol{S}}$. However, the emphasis is on the time dynamics of the polarization. We recall that time dynamics in quantum mechanics is given with first-order differential equations in time (similar to the Schrödinger equation, for example).

Thus, let us begin with

$$i\hbar\frac{d\boldsymbol{P}(t)}{dt} = i\hbar\frac{d}{dt}Tr(\widehat{\sigma}\widehat{\rho}(t)) = Tr\left(\widehat{\sigma}i\hbar\frac{d\widehat{\rho}(t)}{dt}\right).$$

We recall the Liouville–von Neumann equation of motion for the density matrix, $i\hbar\frac{d\widehat{\rho}(t)}{dt} = \left[\widehat{H},\widehat{\rho}(t)\right]$. For the ith component of the polarization vector, we obtain

$$\begin{aligned} i\hbar\frac{dP_a(t)}{dt} &= Tr\left(\widehat{\sigma}_a i\hbar\frac{d\widehat{\rho}(t)}{dt}\right) = Tr\left(\widehat{\sigma}_a\left[\widehat{H},\widehat{\rho}(t)\right]\right) = -\frac{\hbar\gamma}{2}Tr(\widehat{\sigma}_a[\boldsymbol{B}\cdot\widehat{\boldsymbol{\sigma}},\widehat{\rho}(t)]) \\ &= -\frac{\hbar\gamma B^b}{2}Tr(\widehat{\sigma}_a[\widehat{\sigma}_b,\widehat{\rho}(t)]) \\ &= -\frac{\hbar\gamma B^b}{2}Tr([\widehat{\sigma}_a,\widehat{\sigma}_b]\widehat{\rho}(t)) = -\frac{\hbar\gamma B^b}{2}2i\varepsilon_{abc}Tr(\widehat{\sigma}^c\widehat{\rho}(t)) = -i\hbar\gamma Tr\left(\varepsilon_{abc}B^b\widehat{\sigma}^c\widehat{\rho}(t)\right). \end{aligned}$$

In the preceding relation we have used the fact that $[\widehat{\sigma}_a,\widehat{\sigma}_b] = 2i\varepsilon_{abc}\widehat{\sigma}^c$. After some simple algebra, we find

$$\frac{dP_a(t)}{dt} = -\gamma Tr\left(\varepsilon_{abc}B^b\widehat{\sigma}^c\widehat{\rho}(t)\right) = -\gamma\varepsilon_{abc}B^b Tr(\widehat{\sigma}^c\widehat{\rho}(t)) = -\gamma\varepsilon_{abc}B^b P^c(t).$$

Therefore, in a vector form, we find

$$\frac{d\boldsymbol{P}(t)}{dt} = \gamma\boldsymbol{P}(t)\times\boldsymbol{B}.$$

This is an expected outcome classically. The result simply indicates that the polarization vector $\boldsymbol{P}(t)$ has a fixed magnitude that precesses in time around the magnetic field $\boldsymbol{B}$ with frequency $\omega_0 = \gamma B$.

Problem 4.9

The most general density matrix for a spin ½ particle is given by $\widehat{\rho} = \frac{1}{2}(\widehat{\mathbb{1}} + \boldsymbol{P}\cdot\widehat{\boldsymbol{\sigma}})$.

a) What conditions must the polarization $\boldsymbol{P}$ satisfy so that $\widehat{\rho}$ is a density matrix operator?
b) Calculate the von Neumann entropy.

---

a) We see right away that $Tr(\widehat{\rho}) = 1$ since the Pauli matrices $\boldsymbol{\sigma}$ are traceless. We also easily find that

$$Det(\widehat{\rho}) = \frac{1}{4}(1-\boldsymbol{P}^2).$$

Given that the $Det(\widehat{\rho})$ must be nonnegative, we determine that $\boldsymbol{P}^2 \leq 1$. Another way to see this is to examine the purity $\mathcal{P}(\widehat{\rho})$ for the given density operator. Specifically,

$$0 \leq \mathcal{P}(\widehat{\rho}) = Tr(\widehat{\rho}^2) = \frac{1}{2}(1+\boldsymbol{P}^2) \leq 1.$$

Given that $\boldsymbol{P}^2 \geq 0$, we determine that $0 \leq \boldsymbol{P}^2 \leq 1$. This condition essentially determines the interior of the so-called *Bloch sphere*. In fact, when $\boldsymbol{P}^2 < 1$, then the density matrix corresponds to a mixed state, while $\boldsymbol{P}^2 = 1$ determines the density matrix for a pure state.

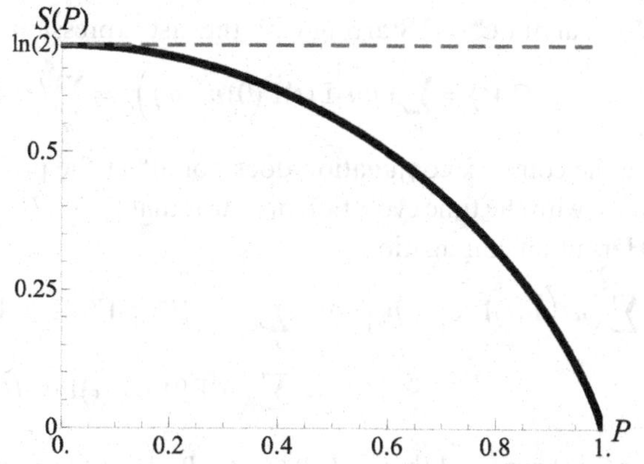

Figure 4.1 A graphical representation of the von Neumann entropy as a function of the polarization magnitude.

b) We remember that the von Neumann entropy is given as

$$S_N(\widehat{\rho}) = -\sum_n \lambda_n \ln(\lambda_n),$$

where λ_i are the eigenvalues of the density matrix. It is easy to find that the eigenvalues for the given ρ are $\lambda_{1,2} = \frac{1 \pm P}{2}$, where $P = |\boldsymbol{P}|$. Thus,

$$S_N(\widehat{\rho}) = -\lambda_1 \ln(\lambda_1) - \lambda_2 \ln(\lambda_2) = -\frac{1+P}{2} \ln\left(\frac{1+P}{2}\right) - \frac{1-P}{2} \ln\left(\frac{1-P}{2}\right)$$
$$= -\frac{1}{2}\left[\ln(1-P^2) - \ln(4)\right] - \frac{P}{2} \ln\left(\frac{1+P}{1-P}\right) = \ln(2) - \frac{1}{2}\ln(1-P^2) - P\operatorname{arctanh}(P).$$

To get an idea of how the entropy evolves as a function of the polarization, we give a graphical representation here (Figure 4.1) showing a smooth decline of $S(P)$ between its highest $\ln(2)$ and lowest 0 values.

Problem 4.10

Often it is necessary to define time-dependent correlation functions for operators. Let's consider the correlation function for some operator $\widehat{\Gamma}(t)$ given as $C(t) = \langle \widehat{\Gamma}(t)\widehat{\Gamma}(0)\rangle = Tr(\widehat{\Gamma}(t)\widehat{\Gamma}(0)\widehat{\rho})$. Here $\widehat{\rho}$ is the equilibrium density operator corresponding to states (basis) that are time-independent as expected from the Schrödinger picture. Specifically for eigenstates $|n\rangle$, the density operator is given as $\widehat{\rho} = \sum_n p_n |n\rangle\langle n|$, where p_n are the probabilities. Evaluate the imaginary part of this correlation function given as $Im(C(t)) = \frac{1}{2i}(C(t) - C^*(t))$.

Let's first see what $C(t)$ looks like explicitly in the basis $|n\rangle$. We can write

$$C(t) = Tr(\widehat{\Gamma}(t)\widehat{\Gamma}(0)\widehat{\rho}) = \sum_m \langle m|\widehat{\Gamma}(t)\widehat{\Gamma}(0)\widehat{\rho}|m\rangle = \sum_{m,n} \langle m|\widehat{\Gamma}(t)\widehat{\Gamma}(0)p_n|n\rangle\langle n|m\rangle$$
$$= \sum_n \langle n|\widehat{\Gamma}(t)\widehat{\Gamma}(0)p_n|n\rangle.$$

What about $C^*(t)$? Starting with the last expression, we find

$$C^*(t) = \sum_n \left(\left\langle n \left| \widehat{\Gamma}(t)\widehat{\Gamma}(0)p_n \right| n \right\rangle\right)^* = \sum_n \left\langle n \left| \left(\widehat{\Gamma}(t)\widehat{\Gamma}(0)\right)^* p_n \right| n \right\rangle,$$

since the complex conjugation does not affect the probabilities p_n. We make the connection with the time evolution operator that $\widehat{\Gamma}(t) = \widehat{U}^+(t)\widehat{\Gamma}(0)\widehat{U}(t)$ and take advantage of Hermitian conjugation,

$$\begin{aligned}\sum_n \left\langle n \left| \left(\widehat{\Gamma}(t)\widehat{\Gamma}(0)\right)^* p_n \right| n \right\rangle &= \sum_n \left\langle n \left| \left(\widehat{U}^+(t)\widehat{\Gamma}(0)\widehat{U}(t)\widehat{\Gamma}(0)\right)^* p_n \right| n \right\rangle \\ &= \sum_n \left\langle n \left| \widehat{\Gamma}(0)\widehat{U}^+(t)\widehat{\Gamma}(0)\widehat{U}(t)p_n \right| n \right\rangle = Tr\left(\widehat{\Gamma}(0)\widehat{\Gamma}(t)\widehat{\rho}\right),\end{aligned}$$

where we have used that $(ABC)^+ = C^+B^+A^+$. Finally, putting everything together,

$$\begin{aligned}Im(C(t)) &= \frac{1}{2i}(C(t)) - C^*(t)) = \frac{1}{2i}\left[Tr\left(\widehat{\Gamma}(t)\widehat{\Gamma}(0)\widehat{\rho}\right) - \left(Tr\left(\widehat{\Gamma}(t)\widehat{\Gamma}(0)\widehat{\rho}\right)\right)^*\right] \\ &= \frac{1}{2i}\left[Tr\left(\widehat{\Gamma}(t)\widehat{\Gamma}(0)\widehat{\rho}\right) - Tr\left(\widehat{\Gamma}(0)\widehat{\Gamma}(t)\widehat{\rho}\right)\right] = \frac{1}{2i}Tr\left(\left[\widehat{\Gamma}(t),\widehat{\Gamma}(0)\right]\widehat{\rho}\right) \\ &= \frac{1}{2i}\left\langle \left[\widehat{\Gamma}(t),\widehat{\Gamma}(0)\right] \right\rangle.\end{aligned}$$

In other words, the imaginary part of the so-defined correlation function is related to the commutator of the time-dependent operator and the operator at $t = 0$.

Problem 4.11

Consider a system for spin ½ particles whose pure state is given as

$$|\Psi\rangle = \frac{1}{\sqrt{1+a^2}}(|\uparrow_A, \uparrow_B\rangle + ae^{i\theta}|\downarrow_A, \downarrow_B\rangle),$$

with $a \geq 0$. What would one record if the spin of particle A is measured? Obtain your answer by

a) using the operator $\widehat{\sigma}_z \otimes \widehat{\mathbb{I}}$ and evaluating $\langle\widehat{O}\rangle = \langle\Psi|\widehat{O}|\Psi\rangle$;
b) using the operator $\widehat{\sigma}_z \otimes \widehat{\mathbb{I}}$ and evaluating $Tr\left(\left[\widehat{\sigma}_z \otimes \widehat{\mathbb{I}}\right] \cdot \widehat{\rho}_{AB}\right)$;
c) using the reduced density $\widehat{\rho}_A$ and evaluating $Tr(\widehat{\sigma}_z\widehat{\rho}_A)$.

This problem is an opportunity to practice the different ways to represent measurements of local observables in composite systems.

a) Here we obtain the result as usually done in quantum mechanics,

$$\begin{aligned}\langle\widehat{\sigma}_z\rangle_A &= \langle\Psi|\widehat{\sigma}_z \otimes \widehat{\mathbb{I}}|\Psi\rangle \\ &= \frac{1}{1+a^2}\left(\langle\uparrow_A|\langle\uparrow_B| + ae^{-i\theta}\langle\downarrow_A|\langle\downarrow_B|\right)\left[\widehat{\sigma}_z \otimes \widehat{\mathbb{I}}\right]\left(|\uparrow_A\rangle|\uparrow_B\rangle + ae^{i\theta}|\downarrow_A\rangle|\downarrow_B\rangle\right) \\ &= \frac{1}{1+a^2}\Big(\langle\uparrow_A|\widehat{\sigma}_z|\uparrow_A\rangle\langle\uparrow_B|\widehat{\mathbb{I}}|\uparrow_B\rangle + ae^{-i\theta}\langle\downarrow_A|\widehat{\sigma}_z|\uparrow_A\rangle\langle\downarrow_B|\widehat{\mathbb{I}}|\uparrow_B\rangle \\ &\quad + ae^{i\theta}\langle\uparrow_A|\widehat{\sigma}_z|\downarrow_A\rangle\langle\uparrow_B|\widehat{\mathbb{I}}|\downarrow_B\rangle + a^2\langle\downarrow_A|\widehat{\sigma}_z|\downarrow_A\rangle\langle\downarrow_B|\widehat{\mathbb{I}}|\downarrow_B\rangle\Big).\end{aligned}$$

Now, using $\widehat{\sigma}_z|\uparrow\rangle = |\uparrow\rangle$ and $\widehat{\sigma}_z|\downarrow\rangle = -|\downarrow\rangle$, we obtain

$$\langle\widehat{\sigma}_z\rangle_A = \langle\Psi|\widehat{\sigma}_z\otimes\widehat{\mathbb{1}}|\Psi\rangle = \frac{1}{1+a^2}\left(\langle\uparrow_A|\widehat{\sigma}_z|\uparrow_A\rangle + 0 + 0 + a^2\langle\downarrow_A|\widehat{\sigma}_z|\downarrow_A\rangle\right) = \frac{1-a^2}{1+a^2}.$$

b) Measurements can also be defined using the ensemble average over quantum mechanical states using the density operator. Since this is a composite system, we have

$$\langle\widehat{\sigma}_z\rangle_A = Tr\left(\left[\widehat{\sigma}_z\otimes\widehat{\mathbb{1}}\right]\cdot\widehat{\rho}_{AB}\right),$$

$$\widehat{\rho}_{AB} = |\Psi\rangle\langle\Psi| = \frac{1}{1+a^2}(|\uparrow_A\rangle|\uparrow_B\rangle + ae^{i\theta}|\downarrow_A\rangle|\downarrow_B\rangle)(\langle\uparrow_A|\langle\uparrow_B| + ae^{-i\theta}\langle\downarrow_A|\langle\downarrow_B|)$$

$$= \frac{1}{1+a^2}\left(|\uparrow_A\rangle|\uparrow_B\rangle\langle\uparrow_A|\langle\uparrow_B| + ae^{-i\theta}|\uparrow_A\rangle|\uparrow_B\rangle\langle\downarrow_A|\langle\downarrow_B|\right.$$
$$\left. + ae^{i\theta}|\downarrow_A\rangle|\downarrow_B\rangle\langle\uparrow_A|\langle\uparrow_B| + a^2|\downarrow_A\rangle|\downarrow_B\rangle\langle\downarrow_A|\langle\downarrow_B|\right).$$

In the basis $\{|\uparrow_A\rangle|\uparrow_B\rangle, |\uparrow_A\rangle|\downarrow_B\rangle, |\downarrow_A\rangle|\uparrow_B\rangle, |\downarrow_A\rangle|\downarrow_B\rangle\}$, we have

$$\rho_{AB} = \frac{1}{1+a^2}\begin{pmatrix} 1 & 0 & 0 & ae^{-i\theta} \\ 0 & 0 & 0 & 0 \\ 0 & 0 & 0 & 0 \\ ae^{i\theta} & 0 & 0 & a^2 \end{pmatrix},$$

$$\left[\widehat{\sigma}_z\otimes\widehat{\mathbb{1}}\right] = \begin{pmatrix} 1 & 0 \\ 0 & -1 \end{pmatrix}\otimes\begin{pmatrix} 1 & 0 \\ 0 & 1 \end{pmatrix} = \begin{pmatrix} +\mathbb{1} & 0\mathbb{1} \\ 0\mathbb{1} & -\mathbb{1} \end{pmatrix} = \begin{pmatrix} 1 & 0 & 0 & 0 \\ 0 & 1 & 0 & 0 \\ 0 & 0 & -1 & 0 \\ 0 & 0 & 0 & -1 \end{pmatrix}.$$

Multiplying the two matrices and finding the trace afterwards yields

$$\langle\widehat{\sigma}_z\rangle_A = Tr\left(\left[\widehat{\sigma}_z\otimes\widehat{\mathbb{1}}\right]\cdot\widehat{\rho}_{AB}\right) = \frac{1-a^2}{1+a^2}.$$

c) Measuring the spin of particle A as part of the two-particle composite system can be defined using the relation $\langle\widehat{\sigma}_z\rangle_A = Tr(\widehat{\sigma}_z\widehat{\rho}_A)$, where $\widehat{\rho}_A$ is the reduced density operator. For this purpose, we first find the reduced density $\widehat{\rho}_A$:

$$\widehat{\rho}_A = Tr_B(\widehat{\rho}_{AB}) = \langle\uparrow_B|\widehat{\rho}_{AB}|\uparrow_B\rangle + \langle\downarrow_B|\widehat{\rho}_{AB}|\downarrow_B\rangle$$

$$= \frac{1}{1+a^2}(|\uparrow_A\rangle\langle\uparrow_B|\uparrow_B\rangle\langle\uparrow_A|\langle\uparrow_B|\uparrow_B\rangle + ae^{-i\theta}|\uparrow_A\rangle\langle\uparrow_B|\uparrow_B\rangle\langle\downarrow_A|\langle\downarrow_B|\uparrow_B\rangle$$
$$+ ae^{i\theta}|\downarrow_A\rangle\langle\uparrow_B|\downarrow_B\rangle\langle\uparrow_A|\langle\uparrow_B|\uparrow_B\rangle + a^2|\downarrow_A\rangle\langle\uparrow_B|\downarrow_B\rangle\langle\downarrow_A|\langle\downarrow_B|\uparrow_B\rangle)$$
$$+ \frac{1}{1+a^2}(|\uparrow_A\rangle\langle\downarrow_B|\uparrow_B\rangle\langle\uparrow_A|\langle\uparrow_B|\downarrow_B\rangle + ae^{-i\theta}|\uparrow_A\rangle\langle\downarrow_B|\uparrow_B\rangle\langle\downarrow_A|\langle\downarrow_B|\downarrow_B\rangle$$
$$+ ae^{i\theta}|\downarrow_A\rangle\langle\downarrow_B|\downarrow_B\rangle\langle\uparrow_A|\langle\uparrow_B|\downarrow_B\rangle + a^2|\downarrow_A\rangle\langle\downarrow_B|\downarrow_B\rangle\langle\downarrow_A|\langle\downarrow_B|\downarrow_B\rangle)$$

$$= \frac{1}{1+a^2}(|\uparrow_A\rangle\langle\uparrow_A| + a^2|\downarrow_A\rangle\langle\downarrow_A|)$$

$$= \frac{1}{1+a^2}\begin{pmatrix} 1 & 0 \\ 0 & a^2 \end{pmatrix}.$$

Therefore,

$$\langle\widehat{\sigma}_z\rangle_A = Tr(\widehat{\sigma}_z\widehat{\rho}_A) = \frac{1-a^2}{1+a^2}.$$

Food for thought: We have given three definitions for measurements in quantum mechanics. For the case of a pure state, all outcomes yield the same result. Can you design a problem for a density operator corresponding to a maximally mixed state and see if the outcomes from these definitions are the same?

Problem 4.12

Consider one electron spin in the state $|\chi_1\rangle = \alpha|\uparrow\rangle + \beta|\downarrow\rangle$ and a second electron in the state $|\chi_2\rangle = \alpha|\downarrow\rangle + \beta|\uparrow\rangle$, where α and β are real. The composite state of the system is $|\chi\rangle = |\chi_1\chi_2\rangle = |\chi_1\rangle\otimes|\chi_2\rangle$.

Construct the operator $\widehat{A} = \widehat{\sigma}_z^{(1)}\otimes\widehat{\sigma}_y^{(2)}$ in the basis $|\uparrow\uparrow\rangle, |\uparrow\downarrow\rangle, |\downarrow\uparrow\rangle, |\downarrow\downarrow\rangle$, and calculate the ensemble average $\langle\widehat{A}\rangle = Tr(\widehat{A}\widehat{\rho})$, where $\widehat{\rho}$ is the density operator corresponding to the pure state $|\chi\rangle$.

This problem concerns the construction of operators as tensors (outer products) of operators corresponding to individual states. The composite operator then will have a larger basis of the constituent subsystems. So, in order to calculate $\langle\widehat{A}\rangle = Tr(\widehat{A}\widehat{\rho})$, we need to know $\widehat{A}$ and $\widehat{\rho}$ in the $\{|\uparrow\uparrow\rangle, |\uparrow\downarrow\rangle, |\downarrow\uparrow\rangle, |\downarrow\downarrow\rangle\}$ basis.

Let's calculate the matrix elements for $\widehat{A}$ first. Start with

$$\langle\uparrow\uparrow|\widehat{A}|\uparrow\uparrow\rangle = \langle\uparrow\uparrow|\widehat{\sigma}_z^{(1)}\otimes\widehat{\sigma}_z^{(2)}|\uparrow\uparrow\rangle = \langle\uparrow|\widehat{\sigma}_z^{(1)}|\uparrow\rangle\otimes\langle\uparrow|\widehat{\sigma}_y^{(2)}|\uparrow\rangle = 0.$$

Here, we remember that $\widehat{\sigma}_y^{(2)}|\uparrow\uparrow\rangle$ operates on the second spin, while $\widehat{\sigma}_z^{(1)}|\uparrow\uparrow\rangle$ operates on the first spin. Since $\langle\uparrow|\widehat{\sigma}_y|\uparrow\rangle = 0$, the whole matrix element is zero. Similarly all matrix elements can be evaluated,

$$\widehat{A} = \begin{pmatrix} 0 & -i & 0 & 0 \\ i & 0 & 0 & 0 \\ 0 & 0 & 0 & i \\ 0 & 0 & -i & 0 \end{pmatrix}.$$

For the density matrix, we use its general expression for a pure state, $\widehat{\rho} = |\chi\rangle\langle\chi|$, which implies that we need to construct $|\chi\rangle$ first. The normalized composite state is then

$$\begin{aligned} |\chi\rangle &= \frac{1}{\alpha^2+\beta^2}((a|\uparrow_1\rangle + \beta|\downarrow_1\rangle)(\alpha|\downarrow_2\rangle + \beta|\uparrow_2\rangle)) \\ &= \frac{1}{\alpha^2+\beta^2}(\alpha\beta|\uparrow_1\uparrow_2\rangle + \alpha^2|\uparrow_1\downarrow_2\rangle + \beta^2|\downarrow_1\uparrow_2\rangle + \alpha\beta|\downarrow_1\downarrow_2\rangle). \end{aligned}$$

From here, we find

$$\hat{\rho} = |\chi\rangle\langle\chi| = \frac{1}{(\alpha^2+\beta^2)^4}\begin{pmatrix} \alpha^2\beta^2 & \alpha^3\beta & \alpha\beta^3 & \alpha^2\beta^2 \\ \alpha^3\beta & \alpha^4 & \alpha^2\beta^2 & \alpha^3\beta \\ \alpha\beta^3 & \alpha^2\beta^2 & \beta^4 & \alpha\beta^3 \\ \alpha^2\beta^2 & \alpha^3\beta & \alpha\beta^3 & \alpha^2\beta^2 \end{pmatrix}.$$

The ensemble average is then

$$\langle\hat{A}\rangle = Tr(\hat{A}\hat{\rho}) = \frac{1}{(\alpha^2+\beta^2)^4} Tr\left[\begin{pmatrix} 0 & -i & 0 & 0 \\ i & 0 & 0 & 0 \\ 0 & 0 & 0 & i \\ 0 & 0 & -i & 0 \end{pmatrix} \cdot \begin{pmatrix} \alpha^2\beta^2 & \alpha^3\beta & \alpha\beta^3 & \alpha^2\beta^2 \\ \alpha^3\beta & \alpha^4 & \alpha^2\beta^2 & \alpha^3\beta \\ \alpha\beta^3 & \alpha^2\beta^2 & \beta^4 & \alpha\beta^3 \\ \alpha^2\beta^2 & \alpha^3\beta & \alpha\beta^3 & \alpha^2\beta^2 \end{pmatrix}\right] = 0.$$

While this solution provides the answer to the question at hand, often it is not necessary to resort to matrix multiplication. Rather, one can work directly with the tensor product, and here we give the solution for extra practice:

$$\begin{aligned}
\langle\hat{A}\rangle &= Tr(\hat{A}\hat{\rho}) = Tr\left(\hat{\sigma}_z^{(1)} \otimes \hat{\sigma}_y^{(2)} |\chi\rangle\langle\chi|\right) = Tr\left(\langle\chi|\hat{\sigma}_z^{(1)} \otimes \hat{\sigma}_y^{(2)} |\chi\rangle\right) \\
&= \left[\frac{1}{\alpha^2+\beta^2}\left(\alpha\beta\langle\uparrow_1\uparrow_2| + \alpha^2\langle\uparrow_1\downarrow_2| + \beta^2\langle\downarrow_1\uparrow_2| + \alpha\beta\langle\downarrow_1\downarrow_2|\right)\right] \\
&\times \left[\sigma_z^{(1)} \otimes \sigma_y^{(2)}\right]\left[\frac{1}{\alpha^2+\beta^2}\left(\alpha\beta|\uparrow_1\uparrow_2\rangle + \alpha^2|\uparrow_1\downarrow_2\rangle + \beta^2|\downarrow_1\uparrow_2\rangle + \alpha\beta|\downarrow_1\downarrow_2\rangle\right)\right] \\
&= \frac{1}{(\alpha^2+\beta^2)^2}\Big[\alpha^2\beta^2\langle\uparrow_1\uparrow_2|\left[\sigma_z^{(1)} \otimes \sigma_y^{(2)}\right]|\uparrow_1\uparrow_2\rangle + \alpha^3\beta\langle\uparrow_1\uparrow_2|\left[\sigma_z^{(1)} \otimes \sigma_y^{(2)}\right]|\uparrow_1\downarrow_2\rangle \\
&+ \alpha\beta^3\langle\uparrow_1\uparrow_2|\left[\sigma_z^{(1)} \otimes \sigma_y^{(2)}\right]|\uparrow_1\downarrow_2\rangle + \alpha^2\beta^2\langle\uparrow_1\uparrow_2|\left[\sigma_z^{(1)} \otimes \sigma_y^{(2)}\right]|\downarrow_1\downarrow_2\rangle \\
&+ \alpha^3\beta\langle\uparrow_1\downarrow_2|\left[\sigma_z^{(1)} \otimes \sigma_y^{(2)}\right]|\uparrow_1\uparrow_2\rangle + \alpha^4\langle\uparrow_1\downarrow_2|\left[\sigma_z^{(1)} \otimes \sigma_y^{(2)}\right]|\uparrow_1\downarrow_2\rangle \\
&+ \alpha^2\beta^2\langle\uparrow_1\downarrow_2|\left[\sigma_z^{(1)} \otimes \sigma_y^{(2)}\right]|\downarrow_1\uparrow_2\rangle + \alpha^3\beta\langle\uparrow_1\downarrow_2|\left[\sigma_z^{(1)} \otimes \sigma_y^{(2)}\right]|\downarrow_1\downarrow_2\rangle \\
&+ \alpha\beta^3\langle\downarrow_1\uparrow_2|\left[\sigma_z^{(1)} \otimes \sigma_y^{(2)}\right]|\uparrow_1\uparrow_2\rangle + \alpha^2\beta^2\langle\downarrow_1\uparrow_2|\left[\sigma_z^{(1)} \otimes \sigma_y^{(2)}\right]|\uparrow_1\downarrow_2\rangle \\
&+ \beta^4\langle\downarrow_1\uparrow_2|\left[\sigma_z^{(1)} \otimes \sigma_y^{(2)}\right]|\downarrow_1\uparrow_2\rangle + \alpha\beta^3\langle\downarrow_1\uparrow_2|\left[\sigma_z^{(1)} \otimes \sigma_y^{(2)}\right]|\downarrow_1\downarrow_2\rangle \\
&+ \alpha^2\beta^2\langle\downarrow_1\downarrow_2|\left[\sigma_z^{(1)} \otimes \sigma_y^{(2)}\right]|\uparrow_1\uparrow_2\rangle + \alpha^3\beta\langle\downarrow_1\downarrow_2|\left[\sigma_z^{(1)} \otimes \sigma_y^{(2)}\right]|\uparrow_1\downarrow_2\rangle \\
&+ \alpha\beta^2\langle\downarrow_1\downarrow_2|\left[\sigma_z^{(1)} \otimes \sigma_y^{(2)}\right]|\downarrow_1\uparrow_2\rangle + \alpha^2\beta^2\langle\downarrow_1\downarrow_2|\left[\sigma_z^{(1)} \otimes \sigma_y^{(2)}\right]|\downarrow_1\uparrow_2\rangle\Big].
\end{aligned}$$

After taking into account the operation of the Pauli matrices, as explained at the beginning of the solution for one case, we find that $\langle\hat{A}\rangle = 0$.

Problem 4.13

Consider a state of the form $|\Psi\rangle = \cos(\theta)|\uparrow\downarrow\rangle + \sin(\theta)|\downarrow\uparrow\rangle$. a) Find the entanglement entropy for this system. b) What values of θ correspond to a Bell state and a product biparticle state? Comment on the entropy entanglement in each case.

a) To find the entanglement entropy, we have to obtain the partial trace of the state $|\Psi\rangle$. Therefore, we need the density matrix operator, found to be

$$\widehat{\rho} = |\psi\rangle\langle\psi| = \begin{pmatrix} 0 & 0 & 0 & 0 \\ 0 & \cos^2(\theta) & \cos(\theta)\sin(\theta) & 0 \\ 0 & \cos(\theta)\sin(\theta) & \sin^2(\theta) & 0 \\ 0 & 0 & 0 & 0 \end{pmatrix}.$$

The reduced density operator for particle B (second spin in the composite spinor) is

$$\widehat{\rho}_B = Tr_A(\widehat{\rho}_{AB}) = \sum_{a=1} \langle a_\alpha|\widehat{\rho}_{AB}|a_a\rangle = \begin{pmatrix} \cos(\theta)^2 & 0 \\ 0 & \sin(\theta)^2 \end{pmatrix}.$$

The entanglement entropy is

$$E = -Tr_B\left[\widehat{\rho}_B \ln(\widehat{\rho}_B)\right] = S(\widehat{\rho}_B),$$
$$E = -\sin^2(\theta)\ln(\sin^2(\theta)) - \cos^2(\theta)\ln(\cos^2(\theta)).$$

The same entanglement entropy with the same reduced density is found for subsystem A (first spin in the composite spinor).

b) The entropy is an oscillatory function of the angle θ, as shown in Figure 4.2. It can be zero when $\theta = n\pi$, with $|\Psi\rangle = |\uparrow\downarrow\rangle = |\uparrow_A\rangle\otimes|\downarrow_B\rangle$ being the noncorrelated pure state. The entropy can also be zero when $\theta = \pi\left(n+\frac{1}{2}\right)$ with $|\Psi\rangle = |\downarrow\uparrow\rangle = |\downarrow_A\rangle\otimes|\uparrow_B\rangle$ being the pure noncorrelated state (n – integer).

The given graphical representation of $E(\theta)$ shows how the entropy changes as a function of the angle θ. In the case of $\theta = \pi\left(n+\frac{1}{4}\right)$ or $\theta = \pi\left(n+\frac{3}{4}\right)$ for integer n, $E = \ln(2)$ reaches its maximum value, which corresponds to a maximally entangled state. In this case, those are the Bell states

$$|a,+\rangle = \frac{1}{\sqrt{2}}(|\uparrow\downarrow\rangle + |\downarrow\uparrow\rangle), \text{ for } \theta = \pi\left(n+\frac{1}{4}\right),$$
$$|a,-\rangle = \frac{1}{\sqrt{2}}(|\uparrow\downarrow\rangle - |\downarrow\uparrow\rangle), \text{ for } \theta = \pi\left(n+\frac{3}{4}\right).$$

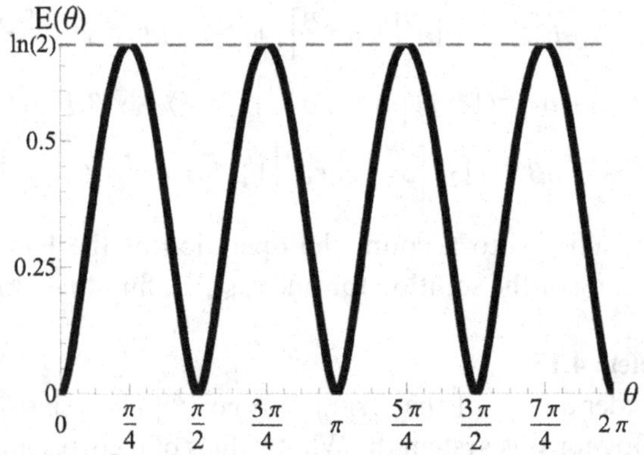

Figure 4.2 A graphical representation of the entanglement entropy E as a function of the angle θ.

Problem 4.14
Show that the Bell states can be transformed to each other by using Pauli matrices. As an example, consider $|c,+\rangle = \widehat{\sigma}_1 \otimes \widehat{\mathbb{I}}_2 |a,+\rangle$ and $|c,+\rangle = -i\widehat{\sigma}_2 \otimes \widehat{\mathbb{I}}_2 |a,-\rangle$.

We start with

$$
\begin{aligned}
|a,+\rangle &= \frac{1}{\sqrt{2}}(|\uparrow_1\rangle \otimes |\downarrow_2\rangle + |\downarrow_1\rangle \otimes |\uparrow_2\rangle),\\
\widehat{\sigma}_1 \otimes \widehat{\mathbb{I}}_2 |a,+\rangle &= \frac{1}{\sqrt{2}}\left(\widehat{\sigma}_1 |\uparrow_1\rangle \otimes \widehat{\mathbb{I}}_2 |\downarrow_2\rangle + \widehat{\sigma}_1 |\downarrow_1\rangle \otimes \widehat{\mathbb{I}}_2 |\uparrow_2\rangle\right)\\
&= \frac{1}{\sqrt{2}}(|\downarrow_1\rangle \otimes |\downarrow_2\rangle + |\uparrow_1\rangle \otimes |\uparrow_2\rangle) = |c,+\rangle.
\end{aligned}
$$

Also,

$$
\begin{aligned}
|a,-\rangle &= \frac{1}{\sqrt{2}}(|\uparrow_1\rangle \otimes |\downarrow_2\rangle - |\downarrow_1\rangle \otimes |\uparrow_2\rangle),\\
i\widehat{\sigma}_2 \otimes \widehat{\mathbb{I}}_2 |a,-\rangle &= \frac{1}{\sqrt{2}}\left(i\widehat{\sigma}_2 |\uparrow_1\rangle \otimes \widehat{\mathbb{I}}_2 |\downarrow_2\rangle - i\widehat{\sigma}_2 |\downarrow_1\rangle \otimes \widehat{\mathbb{I}}_2 |\uparrow_2\rangle\right)\\
&= \frac{1}{\sqrt{2}}(-|\downarrow_1\rangle \otimes |\downarrow_2\rangle - |\uparrow_1\rangle \otimes |\uparrow_2\rangle) = -|c,+\rangle.
\end{aligned}
$$

> *Food for thought:* For extra practice, the reader can show similar relations expressing $|a,\pm\rangle$ in terms of $|c,\pm\rangle$.

Problem 4.15
Consider the Hamiltonian for the Ising model for a system of two spin ½ particles,

$$\widehat{H} = -J\widehat{\sigma}_x^{(1)} \otimes \widehat{\sigma}_x^{(2)} - B\widehat{\sigma}_z^{(1)} \otimes \widehat{\sigma}_z^{(2)},$$

where the $\widehat{\sigma}_{x,y,z}^{(1,2)}$ are the Pauli matrices for the two-particle systems. Also, J, B are positive real constants.

a) Write the density operator $\widehat{\rho}$ for the pure state of the ground state of $\widehat{H}$ in the basis set $|\uparrow\uparrow\rangle, |\uparrow\downarrow\rangle, |\downarrow\uparrow\rangle, |\downarrow\downarrow\rangle$.
b) Evaluate the entanglement entropy by using reduced density operators $\widehat{\rho}_1$, $\widehat{\rho}_2$ for the two subsystems.
c) Check to see if $\widehat{\rho}_1 \otimes \widehat{\rho}_2$ is the same as $\widehat{\rho}$.

---

a) The Hamiltonian for the Ising model appears often in physics and, in this problem, we are asked to write the density operator for the pure ground state of $\widehat{H}$. This implies that the eigenvalues and eigenfunctions must be found beforehand. The problem also gives an opportunity to practice partial traces in the context of entanglement entropy.

From the tensor product of the Pauli matrices, we obtain

$$\widehat{H} = -J\begin{pmatrix} 0 & 0 & 0 & 1 \\ 0 & 0 & 1 & 0 \\ 0 & 1 & 0 & 0 \\ 1 & 0 & 0 & 0 \end{pmatrix} - B\begin{pmatrix} 1 & 0 & 0 & 0 \\ 0 & -1 & 0 & 0 \\ 0 & 0 & -1 & 0 \\ 0 & 0 & 0 & 1 \end{pmatrix} = \begin{pmatrix} -B & 0 & 0 & -J \\ 0 & B & -J & 0 \\ 0 & -J & B & 0 \\ -J & 0 & 0 & -B \end{pmatrix}.$$

The eigenvalues and corresponding eigenstates are found as

$$E_1 = B + J; \quad E_2 = B - J; \quad E_3 = -B + J; \quad E_4 = -B - J,$$

$$|\chi_1\rangle = \frac{1}{\sqrt{2}}(|\uparrow_1\downarrow_2\rangle - |\downarrow_1\uparrow_2\rangle); \quad |\chi_2\rangle\frac{1}{\sqrt{2}}(|\uparrow_1\downarrow_2\rangle - |\downarrow_1\uparrow_2\rangle);$$

$$|\chi_3\rangle = \frac{1}{\sqrt{2}}(|\uparrow_1\uparrow_2\rangle - |\downarrow_1\downarrow_2\rangle); \quad |\chi_4\rangle = \frac{1}{\sqrt{2}}(|\uparrow_1\uparrow_2\rangle + |\downarrow_1\downarrow_2\rangle).$$

In the preceding, we introduced the subscripts 1,2 to explicitly denote particles 1,2. Note that the eigenstates for this Hamiltonian are the four Bell states for anticorrelated and correlated pairs of spins, as shown earlier.

The ground state energy is $E_4 = -B - J$ with the corresponding eigenstate $|\chi_4\rangle$. The density matrix for the case of the pure ground state then becomes

$$\widehat{\rho} = |\chi_4\rangle\langle\chi_4| = \frac{1}{2}(|\uparrow_1\uparrow_2\rangle\langle\uparrow_2\uparrow_2| + |\uparrow_1\uparrow_2\rangle\langle\downarrow_1\downarrow_2| + |\downarrow_1\downarrow_2\rangle\langle\uparrow_1\uparrow_2| + |\downarrow_1\downarrow_2\rangle\langle\downarrow_1\downarrow_2|)$$

$$= \frac{1}{2}\begin{pmatrix} 1 & 0 & 0 & 1 \\ 0 & 0 & 0 & 0 \\ 0 & 0 & 0 & 0 \\ 1 & 0 & 0 & 1 \end{pmatrix}.$$

b) Continuing further, the entanglement entropy is given by $E = -Tr[\widehat{\rho}_1 \ln(\widehat{\rho}_1)]$ and $E = -Tr[\widehat{\rho}_2 \ln(\widehat{\rho}_2)]$ in terms of the reduced matrices. Notice that no matter which reduced density matrix is going to be used in E, the result is going to be the same.

The partial density matrices are evaluated in the usual manner, and as expected for a pure state,

$$\widehat{\rho}_2 = Tr_1(\widehat{\rho})$$

$$= \langle\uparrow_1|\widehat{\rho}|\uparrow_1\rangle + \langle\downarrow_1|\widehat{\rho}|\downarrow_1\rangle = \frac{1}{2}(\langle\uparrow_1|\uparrow_1\uparrow_2\rangle\langle\uparrow_1\uparrow_2|\uparrow_1\rangle + \langle\uparrow_1|\uparrow_1\uparrow_2\rangle\langle\downarrow_1\downarrow_2|\uparrow_1\rangle$$

$$+ \langle\uparrow_1|\downarrow_1\downarrow_2\rangle\langle\uparrow_1\uparrow_2|\uparrow_1\rangle + \langle\uparrow_1|\downarrow_1\downarrow_2\rangle\langle\downarrow_1\downarrow_2|\uparrow_1\rangle + \langle\downarrow_1|\uparrow_1\uparrow_2\rangle\langle\uparrow_1\uparrow_2|\downarrow_3\rangle$$

$$+ \langle\downarrow_1|\uparrow_1\uparrow_2\rangle\langle\downarrow_1\downarrow_2|\downarrow_1\rangle + \langle\downarrow_1|\downarrow_1\downarrow_2\rangle\langle\uparrow_1\uparrow_2|\downarrow_1\rangle + \langle\downarrow_1|\downarrow_1\downarrow_2\rangle\langle\downarrow_1\downarrow_2|\downarrow_1\rangle)$$

$$= \frac{1}{2}(|\uparrow_2\rangle\langle\uparrow_2| + |\downarrow_2\rangle\langle\downarrow_2|) = \frac{1}{2}\begin{pmatrix} 1 & 0 \\ 0 & 1 \end{pmatrix}.$$

$$\widehat{\rho}_1 = Tr_2(\widehat{\rho}) = \langle\uparrow_2|\widehat{\rho}|\uparrow_2\rangle + \langle\downarrow_2|\widehat{\rho}|\downarrow_2\rangle = \frac{1}{2}(|\uparrow_1\rangle\langle\uparrow_1| + |\downarrow_1\rangle\langle\downarrow_1|) = \frac{1}{2}\begin{pmatrix} 1 & 0 \\ 0 & 1 \end{pmatrix}.$$

In the preceding, the subindices (1,2) for the up and down states are kept explicitly for a more transparent calculation. Therefore, the entanglement entropy is

$$E = -Tr\left[\widehat{\rho}_1 \ln(\widehat{\rho}_1)\right] = -Tr\left[\frac{1}{2}\begin{pmatrix} 1 & 0 \\ 0 & 1 \end{pmatrix} \ln\left(\frac{1}{2}\begin{pmatrix} 1 & 0 \\ 0 & 1 \end{pmatrix}\right)\right]$$

$$= -Tr\left[\frac{\ln\left(\frac{1}{2}\right)}{2}\begin{pmatrix} 1 & 0 \\ 0 & 1 \end{pmatrix}\right] = \ln(2).$$

For the preceding relation, we have used the fact that, for a diagonal matrix,

$$\log_b(A) = \log_b\begin{pmatrix} a_1 & \cdots & 0 \\ \vdots & \ddots & \vdots \\ 0 & \cdots & a_n \end{pmatrix} = \begin{pmatrix} \log_b(a_1) & \cdots & 0 \\ \vdots & \ddots & \vdots \\ 0 & \cdots & \log_b(a_n) \end{pmatrix}.$$

c) Finally, for the last question, we check that

$$\widehat{\rho}_1 \otimes \widehat{\rho}_2 = \frac{1}{4}\begin{pmatrix} 1 & 0 \\ 0 & 1 \end{pmatrix} \otimes \begin{pmatrix} 1 & 0 \\ 0 & 1 \end{pmatrix} = \frac{1}{2}\begin{pmatrix} 1 & 0 & 0 & 0 \\ 0 & 1 & 0 & 0 \\ 0 & 0 & 1 & 0 \\ 0 & 0 & 0 & 1 \end{pmatrix} \neq \widehat{\rho}.$$

Problem 4.16

Consider the Hamiltonian for the Ising model for a system of two spin 1/2 particles,

$$\widehat{H} = -J\widehat{\sigma}_x^{(1)} \otimes \widehat{\sigma}_x^{(2)} - B\left[\widehat{\sigma}_z^{(1)} \otimes \widehat{\mathbb{I}}^{(2)} + \widehat{\mathbb{I}}^{(1)} \otimes \widehat{\sigma}_z^{(2)}\right],$$

where the $\widehat{\sigma}_{x,y,z}^{(1,2)}$ are the Pauli matrices for the two particle systems and $\widehat{\mathbb{I}}^{(1,2)}$ are the corresponding identity operators. Also, J, B are positive real constants.

a) Write the density operator $\widehat{\rho}$ for the pure state of the ground state of $\widehat{H}$ in the basis set $|\uparrow\uparrow\rangle, |\uparrow\downarrow\rangle, |\downarrow\uparrow\rangle, |\downarrow\downarrow\rangle$.
b) Evaluate the entanglement entropies for the two subsystems.
c) Check to see if $\widehat{\rho}_1 \otimes \widehat{\rho}_2$ is the same as $\widehat{\rho}$.

---

a) The Hamiltonian for the Ising model appears often in physics and, in this problem, we are asked to write the density operator for the pure ground state of $\widehat{H}$. This implies that the eigenvalues and eigenfunctions must be found beforehand. The problem also gives an opportunity to practice partial traces in the context of entanglement entropy.

From the tensor product of the Pauli matrices, we obtain

$$\widehat{H} = -J\begin{pmatrix} 0 & 0 & 0 & 1 \\ 0 & 0 & 1 & 0 \\ 0 & 1 & 0 & 0 \\ 1 & 0 & 0 & 0 \end{pmatrix} - B\begin{pmatrix} 2 & 0 & 0 & 0 \\ 0 & 0 & 0 & 0 \\ 0 & 0 & 0 & 0 \\ 0 & 0 & 0 & -2 \end{pmatrix} = -\begin{pmatrix} 2B & 0 & 0 & J \\ 0 & 0 & J & 0 \\ 0 & J & 0 & 0 \\ J & 0 & 0 & -2B \end{pmatrix}.$$

The eigenvalues and corresponding eigenstates are found as (with $R = \sqrt{J^2+4B^2}$)

$$E_1 = -R; \quad E_2 = -J; \quad E_3 = +J; \quad E_4 = +R.$$

$$|\chi_1\rangle = \frac{1}{\sqrt{N_+}}\left[(2B+R)|\uparrow_1\uparrow_2\rangle + J|\downarrow_1\downarrow_2\rangle\right]; \quad |\chi_2\rangle = \frac{1}{\sqrt{2}}(|\uparrow_1\downarrow_2\rangle + |\downarrow_1\uparrow_2\rangle);$$

$$|\chi_3\rangle = \frac{1}{\sqrt{2}}(|\uparrow_1\downarrow_2\rangle - |\downarrow_1\uparrow_2\rangle); \quad |\chi_4\rangle = \frac{1}{\sqrt{N_-}}\left[(2B-R)|\uparrow_1\uparrow_2\rangle + J|\downarrow_1\downarrow_2\rangle\right].$$

Here, we introduced the subscripts 1,2 to explicitly denote particles 1,2 and $N_\pm = 2R(R \pm 2B)$.

The ground state energy is $E_1 = -R$, with the corresponding eigenstate $|\chi_1\rangle$.

The density matrix for the case of the pure ground state then becomes

$$\begin{aligned}
\widehat{\rho} &= |\chi_1\rangle\langle\chi_1| \\
&= \frac{1}{N_+}\left[(2B+R)^2|\uparrow_1\uparrow_2\rangle\langle\uparrow_1\uparrow_2| + J(2B+R)|\uparrow_1\uparrow_2\rangle\langle\downarrow_1\downarrow_2|\right. \\
&\qquad \left. + J(2B+R)|\downarrow_1\downarrow_2\rangle\langle\uparrow_1\uparrow_2| + J^2|\downarrow_1\downarrow_2\rangle\langle\downarrow_1\downarrow_2|\right] \\
&= \frac{1}{2R}\begin{pmatrix} 2B+R & 0 & 0 & J \\ 0 & 0 & 0 & 0 \\ 0 & 0 & 0 & 0 \\ J & 0 & 0 & \frac{J^2}{2B+R} \end{pmatrix}.
\end{aligned}$$

b) Continuing further, the entanglement entropy is given by $E = -Tr[\widehat{\rho}_1 \ln(\widehat{\rho}_1)]$ and $E = -Tr[\widehat{\rho}_2 \ln(\widehat{\rho}_2)]$ in terms of the reduced matrices. Notice that no matter which reduced density matrix is going to be used in E, the result is going to be the same.

The partial density matrices are evaluated in the usual manner, and as expected for a pure state,

$$\widehat{\rho}_1 = Tr_2(\widehat{\rho}) = \frac{1}{2R}\left[(2B+R)|\uparrow_1\rangle\langle\uparrow_1| + \frac{J^2}{2B+R}|\downarrow_1\rangle\langle\downarrow_1|\right] = \frac{1}{2R}\begin{pmatrix} 2B+R & 0 \\ 0 & \frac{J^2}{2B+R}\end{pmatrix},$$

$$\widehat{\rho}_2 = Tr_1(\widehat{\rho}) = \frac{1}{2R}\left[(2B+R)|\uparrow_2\rangle\langle\uparrow_2| + \frac{J^2}{2B+R}|\downarrow_2\rangle\langle\downarrow_2|\right] = \frac{1}{2R}\begin{pmatrix} 2B+R & 0 \\ 0 & \frac{J^2}{2B+R}\end{pmatrix}.$$

Therefore, using that the eigenvalues of those density matrices are $\lambda_\pm = \frac{1}{2} \pm \frac{B}{R}$, the entanglement entropy is

$$\begin{aligned}
E = -Tr[\widehat{\rho}_1 \ln(\widehat{\rho}_1)] &= -\left(\frac{1}{2} + \frac{B}{R}\right)\ln\left(\frac{1}{2} + \frac{B}{R}\right) - \left(\frac{1}{2} - \frac{B}{R}\right)\ln\left(\frac{1}{2} - \frac{B}{R}\right) \\
&= -\frac{1}{2}\ln\left(\frac{1}{4} - \frac{B^2}{R^2}\right) - \frac{2B}{R}\operatorname{arctanh}\left(\frac{2B}{R}\right),
\end{aligned}$$

which is obtained by noting that for a diagonal matrix:

$$\log_b(A) = \log_b \begin{pmatrix} a_1 & \cdots & 0 \\ \vdots & \ddots & \vdots \\ 0 & \cdots & a_n \end{pmatrix} = \begin{pmatrix} \log_b(a_1) & \cdots & 0 \\ \vdots & \ddots & \vdots \\ 0 & \cdots & \log_b(a_n) \end{pmatrix}.$$

c) Finally, for the last question, we check that

$$\widehat{\rho}_1 \otimes \widehat{\rho}_2 = \frac{1}{4R^2} \begin{pmatrix} 2B+R & 0 \\ 0 & \frac{J^2}{2B+R} \end{pmatrix} \otimes \begin{pmatrix} 2B+R & 0 \\ 0 & \frac{J^2}{2B+R} \end{pmatrix}$$

$$= \frac{1}{4R^2} \begin{pmatrix} (2B+R)^2 & 0 & 0 & 0 \\ 0 & J^2 & 0 & 0 \\ 0 & 0 & J^2 & 0 \\ 0 & 0 & 0 & \frac{J^4}{(2B+R)^2} \end{pmatrix} \neq \widehat{\rho}.$$

Problem 4.17

Repeat the previous problem, but for the state with the highest energy for the XY Hamiltonian, given as

$$\widehat{H} = -J\left(\widehat{\sigma}_x^{(1)} \otimes \widehat{\sigma}_x^{(2)} + \widehat{\sigma}_y^{(1)} \otimes \widehat{\sigma}_y^{(2)}\right) - B\widehat{\sigma}_z^{(1)} \otimes \widehat{\sigma}_x^{(2)},$$

where J, B are positive real constants.

---

a) Here we repeat the same steps as in the previous problem, starting with finding the eigenvalues and eigenstates of the Hamiltonian in the $|\uparrow_1\uparrow_2\rangle, |\uparrow_1\downarrow_2\rangle, |\downarrow_1\uparrow_2\rangle, |\downarrow_1\downarrow_2\rangle$ basis.

$$\widehat{H} = -2J \begin{pmatrix} 0 & 0 & 0 & 0 \\ 0 & 0 & 1 & 0 \\ 0 & 1 & 0 & 0 \\ 0 & 0 & 0 & 0 \end{pmatrix} - B \begin{pmatrix} 1 & 0 & 0 & 0 \\ 0 & -1 & 0 & 0 \\ 0 & 0 & -1 & 0 \\ 0 & 0 & 0 & 1 \end{pmatrix} = \begin{pmatrix} -B & 0 & 0 & 0 \\ 0 & B & -2J & 0 \\ 0 & -2J & B & 0 \\ 0 & 0 & 0 & -B \end{pmatrix}.$$

The eigenvalues and corresponding eigenstates are

$$E_1 = -B; \quad E_2 = B - 2J; \quad E_3 = B + 2J; \quad E_4 = -B,$$

$$|\chi_1\rangle = |\uparrow_1\uparrow_2\rangle; \; |\chi_2\rangle = \frac{1}{\sqrt{2}}(|\uparrow_1\downarrow_2\rangle + |\downarrow_1\uparrow_2\rangle); \; |\chi_3\rangle = \frac{1}{\sqrt{2}}(|\uparrow_1\downarrow_2\rangle - |\downarrow_1\uparrow_2\rangle); \; |\chi_4\rangle = |\downarrow_1\downarrow_2\rangle.$$

The highest energy$E_3 = B + 2J$with the corresponding state $|\chi_3\rangle = \frac{1}{\sqrt{2}}(|\uparrow_1\downarrow_2\rangle - |\downarrow_1\uparrow_2\rangle)$.
The density matrix is found as

$$\widehat{\rho} = |\chi_3\rangle\langle\chi_3| = \frac{1}{2}(|\uparrow_1\downarrow_2\rangle\langle\uparrow_1\downarrow_2| - |\uparrow_1\downarrow_2\rangle\langle\downarrow_1\uparrow_2| - |\downarrow_1\uparrow_2\rangle\langle\uparrow_1\downarrow_2| + |\downarrow_1\uparrow_2\rangle\langle\downarrow_1\uparrow_2|)$$

$$= \frac{1}{2} \begin{pmatrix} 0 & 0 & 0 & 0 \\ 0 & 1 & -1 & 0 \\ 0 & -1 & 1 & 0 \\ 0 & 0 & 0 & 0 \end{pmatrix}.$$

b) The partial densities are

$$\widehat{\rho}_2 = Tr_1(\widehat{\rho}) = \langle\uparrow_1|\widehat{\rho}|\uparrow_1\rangle + \langle\downarrow_1|\widehat{\rho}|\downarrow_1\rangle$$
$$= \frac{1}{2}\big(\langle\uparrow_1|\uparrow_1\downarrow_2\rangle\langle\uparrow_1\downarrow_2|\uparrow_1\rangle + \langle\uparrow_1|\uparrow_1\downarrow_2\rangle\langle\downarrow_1\uparrow_2|\uparrow_1\rangle + \langle\uparrow_1|\downarrow_1\uparrow_2\rangle\langle\uparrow_1\downarrow_2|\uparrow_1\rangle$$
$$+ \langle\uparrow_1|\downarrow_1\uparrow_2\rangle\langle\downarrow_1\uparrow_2|\uparrow_1\rangle + \langle\downarrow_1|\uparrow_1\downarrow_2\rangle\langle\uparrow_1\downarrow_2|\downarrow_1\rangle + \langle\downarrow_1|\uparrow_1\downarrow_2\rangle\langle\downarrow_1\uparrow_2|\downarrow_1\rangle$$
$$+ \langle\downarrow_1|\downarrow_1\uparrow_2\rangle\langle\uparrow_1\downarrow_2|\downarrow_1\rangle + \langle\downarrow_1|\downarrow_1\uparrow_2\rangle\langle\downarrow_1\uparrow_2|\downarrow_1\rangle\big)$$
$$= \frac{1}{2}\big(\langle\uparrow_1|\uparrow_1\downarrow_2\rangle\langle\uparrow_1\downarrow_2|\uparrow_1\rangle + \langle\downarrow_1|\downarrow_1\uparrow_2\rangle\langle\downarrow_1\uparrow_2|\downarrow_1\rangle\big)$$
$$= \frac{1}{2}(|\uparrow_2\rangle\langle\uparrow_2| + |\downarrow_2\rangle\langle\downarrow_2|) = \frac{1}{2}\begin{pmatrix} 1 & 0 \\ 0 & 1 \end{pmatrix},$$

$$\widehat{\rho}_1 = Tr_2(\widehat{\rho}) = \langle\uparrow_2|\widehat{\rho}|\uparrow_2\rangle + \langle\downarrow_2|\widehat{\rho}|\downarrow_2\rangle = \frac{1}{2}(|\uparrow_1\rangle\langle\uparrow_1| + |\downarrow_1\rangle\langle\downarrow_1|) = \frac{1}{2}\begin{pmatrix} 1 & 0 \\ 0 & 1 \end{pmatrix},$$

$$\widehat{\rho}_1 = \frac{1}{2}\begin{pmatrix} 1 & 0 \\ 0 & 1 \end{pmatrix} = \widehat{\rho}_2.$$

Therefore, the entanglement entropy is

$$E = \ln(2).$$

c) Finally, for the last question, we check that

$$\widehat{\rho}_1 \otimes \widehat{\rho}_2 = \frac{1}{4}\begin{pmatrix} 1 & 0 \\ 0 & 1 \end{pmatrix} \otimes \begin{pmatrix} 1 & 0 \\ 0 & 1 \end{pmatrix} = \frac{1}{2}\begin{pmatrix} 1 & 0 & 0 & 0 \\ 0 & 1 & 0 & 0 \\ 0 & 0 & 1 & 0 \\ 0 & 0 & 0 & 1 \end{pmatrix} \neq \widehat{\rho}.$$

Problem 4.18

Repeat the previous problem, but for the state with the highest energy for the XY Hamiltonian, given as

$$\widehat{H} = -J\left(\widehat{\sigma}_x^{(1)} \otimes \widehat{\sigma}_x^{(2)} + \widehat{\sigma}_y^{(1)} \otimes \widehat{\sigma}_y^{(2)}\right) - B\left[\widehat{\sigma}_z^{(1)} \otimes \widehat{\mathbb{I}}^{(2)} + \widehat{\mathbb{I}}^{(1)} \otimes \widehat{\sigma}_z^{(2)}\right],$$

where J, B are positive real constants and $B < J$.

---

a) Here we repeat the same steps as in the previous problem, starting with finding the eigenvalues and eigenstates of the Hamiltonian in the $|\uparrow_1\uparrow_2\rangle, |\uparrow_1\downarrow_2\rangle, |\downarrow_1\uparrow_2\rangle, |\downarrow_1\downarrow_2\rangle$ basis.

$$\widehat{H} = -2J\begin{pmatrix} 0 & 0 & 0 & 0 \\ 0 & 0 & 1 & 0 \\ 0 & 1 & 0 & 0 \\ 0 & 0 & 0 & 0 \end{pmatrix} - B\begin{pmatrix} 2 & 0 & 0 & 0 \\ 0 & 0 & 0 & 0 \\ 0 & 0 & 0 & 0 \\ 0 & 0 & 0 & -2 \end{pmatrix} = -2\begin{pmatrix} B & 0 & 0 & 0 \\ 0 & 0 & J & 0 \\ 0 & J & 0 & 0 \\ 0 & 0 & 0 & -B \end{pmatrix}.$$

The eigenvalues and corresponding eigenstates are

$$E_1 = -2B; \quad E_2 = -2J; \quad E_3 = 2J; \quad E_4 = 2B,$$

$$|\chi_1\rangle = |\uparrow_1\uparrow_2\rangle; \ |\chi_2\rangle = \frac{1}{\sqrt{2}}(|\uparrow_1\downarrow_2\rangle + |\downarrow_1\uparrow_2\rangle); \ |\chi_3\rangle = \frac{1}{\sqrt{2}}(|\uparrow_1\downarrow_2\rangle - |\downarrow_1\uparrow_2\rangle); \ |\chi_4\rangle = |\downarrow_1\downarrow_2\rangle.$$

The highest energy $E_3=2J$ with the corresponding state $|\chi_3\rangle=\frac{1}{\sqrt{2}}(|\uparrow_1\downarrow_2\rangle-|\downarrow_1\uparrow_2\rangle)$. The density matrix is found as

$$\widehat{\rho}=|\chi_3\rangle\langle\chi_3|=\frac{1}{2}(|\uparrow_1\downarrow_2\rangle\langle\uparrow_1\downarrow_2|-|\uparrow_1\downarrow_2\rangle\langle\downarrow_1\uparrow_2|-|\downarrow_1\uparrow_2\rangle\langle\uparrow_1\downarrow_2|+|\downarrow_1\uparrow_2\rangle\langle\downarrow_1\uparrow_2|)$$

$$=\frac{1}{2}\begin{pmatrix} 0 & 0 & 0 & 0 \\ 0 & 1 & -1 & 0 \\ 0 & -1 & 1 & 0 \\ 0 & 0 & 0 & 0 \end{pmatrix}.$$

b) The partial densities are

$$\widehat{\rho}_2=Tr_1(\widehat{\rho})=\langle\uparrow_1|\widehat{\rho}|\uparrow_1\rangle+\langle\downarrow_1|\widehat{\rho}|\downarrow_1\rangle$$

$$=\frac{1}{2}(\langle\uparrow_1|\uparrow_1\downarrow_2\rangle\langle\uparrow_1\downarrow_2|\uparrow_1\rangle+\langle\uparrow_1|\uparrow_1\downarrow_2\rangle\langle\downarrow_1\uparrow_2|\uparrow_1\rangle+\langle\uparrow_1|\downarrow_1\uparrow_2\rangle\langle\uparrow_1\downarrow_2|\uparrow_1\rangle$$
$$+\langle\uparrow_1|\downarrow_1\uparrow_2\rangle\langle\downarrow_1\uparrow_2|\uparrow_1\rangle+\langle\downarrow_1|\uparrow_1\downarrow_2\rangle\langle\uparrow_1\downarrow_2|\downarrow_1\rangle+\langle\downarrow_1|\uparrow_1\downarrow_2\rangle\langle\downarrow_1\uparrow_2|\downarrow_1\rangle$$
$$+\langle\downarrow_1|\downarrow_1\uparrow_2\rangle\langle\uparrow_1\downarrow_2|\downarrow_1\rangle+\langle\downarrow_1|\downarrow_1\uparrow_2\rangle\langle\downarrow_1\uparrow_2|\downarrow_1\rangle)$$

$$=\frac{1}{2}(\langle\uparrow_1|\uparrow_1\downarrow_2\rangle\langle\uparrow_1\downarrow_2|\uparrow_1\rangle+0+0+0+0+0+0+\langle\downarrow_1|\downarrow_1\uparrow_2\rangle\langle\downarrow_1|\uparrow_2\downarrow_1\rangle)$$

$$=\frac{1}{2}(|\uparrow_2\rangle\langle\uparrow_2|+|\downarrow_2\rangle\langle\downarrow_2|)=\frac{1}{2}\begin{pmatrix} 1 & 0 \\ 0 & 1 \end{pmatrix},$$

$$\widehat{\rho}_1=Tr_2(\widehat{\rho})=\langle\uparrow_2|\widehat{\rho}|\uparrow_2\rangle+\langle\downarrow_2|\widehat{\rho}|\downarrow_2\rangle=\frac{1}{2}(|\uparrow_1\rangle\langle\uparrow_1|+|\downarrow_1\rangle\langle\downarrow_1|)=\frac{1}{2}\begin{pmatrix} 1 & 0 \\ 0 & 1 \end{pmatrix},$$

$$\widehat{\rho}_1=\frac{1}{2}\begin{pmatrix} 1 & 0 \\ 0 & 1 \end{pmatrix}=\widehat{\rho}_2.$$

Therefore, the entanglement entropy is

$$E=\ln(2).$$

c) Finally, for the last question, we check that

$$\widehat{\rho}_1\otimes\widehat{\rho}_2=\frac{1}{4}\begin{pmatrix} 1 & 0 \\ 0 & 1 \end{pmatrix}\otimes\begin{pmatrix} 1 & 0 \\ 0 & 1 \end{pmatrix}=\frac{1}{2}\begin{pmatrix} 1 & 0 & 0 & 0 \\ 0 & 1 & 0 & 0 \\ 0 & 0 & 1 & 0 \\ 0 & 0 & 0 & 1 \end{pmatrix}\neq\widehat{\rho}.$$

Problem 4.19

Consider the Hamiltonian for a two-particle system,

$$\widehat{H}=-J\left(\widehat{\sigma}_x^{(1)}\otimes\widehat{\mathbb{I}}^{(2)}+\widehat{\mathbb{I}}^{(1)}\otimes\widehat{\sigma}_x^{(2)}\right)-B\widehat{\sigma}_z^{(1)}\otimes\widehat{\sigma}_z^{(2)},$$

where the $\widehat{\sigma}_{x,y,z}^{(1,2)}$ are the Pauli matrices for each particle and $\widehat{\mathbb{I}}^{(1,2)}$ are the corresponding identity operators. Also, J,B are real positive constants.

a) Write the density operator $\widehat{\rho}_{12}$ in the basis set $|\uparrow\uparrow\rangle,|\uparrow\downarrow\rangle,|\downarrow\uparrow\rangle,|\downarrow\downarrow\rangle$ when the composite system is in the ground state of $\widehat{H}$. b) What is the entanglement entropy?

a) Let's begin by writing $\widehat{H}$ explicitly in the given basis,

$$\widehat{H} = -J\begin{pmatrix} 0 & 1 & 1 & 0 \\ 1 & 0 & 0 & 1 \\ 1 & 0 & 0 & 1 \\ 0 & 1 & 1 & 0 \end{pmatrix} - B\begin{pmatrix} 1 & 0 & 0 & 0 \\ 0 & -1 & 0 & 0 \\ 0 & 0 & -1 & 0 \\ 0 & 0 & 0 & 1 \end{pmatrix} = \begin{pmatrix} -B & -J & -J & 0 \\ -J & B & 0 & -J \\ -J & 0 & B & -J \\ 0 & -J & -J & -B \end{pmatrix}.$$

The eigenvalues and corresponding eigenstates are

$$E_1 = B; \quad E_2 = -B; \quad E_3 = -\sqrt{B^2+4J^2}; \quad E_4 = \sqrt{B^2+4J^2},$$

$$|\chi_1\rangle = \frac{1}{\sqrt{2}}(|\uparrow_1\downarrow_2\rangle - |\downarrow_1\uparrow_2\rangle); \quad |\chi_2\rangle = \frac{1}{\sqrt{2}}(|\uparrow_1\uparrow_2\rangle - |\downarrow_1\downarrow_2\rangle);$$

$$|\chi_3\rangle = \frac{1}{2\sqrt{\sqrt{1+\alpha^2}\left(\sqrt{1+\alpha^2}-a\right)}}\left[|\uparrow_1\uparrow_2\rangle + |\downarrow_1\downarrow_2\rangle + (\sqrt{1+\alpha^2}-\alpha)(|\uparrow_1\downarrow_2\rangle + |\downarrow_1\uparrow_2\rangle)\right];$$

$$|\chi_4\rangle = \frac{1}{2\sqrt{\sqrt{1+\alpha^2}\left(\sqrt{1+\alpha^2}+a\right)}}\left[|\uparrow_1\uparrow_2\rangle + |\downarrow_1\downarrow_2\rangle - (\sqrt{1+\alpha^2}+\alpha)(|\uparrow_1\downarrow_2\rangle + |\downarrow_1\uparrow_2\rangle)\right],$$

where $\alpha = \frac{B}{2J}$. Clearly the ground state energy is E_3 with the corresponding ground state $|\chi_3\rangle$.

The density operator for the pure ground state is then

$$\widehat{\rho}_{12} = |\chi_3\rangle\langle\chi_3| = \frac{1}{4\sqrt{1+\alpha^2}}\begin{pmatrix} \frac{1}{\sqrt{1+\alpha^2}-\alpha} & 1 & 1 & \frac{1}{\sqrt{1+\alpha^2}-\alpha} \\ 1 & \sqrt{1+\alpha^2}-\alpha & \sqrt{1+\alpha^2}-\alpha & 1 \\ 1 & \sqrt{1+\alpha^2}-\alpha & \sqrt{1+\alpha^2}-\alpha & 1 \\ \frac{1}{\sqrt{1+\alpha^2}-\alpha} & 1 & 1 & \frac{1}{\sqrt{1+\alpha^2}-\alpha} \end{pmatrix}.$$

b) The eigenvalues of this $\widehat{\rho}$ are $\{0,0,0,1\}$; therefore, the von Neumann entropy is given as

$$S_N(\widehat{\rho}_{12}) = -\sum_i \lambda_i \ln(\lambda_i) = 0.$$

Then we conclude that $|\chi_3\rangle$ is a pure state, since its purity property is $\mathscr{P}(\widehat{\rho}_{12}) = Tr(\widehat{\rho}_{12}^2) = Tr(|\chi_3\rangle\langle\chi_3|\chi_3\rangle\langle\chi_3|) = Tr(|\chi_3\rangle\langle\chi_3|) = Tr(\widehat{\rho}_{12}) = 1$.

For the entanglement entropy, we need to find partial density operators $\widehat{\rho}_1 = Tr_2(\widehat{\rho}_{12}) = \widehat{\rho}_2 = Tr_1(\widehat{\rho}_{12})$ and their eigenvalues,

$$\widehat{\rho}_2 = Tr_1(\widehat{\rho}_{12}) = \langle\uparrow_1|\widehat{\rho}_{12}|\uparrow_1\rangle + \langle\downarrow_1\widehat{\rho}_{12}|\downarrow_1\rangle = \frac{1}{2\sqrt{1+\alpha^2}}\begin{pmatrix} \sqrt{1+\alpha^2} & 1 \\ 1 & \sqrt{1+\alpha^2} \end{pmatrix},$$

$$\widehat{\rho}_1 = Tr_2(\widehat{\rho}_{12}) = \langle\uparrow_2|\widehat{\rho}_{12}|\uparrow_2\rangle + \langle\downarrow_2|\widehat{\rho}_{12}|\downarrow_2\rangle = \frac{1}{2\sqrt{1+\alpha^2}}\begin{pmatrix} \sqrt{1+\alpha^2} & 1 \\ 1 & \sqrt{1+\alpha^2} \end{pmatrix}.$$

The somewhat lengthy algebra here is left for extra practice. The eigenvalues of the partial density operators are then obtained as

$$\lambda_\pm = \frac{1}{2} \pm \frac{1}{2\sqrt{1+\alpha^2}}.$$

Finally, the entanglement entropy is

$$E = -\left(\frac{1}{2} - \frac{1}{2\sqrt{1+\alpha^2}}\right)\ln\left(\frac{1}{2} - \frac{1}{2\sqrt{1+\alpha^2}}\right) - \left(\frac{1}{2} + \frac{1}{2\sqrt{1+\alpha^2}}\right)\ln\left(\frac{1}{2} + \frac{1}{2\sqrt{1+\alpha^2}}\right)$$
$$= \ln(2) - \frac{1}{2}\ln\left(1 - \frac{1}{1+\alpha^2}\right) - \frac{\operatorname{arcoth}(\sqrt{1+\alpha^2})}{\sqrt{1+\alpha^2}}.$$

Problem 4.20

For a biparticle system, the Hamiltonian is written as $\widehat{H} = \widehat{H}_1 \otimes \widehat{H}_2$, where the Hamiltonians of the individual particles 1,2 are $\widehat{H}_1$ with dimension n_1 and $\widehat{H}_2$ with dimension $n_2 \geq n_1$, respectively. Show that any given quantum mechanical state $|\Psi\rangle$ associated with $\widehat{H}$ can always be represented using the *Schmidt decomposition* as a product of orthonormal states $|\psi_i^{(1)}\rangle$ and $|\psi_i^{(2)}\rangle$,

$$|\Psi\rangle = \sum_{i=1}^{n_1} g_i |\psi_i^{(1)}\rangle \otimes |\psi_i^{(2)}\rangle,$$

where g_i are nonnegative numbers with $\sum_{i=1}^{n_1} |g_i|^2 = 1$.

We can easily identify what $|\psi_i^{(1)}\rangle$ and $|\psi_i^{(2)}\rangle$ are in the context of the density operator for the pure state $|\Psi\rangle$, with $\widehat{\rho} = |\Psi\rangle\langle\Psi|$.

Let's find the partial traces over system 1 and system 2 and obtain the reduced density matrices,

$$\widehat{\rho}_1 = Tr_2(|\Psi\rangle\langle\Psi|)$$
$$= Tr_2\left(\sum_{i,j=1}^{n_1} g_i g_j^* |\psi_i^{(1)}\rangle|\psi_i^{(2)}\rangle\langle\psi_j^{(2)}|\langle\psi_j^{(1)}|\right)$$
$$= \sum_{i,j=1}^{n_1} g_i g_j^* |\psi_i^{(1)}\rangle\langle\psi_j^{(1)}| Tr_2\left(|\psi_i^{(2)}\rangle\langle\psi_j^{(2)}|\right) = \sum_{i,j=1}^{n_1} g_i g_j^* |\psi_i^{(1)}\rangle\langle\psi_j^{(1)}|\delta_{ij}$$
$$= \sum_{i=1}^{n_1} |g_i|^2 |\psi_i^{(1)}\rangle\langle\psi_i^{(1)}|,$$

$$\widehat{\rho}_2 = Tr_1(|\Psi\rangle\langle\Psi|) = Tr_1\left(\sum_{i,j=1}^{n_1} g_i g_j^* |\psi_i^{(1)}\rangle|\psi_i^{(2)}\rangle\langle\psi_j^{(2)}|\langle\psi_j^{(1)}|\right)$$
$$= \sum_{i,j=1}^{n_1} g_i g_j^* Tr_1\left(|\psi_i^{(1)}\rangle|\psi_j^{(1)}|\right) |\psi_i^{(2)}\rangle\langle\psi_j^{(2)}|$$
$$= \sum_{i,j=1}^{n_1} g_i g_j^* \delta_{ij} |\psi_i^{(2)}\rangle\langle\psi_j^{(2)}| = \sum_{i=1}^{n_1} |g_i|^2 |\psi_i^{(2)}\rangle\langle\psi_i^{(2)}|.$$

The preceding relations show that $|\psi_i^{(1)}\rangle$ are the eigenstates corresponding to the partial density matrix for system 1, while $|\psi_i^{(2)}\rangle$ are the eigenstates for the partial density matrix for system 2. It is also clear that in both cases, $|g_i|^2$ are the eigenvalues of both $\widehat{\rho}_1$ and $\widehat{\rho}_2$. Note further that in both cases, the eigenvalues are the same.

Problem 4.21

Consider a system for spin ½ particles whose pure state is given as

$$|\Psi\rangle = \frac{1}{\sqrt{2}}|\uparrow_1\rangle|\uparrow_2\rangle + \frac{1}{2}|\downarrow_1\rangle(|\uparrow_2\rangle - |\downarrow_2\rangle).$$

Find its Schmidt decomposition.

The pure state $|\Psi\rangle = \frac{1}{\sqrt{2}}|\uparrow_1\rangle|\uparrow_2\rangle + \frac{1}{2}|\downarrow_1\rangle|\uparrow_2\rangle - \frac{1}{2}|\downarrow_1\rangle|\downarrow_2\rangle$ can be written in a matrix form using the equation $|\Psi\rangle = \sum_{n,m} \overline{\Psi}_{n,m}|a_n\rangle|b_m\rangle$, where $|a_n\rangle, |b_m\rangle$ in our case are the up and down spinors for the two particles: $|\uparrow_1\rangle|\uparrow_2\rangle, |\downarrow_1\rangle|\uparrow_2\rangle|, |\uparrow_1\rangle|\downarrow_2\rangle, |\downarrow_1\rangle|\downarrow_2\rangle$. Specifically, we can write

$$|\Psi\rangle = (|\uparrow_1\rangle, |\downarrow_1\rangle)\cdot\begin{pmatrix} \frac{1}{\sqrt{2}} & -\frac{1}{2} \\ \frac{1}{2} & 0 \end{pmatrix}\cdot\begin{pmatrix} |\uparrow_2\rangle \\ |\downarrow_2\rangle \end{pmatrix} \to \overline{\Psi} = \begin{pmatrix} \frac{1}{\sqrt{2}} & -\frac{1}{2} \\ \frac{1}{2} & 0 \end{pmatrix},$$

where the matrix $\overline{\Psi}$ has matrix elements $\overline{\Psi}_{n,m}$ for the given two-particle pure state.

The Schmidt decomposition is basically the *Singular Value Decomposition* of a general matrix (not necessarily square) $A \in \mathbb{C}^{n\times m}$. This matrix can be represented as $A = U\cdot\Sigma\cdot W$, where the rows of the $U \in \mathbb{C}^{n\times n}$ are the left singular vectors of the original matrix, the columns of $W \in \mathbb{C}^{m\times m}$ are the right singular vectors, and $\Sigma \in \mathbb{R}_+^{n\times m}$ is a diagonal matrix containing the singular (nonnegative) values σ_n of the original matrix. U and W are unitary matrices ($U\cdot U^+ = \mathbb{I}$ and $W\cdot W^+ = \mathbb{I}$). Singular Value Decomposition (Johnston, 2021; Steeb and Hardy, 2018) is related but different from the eigenvalue decomposition of a matrix. It can be applied to any nonsquare matrices and, contrary to eigenvalues, the singular values are always nonnegative numbers. The connection between the two types of decompositions is the following: the singular values of A are the (positive) square root of the nonzero eigenvalues of $A\cdot A^+$ and of $A^+\cdot A$, with eigenvectors given by the columns of U and W respectively. In addition to that, both decompositions coincide when A is a semidefinite positive square matrix.

Thus, we find

$$\begin{pmatrix} \frac{1}{\sqrt{2}} & -\frac{1}{2} \\ \frac{1}{2} & 0 \end{pmatrix} = \begin{pmatrix} \frac{-2-\sqrt{3}}{\sqrt{9+5\sqrt{3}}} & \frac{-2+\sqrt{3}}{\sqrt{9-5\sqrt{3}}} \\ \frac{-1-\sqrt{3}}{\sqrt{2}\sqrt{9+5\sqrt{3}}} & \frac{-1+\sqrt{3}}{\sqrt{2}\sqrt{9-5\sqrt{3}}} \end{pmatrix}\cdot\begin{pmatrix} \frac{\sqrt{2+\sqrt{3}}}{2} & 0 \\ 0 & \frac{\sqrt{2-\sqrt{3}}}{2} \end{pmatrix}$$
$$\cdot\begin{pmatrix} \frac{-1-\sqrt{3}}{\sqrt{2}\sqrt{3+\sqrt{3}}} & \frac{1}{\sqrt{3+\sqrt{3}}} \\ \frac{-1+\sqrt{3}}{\sqrt{2}\sqrt{3-\sqrt{3}}} & \frac{1}{\sqrt{3-\sqrt{3}}} \end{pmatrix}.$$

The Schmidt representation is then

$$|\Psi\rangle = \lambda_1|\psi_1^{(1)}\rangle\otimes|\psi_1^{(2)}\rangle + \lambda_2|\psi_2^{(1)}\rangle\otimes|\psi_2^{(2)}\rangle,$$

where

$$\left(|\psi_1^{(1)}\rangle, |\psi_1^{(2)}\rangle\right) = (|\uparrow_1\rangle, |\downarrow_1\rangle)\cdot U$$
$$= \left(\left[\frac{-2-\sqrt{3}}{\sqrt{9+5\sqrt{3}}}|\uparrow_1\rangle + \frac{-1-\sqrt{3}}{\sqrt{2}\sqrt{9+5\sqrt{3}}}|\downarrow_1\rangle\right], \left[\frac{-2+\sqrt{3}}{\sqrt{9-5\sqrt{3}}}|\uparrow_1\rangle + \frac{-1+\sqrt{3}}{\sqrt{2}\sqrt{9-5\sqrt{3}}}|\downarrow_1\rangle\right]\right),$$

$$\begin{pmatrix} |\psi_2^{(1)}\rangle \\ |\psi_2^{(2)}\rangle \end{pmatrix} = W\cdot\begin{pmatrix} |\uparrow_2\rangle \\ |\downarrow_2\rangle \end{pmatrix} = \begin{pmatrix} \left[\frac{-1-\sqrt{3}}{\sqrt{2}\sqrt{3+\sqrt{3}}}|\uparrow_2\rangle + \frac{1}{\sqrt{3+\sqrt{3}}}|\downarrow_2\rangle\right] \\ \left[\frac{-1+\sqrt{3}}{\sqrt{2}\sqrt{3-\sqrt{3}}}|\uparrow_2\rangle + \frac{1}{\sqrt{3-\sqrt{3}}}|\downarrow_2\rangle\right] \end{pmatrix}.$$

The reader can now verify by substitution that indeed the Schmidt decomposition corresponds to the original form of $|\Psi\rangle$.

Problem 4.22

Consider a system for two particles whose pure state is given as

$$|\Psi\rangle = \frac{1}{\sqrt{3}}\left(|0_1, 0_2\rangle + i|0_1, 2_2\rangle + \frac{1}{\sqrt{2}}|1_1, 1_2\rangle - \frac{1}{\sqrt{2}}|1_1, 2_2\rangle\right).$$

where $|0_{1,2}\rangle$, $|1_{1,2}\rangle$, $|2_{1,2}\rangle$ are various ket states for the particles. Find its Schmidt decomposition.

This problem is similar to the previous one; however, here the basis for particle 1 has two states $|0_1\rangle, |1_1\rangle$, while the basis for particle 2 contains three states $|0_2\rangle, |1_2\rangle, |2_2\rangle$.

This pure state $|\Psi\rangle$ can be written in a matrix form using the equation $|\Psi\rangle = \sum_{n,m} \overline{\Psi}_{n,m}|a_n\rangle|b_m\rangle$, where $|a_n\rangle, |b_m\rangle$ in our case are $\{|0_1\rangle|0_2\rangle, |0_1\rangle|1_2\rangle, |0_1\rangle|2_2\rangle, |1_1\rangle|0_2\rangle, |1_1\rangle|1_2\rangle, |1_1\rangle|2_2\rangle\}$. Specifically, we can write

$$|\Psi\rangle = (|0_1\rangle, |1_1\rangle)\cdot\begin{pmatrix} \frac{1}{\sqrt{3}} & 0 & \frac{i}{\sqrt{3}} \\ 0 & \frac{1}{\sqrt{6}} & \frac{-1}{\sqrt{6}} \end{pmatrix}\cdot\begin{pmatrix} |0_2\rangle \\ |1_2\rangle \\ |2_2\rangle \end{pmatrix} \rightarrow \overline{\Psi} = \frac{1}{\sqrt{3}}\begin{pmatrix} 1 & 0 & i \\ 0 & \frac{1}{\sqrt{2}} & \frac{-1}{\sqrt{2}} \end{pmatrix}.$$

Because of the two-state and three-state eigenbasis for particle 1 and particle 2, respectively, the matrix $\overline{\Psi}$ is not square (as it was in the previous problem). We further note that we could exchange the role of rows and columns here without affecting the outcome of our problem.

From the Singular Value Decomposition of the matrix $\overline{\Psi} = U\cdot\Sigma\cdot W$, we write

$$\overline{\Psi} = \begin{pmatrix} -i\frac{\sqrt{3+\sqrt{3}}}{\sqrt{6}} & i\frac{\sqrt{3-\sqrt{3}}}{\sqrt{6}} \\ \frac{\sqrt{3-\sqrt{3}}}{\sqrt{6}} & \frac{\sqrt{3+\sqrt{3}}}{\sqrt{6}} \end{pmatrix}\cdot\begin{pmatrix} \frac{\sqrt{3+\sqrt{3}}}{\sqrt{6}} & 0 & 0 \\ 0 & \frac{\sqrt{3-\sqrt{3}}}{\sqrt{6}} & 0 \end{pmatrix}\begin{pmatrix} \frac{i}{\sqrt{3}} & \frac{3-\sqrt{3}}{6} & \frac{-3-\sqrt{3}}{6} \\ \frac{-i}{\sqrt{3}} & \frac{3+\sqrt{3}}{6} & \frac{-3+\sqrt{3}}{6} \\ \frac{i}{\sqrt{3}} & \frac{1}{\sqrt{3}} & \frac{1}{\sqrt{3}} \end{pmatrix}.$$

The Schmidt representation is then

$$|\Psi\rangle = \lambda_1|\psi_1^{(1)}\rangle\otimes|\psi_1^{(2)}\rangle + \lambda_2|\psi_2^{(1)}\rangle\otimes|\psi_2^{(2)}\rangle,$$

where

$$\lambda_{1,2} = \frac{\sqrt{3\pm\sqrt{3}}}{\sqrt{6}},$$

$$\left(|\psi_1^{(1)}\rangle, |\psi_1^{(2)}\rangle\right) = (|0_1\rangle, |1_1\rangle) \cdot U$$
$$= \left(\left[-i\frac{\sqrt{3+\sqrt{3}}}{\sqrt{6}}|0_1\rangle + \frac{\sqrt{3-\sqrt{3}}}{\sqrt{6}}|1_1\rangle\right], \left[i\frac{\sqrt{3-\sqrt{3}}}{\sqrt{6}}|0_1\rangle + \frac{\sqrt{3+\sqrt{3}}}{\sqrt{6}}|1_1\rangle\right]\right).$$

$$\begin{pmatrix} |\psi_2^{(1)}\rangle \\ |\psi_2^{(2)}\rangle \\ |\psi_2^{(3)}\rangle \end{pmatrix} = W \cdot \begin{pmatrix} |0_2\rangle \\ |1_2\rangle \\ |2_2\rangle \end{pmatrix} = \begin{pmatrix} \left[\frac{i}{\sqrt{3}}|0_2\rangle + \frac{3-\sqrt{3}}{6}|1_2\rangle + \frac{-3-\sqrt{3}}{6}|2_2\rangle\right] \\ \left[\frac{-i}{\sqrt{3}}|0_2\rangle + \frac{3+\sqrt{3}}{6}|1_2\rangle + \frac{-3+\sqrt{3}}{6}|2_2\rangle\right] \\ \left[\frac{i}{\sqrt{3}}|0_2\rangle + \frac{1}{\sqrt{3}}|1_2\rangle + \frac{1}{\sqrt{3}}|2_2\rangle\right] \end{pmatrix}.$$

The reader can now verify by substitution that indeed the Schmidt decomposition corresponds to the original form of $|\Psi\rangle$.

Problem 4.23

Cross (or tensor) products between different states appear very often in the context of density operators. For basic practice, let's consider the state

$$|\psi\rangle = \frac{1}{\sqrt{2}}\left[|0\rangle \otimes |1\rangle \otimes |0\rangle + |1\rangle \otimes |0\rangle \otimes |1\rangle\right],$$

where $|0\rangle = \begin{pmatrix} 1 \\ 0 \end{pmatrix}$ and $|1\rangle = \frac{1}{\sqrt{2}}\begin{pmatrix} 1 \\ 1 \end{pmatrix}$.

a) Write explicitly the density operator in a matrix form corresponding to the pure state $|\psi\rangle$.
b) Calculate the expectation value of $\langle\psi|\widehat{\sigma}_1 \otimes \widehat{\sigma}_2 \otimes \widehat{\sigma}_1|\psi\rangle$ and $\langle\psi|\widehat{\sigma}_3 \otimes \widehat{\sigma}_2 \otimes \widehat{\sigma}_1|\psi\rangle$, where $\widehat{\sigma}_i$ are the Pauli matrices.

---

a) The density operator of the state $\widehat{\rho} = |\psi\rangle\langle\psi|$ can be given in the composite basis corresponding to the three-particle system. Specifically,

$$\widehat{\rho} = |\psi\rangle\langle\psi| = \left(\frac{1}{\sqrt{2}}[|010\rangle + |101\rangle]\right)\left(\frac{1}{\sqrt{2}}[\langle 010| + \langle 101|]\right)$$
$$= \frac{1}{2}(|010\rangle\langle 010| + |101\rangle\langle 010| + |010\rangle\langle 101| + |101\rangle\langle 101|).$$

Writing this in a matrix form, one finds

$$\widehat{\rho} = \begin{pmatrix} 0&0&0&0&0&0&0&0 \\ 0&0&0&0&0&0&0&0 \\ 0&0&1&0&0&1&0&0 \\ 0&0&0&0&0&0&0&0 \\ 0&0&0&0&0&0&0&0 \\ 0&0&1&0&0&1&0&0 \\ 0&0&0&0&0&0&0&0 \\ 0&0&0&0&0&0&0&0 \end{pmatrix} \text{ given in the basis } \begin{pmatrix} |111\rangle \\ |110\rangle \\ |101\rangle \\ |100\rangle \\ |011\rangle \\ |010\rangle \\ |001\rangle \\ |000\rangle \end{pmatrix}.$$

b) Next, using $|0\rangle = \begin{pmatrix} 1 \\ 0 \end{pmatrix}$ and $|1\rangle = \frac{1}{\sqrt{2}}\begin{pmatrix} 1 \\ 1 \end{pmatrix}$, we obtain that:

$$\langle 0|\widehat{\sigma}_1|0\rangle = 0, \quad \langle 0|\widehat{\sigma}_1|1\rangle = \langle 1|\widehat{\sigma}_1|0\rangle = \frac{1}{\sqrt{2}}, \quad \langle 1|\widehat{\sigma}_1|1\rangle = 1,$$

$$\langle 0|\widehat{\sigma}_2|0\rangle = 0, \quad -\langle 0|\widehat{\sigma}_2|1\rangle = \langle 1|\widehat{\sigma}_2|0\rangle = \frac{-i}{\sqrt{2}}, \quad \langle 1|\widehat{\sigma}_2|1\rangle = 0,$$

$$\langle 0|\widehat{\sigma}_3|0\rangle = 1, \quad \langle 0|\widehat{\sigma}_3|1\rangle = \langle 1|\widehat{\sigma}_3|0\rangle = \frac{1}{\sqrt{2}}, \quad \langle 1|\widehat{\sigma}_3|1\rangle = 0.$$

Finally,

$$\begin{aligned}
&\langle\psi|\widehat{\sigma}_1\otimes\widehat{\sigma}_2\otimes\widehat{\sigma}_1|\psi\rangle \\
&= \left(\frac{1}{\sqrt{2}}[\langle 0|\otimes\langle 1|\otimes\langle 0| + \langle 1|\otimes\langle 0|\otimes\langle 1|]\right)(\widehat{\sigma}_1\otimes\widehat{\sigma}_2\otimes\widehat{\sigma}_1)\left(\frac{1}{\sqrt{2}}[|0\rangle\otimes|1\rangle\otimes|0\rangle + |1\rangle\otimes|0\rangle\otimes|1\rangle]\right) \\
&= \frac{1}{2}[\langle 0|\otimes\langle 1|\otimes\langle 0|(\widehat{\sigma}_1\otimes\widehat{\sigma}_2\otimes\widehat{\sigma}_1)|0\rangle\otimes|1\rangle\otimes|0\rangle + \langle 0|\otimes\langle 1|\otimes\langle 0|(\widehat{\sigma}_1\otimes\widehat{\sigma}_2\otimes\widehat{\sigma}_1)|1\rangle\otimes|0\rangle\otimes|1\rangle \\
&\quad + \langle 1|\otimes\langle 0|\otimes\langle 1|(\widehat{\sigma}_1\otimes\widehat{\sigma}_2\otimes\widehat{\sigma}_1)|0\rangle\otimes|1\rangle\otimes|0\rangle + \langle 1|\otimes\langle 0|\otimes\langle 1|(\widehat{\sigma}_1\otimes\widehat{\sigma}_2\otimes\widehat{\sigma}_1)|1\rangle\otimes|0\rangle\otimes|1\rangle] \\
&= \frac{1}{2}[\langle 0|\widehat{\sigma}_1|0\rangle\langle 1|\widehat{\sigma}_2|1\rangle\langle 0|\widehat{\sigma}_1|0\rangle + \langle 0|\widehat{\sigma}_1|1\rangle\langle 1|\widehat{\sigma}_2|0\rangle\langle 0|\widehat{\sigma}_1|1\rangle \\
&\quad + \langle 1|\widehat{\sigma}_1|0\rangle\langle 0|\widehat{\sigma}_2|1\rangle\langle 1|\widehat{\sigma}_1|0\rangle + \langle 1|\widehat{\sigma}_1|1\rangle\langle 0|\widehat{\sigma}_2|0\rangle\langle 1|\widehat{\sigma}_1|1\rangle] \\
&= \frac{1}{2}\left[0 + \frac{i}{2\sqrt{2}} - \frac{i}{2\sqrt{2}} + 0\right] = 0.
\end{aligned}$$

With a similar procedure, we also obtain $\langle\psi\left|\widehat{\sigma}_3\otimes\widehat{\sigma}_2\otimes\widehat{\sigma}_1\right|\psi\rangle = 0$.

Problem 4.24
Let $\widehat{M}$ be a Hermitian $m\times m$ matrix operator and $\widehat{N}$ a Hermitian $n\times n$ matrix operator. The Hamiltonian of a given system can be given as

$$\widehat{H} = \hbar\omega\widehat{M}\otimes\widehat{N}.$$

Calculate the partition function $Z = Tr(e^{-\beta\widehat{H}})$, where β is a real positive constant.

Since $\widehat{M}$ and $\widehat{N}$ are Hermitian, their eigenvalues are real. One can always write $\tilde{M} = U_M^+\widehat{M}U_M = \begin{pmatrix} \lambda_1 & \cdots & 0 \\ \vdots & \ddots & \vdots \\ 0 & \cdots & \lambda_m \end{pmatrix}$, where the diagonal matrix $\tilde{M}$ is composed of the eigenvalues λ_i of matrix M. Similarly, $\tilde{N} = U_N^+\widehat{N}U_N = \begin{pmatrix} \mu_1 & \cdots & 0 \\ \vdots & \ddots & \vdots \\ 0 & \cdots & \mu_n \end{pmatrix}$, where the diagonal matrix $\tilde{N}$ is composed of the eigenvalues μ_i of matrix N. Then,

$$\begin{aligned}
Z &= Tr(e^{-\beta\widehat{H}}) = Tr(e^{-\beta\hbar\omega\widehat{M}\otimes\widehat{N}}) = Tr\left[U_M^+\otimes U_N^+(e^{-\beta\hbar\omega\widehat{M}\otimes\widehat{N}})U_M\otimes U_N\right] \\
&= Tr\left[e^{-\beta\hbar\omega(U_M^+\otimes U_N^+)\widehat{M}\otimes\widehat{N}(U_M\otimes U_N)}\right] = Tr\left[e^{-\beta\hbar\omega(U_M^+\widehat{M}U_M)\otimes(U_N^+\widehat{N}U_N)}\right] \\
&= Tr\left[e^{-\beta\hbar\omega\tilde{M}\otimes\tilde{N}}\right] = \sum_{i=1}^{m}\sum_{j=1}^{n} e^{-\beta\hbar\omega\lambda_i\mu_j}.
\end{aligned}$$

Problem 4.25

Consider the density matrices $\widehat{\rho}_1 = \begin{pmatrix} \frac{3}{4} & ae^{-i\phi} \\ ae^{i\phi} & \frac{1}{4} \end{pmatrix}$ and $\widehat{\rho}_2 = \begin{pmatrix} \frac{3}{4} & 0 \\ 0 & \frac{1}{4} \end{pmatrix}$. Calculate the following properties:

a) $D_{HS}(\widehat{\rho}_1,\widehat{\rho}_2) = \sqrt{Tr\left[(\widehat{\rho}_1 - \widehat{\rho}_2)^2\right]}$ (This is termed *Hilbert–Schmidt distance.*)

b) $D_B(\widehat{\rho}_1,\widehat{\rho}_2) = \sqrt{2\left(1 - Tr\left(\sqrt{\widehat{\rho}_1^{\frac{1}{2}}\widehat{\rho}_2\widehat{\rho}_1^{\frac{1}{2}}}\right)\right)}$ (This is termed *Bures distance.*)

c) $D(\widehat{\rho}_1,\widehat{\rho}_2) = \frac{1}{2}Tr\left[\sqrt{(\widehat{\rho}_1^* - \widehat{\rho}_2^*)(\widehat{\rho}_1 - \widehat{\rho}_2)}\right]$ (This is termed *trace distance.*)

Here we note that since $\widehat{\rho}_1$ is a density matrix, the parameter a must satisfy the purity condition $\mathcal{P}(\widehat{\rho}_1) \leq 1$. Therefore, $Tr\left(\widehat{\rho}_1^2\right) = \frac{5}{8} + 2a^2 \leq 1, a \leq \frac{\sqrt{3}}{4}$.

a) For the Hilbert-Schmidt distance, we first evaluate

$$Tr\left[(\widehat{\rho}_1 - \widehat{\rho}_2)^2\right] = Tr\left[\begin{pmatrix} 0 & ae^{-i\phi} \\ ae^{i\phi} & 0 \end{pmatrix}^2\right] = Tr\left[\begin{pmatrix} a^2 & 0 \\ 0 & a^2 \end{pmatrix}\right] = 2a^2.$$

Therefore, $D_{HS}(\widehat{\rho}_1,\widehat{\rho}_2) = \sqrt{2}a$.

b) For the Bures distance, we must find $\widehat{\rho}_1^{\frac{1}{2}}$ first. However, since $\widehat{\rho}_1$ is not diagonal, we use the relation $\widehat{\rho}_1 = S\Lambda S^{-1}$, where $\Lambda = \begin{pmatrix} \lambda_1 & 0 \\ 0 & \lambda_2 \end{pmatrix}$ and the columns of S contain the eigenvectors corresponding to the $\lambda_{1,2}$ eigenvalues. Also, S^{-1} is the inverse of S. We obtain

$$\lambda_{1,2} = \frac{1}{2}\left[1 \pm \sqrt{1 - 4\left(\frac{3}{16} - a^2\right)}\right]; \quad S = \begin{pmatrix} \frac{-ae^{-i\phi}}{\sqrt{a^2+\left(\frac{3}{4}-\lambda_1\right)^2}} & \frac{-ae^{-i\phi}}{\sqrt{a^2+\left(\frac{3}{4}-\lambda_2\right)^2}} \\ \frac{\frac{3}{4}-\lambda_1}{\sqrt{a^2+\left(\frac{3}{4}-\lambda_1\right)^2}} & \frac{\frac{3}{4}-\lambda_2}{\sqrt{a^2+\left(\frac{3}{4}-\lambda_2\right)^2}} \end{pmatrix}.$$

Using that $\widehat{\rho}_1^{\frac{1}{2}} = S\Lambda^{\frac{1}{2}}S^{-1}$ and after some lengthy, but straightforward algebra (left for the reader's own practice),

$$D_B(\widehat{\rho}_1,\widehat{\rho}_2) = \frac{\sqrt{4 - \sqrt{5 - 4\sqrt{1+3a^2}} - \sqrt{5 + 4\sqrt{1+3a^2}}}}{\sqrt{2}}.$$

c) For the trace distance, we first evaluate

$$Tr\sqrt{(\widehat{\rho}_1^* - \widehat{\rho}_2^*)(\widehat{\rho}_1 - \widehat{\rho}_2)} = Tr\left[\left(\begin{pmatrix} 0 & ae^{i\phi} \\ ae^{-i\phi} & 0 \end{pmatrix}\cdot\begin{pmatrix} 0 & ae^{-i\phi} \\ ae^{i\phi} & 0 \end{pmatrix}\right)^{1/2}\right]$$
$$= Tr\left[\begin{pmatrix} a^2e^{2i\phi} & 0 \\ 0 & a^2e^{-2i\phi} \end{pmatrix}^{1/2}\right] = Tr\left[\begin{pmatrix} ae^{i\phi} & 0 \\ 0 & ae^{-i\phi} \end{pmatrix}\right]$$
$$= 2a\cos(\phi).$$

Therefore, $D(\widehat{\rho}_1,\widehat{\rho}_2) = a\cos(\phi)$.

Food for thought: The distances we have considered here are used in quantum information science as a measure of distinguishability of states. Their geometrical meaning together with other types of distances are widely used for the interpretation of practical experiments.

Problem 4.26

The Hamiltonian for a bosonic particle is given as $\widehat{H} = \hbar\omega\widehat{b}^+\widehat{b}$, where $\widehat{b}^+(\widehat{b})$ are its creation (annihilation) operators.

a) Calculate $Z = Tr(e^{-\beta\widehat{H}})$, where β is a nonzero positive constant.
b) The density operator for this system is defined as $\widehat{\rho} = \frac{e^{-\beta\widehat{H}}}{Tr(e^{-\beta\widehat{H}})}$. Calculate the ensemble average of the Hamiltonian $\langle\widehat{H}\rangle = Tr(\widehat{H}\widehat{\rho})$. Calculate $Tr(\widehat{\rho}\widehat{b})$ and $Tr(\widehat{\rho}\widehat{b}^+)$.
c) Check that $Tr(\widehat{\rho}) = 1$.

---

a) Starting from the definition of the partition function $Z = Tr(e^{-\beta\widehat{H}})$,

$$Z = Tr(e^{-\beta\widehat{H}}) = \sum_{n=0}^{\infty}\langle n|e^{-\beta\widehat{H}}|n\rangle = \sum_{n=0}^{\infty}\langle n|\sum_{k=0}^{\infty}\frac{(-\beta\hbar\omega\widehat{b}^+\widehat{b})^k}{k!}|n\rangle$$
$$= \sum_{n=0}^{\infty}\sum_{k=0}^{\infty}\frac{(-\beta\hbar\omega)^k}{k!}\langle n|(\widehat{b}^+\widehat{b})^k|n\rangle.$$

In the preceding, we use that $\widehat{b}^+\widehat{b}|n\rangle = n|n\rangle$, thus

$$Z = Tr(e^{-\beta\widehat{H}}) = \sum_{n=0}^{\infty}\sum_{k=0}^{\infty}\frac{(-\beta\hbar\omega)^k}{k!}n^k = \sum_{n=0}^{\infty}\sum_{k=0}^{\infty}\frac{(-\beta\hbar\omega n)^k}{k!}$$
$$= \sum_{n=0}^{\infty}e^{-\beta\hbar\omega n} = \frac{e^{\beta\hbar\omega}}{e^{\beta\hbar\omega}-1}.$$

This is the partition function for the Bose–Einstein distribution.

b) To calculate the mean energy, we use $\langle\widehat{H}\rangle = Tr(\widehat{H}\widehat{\rho})$:

$$\langle\widehat{H}\rangle = Tr(\widehat{H}\widehat{\rho}) = Tr\left(\widehat{H}\frac{e^{-\beta\widehat{H}}}{Z}|n\rangle\langle n|\right) = Tr\left(\langle n|\widehat{H}\frac{e^{-\beta\widehat{H}}}{Z}|n\rangle\right)$$
$$= \sum_{n=0}^{\infty}\langle n|\widehat{H}\frac{e^{-\beta\widehat{H}}}{Z}|n\rangle = \frac{1}{Z}\sum_{n=0}^{\infty}\langle n|\hbar\omega\widehat{b}^+\widehat{b}\sum_{k=0}^{\infty}\frac{(-\beta\hbar\omega\widehat{b}^+\widehat{b})^k}{k!}|n\rangle$$
$$= \frac{\hbar\omega}{Z}\sum_{n=0}^{\infty}\sum_{k=0}^{\infty}\frac{(-\beta\hbar\omega)^k}{k!}\langle n|\widehat{b}^+\widehat{b}(\widehat{b}^+\widehat{b})^k|n\rangle = \frac{\hbar\omega}{Z}\sum_{n=0}^{\infty}\sum_{k=0}^{\infty}\frac{(-\beta\hbar\omega)^k}{k!}n^{k+1}$$
$$= \frac{\hbar\omega}{Z}\sum_{n=0}^{\infty}n\sum_{k=0}^{\infty}\frac{(-\beta\hbar\omega n)^k}{k!} = \frac{\hbar\omega}{Z}\sum_{n=0}^{\infty}ne^{-\beta\hbar\omega n} = \frac{\hbar\omega}{Z}\frac{e^{\beta\hbar\omega}}{(e^{\beta\hbar\omega}-1)^2} = \frac{\hbar\omega}{e^{\beta\hbar\omega}-1}.$$

The mean number of bosons is

$$\langle\widehat{b}\rangle = Tr(\widehat{b}\widehat{\rho}) = Tr\left(\widehat{b}\frac{e^{-\beta\widehat{H}}}{Z}|n\rangle\langle n|\right) = Tr\left(\langle n|\widehat{b}\frac{e^{-\beta\widehat{H}}}{Z}|n\rangle\right) = \sum_{n=0}^{\infty}\langle n|\widehat{b}\frac{e^{-\beta\widehat{H}}}{Z}|n\rangle$$
$$= \frac{1}{Z}\sum_{n=0}^{\infty}\langle n|\widehat{b}\sum_{k=0}^{\infty}\frac{(-\beta\hbar\omega\widehat{b}^{+}\widehat{b})^{k}}{k!}|n\rangle = \frac{1}{Z}\sum_{n=0}^{\infty}\sqrt{n+1}\langle n+1|\sum_{k=0}^{\infty}\frac{(-\beta\hbar\omega\widehat{b}^{+}\widehat{b})^{k}}{k!}|n\rangle$$
$$= \frac{1}{Z}\sum_{n=0}^{\infty}\sqrt{n+1}\sum_{k=0}^{\infty}\frac{(-\beta\hbar\omega)^{k}}{k!}\langle n+1|(\widehat{b}^{+}\widehat{b})^{k}|n\rangle.$$

However, since $\langle n+1|n\rangle = 0$, then

$$\langle\widehat{b}\rangle = \frac{1}{Z}\sum_{n=0}^{\infty}\sqrt{n+1}\sum_{k=0}^{\infty}\frac{(-\beta\hbar\omega n)^{k}}{k!}\langle n+1|n\rangle = \frac{1}{Z}\sum_{n=0}^{\infty}\sqrt{n+1}e^{-\beta\hbar\omega n}\langle n+1|n\rangle = 0.$$

The mean number of antibosons is

$$\langle\widehat{b}^{+}\rangle = Tr(\widehat{b}^{+}\widehat{\rho}) = Tr\left(\widehat{b}^{+}\frac{e^{-\beta\widehat{H}}}{Z}|n\rangle\langle n|\right) = Tr\left(\langle n|\widehat{b}^{+}\frac{e^{-\beta\widehat{H}}}{Z}|n\rangle\right)$$
$$= \sum_{n=0}^{\infty}\langle n|\widehat{b}^{+}\frac{e^{-\beta\widehat{H}}}{Z}|n\rangle = \frac{1}{Z}\sum_{n=0}^{\infty}\langle n|\widehat{b}^{+}\sum_{k=0}^{\infty}\frac{(-\beta\hbar\omega\widehat{b}^{+}\widehat{b})^{k}}{k!}|n\rangle$$
$$= \frac{1}{Z}\sum_{n=0}^{\infty}\sqrt{n}\langle n-1|\sum_{k=0}^{\infty}\frac{(-\beta\hbar\omega\widehat{b}^{+}\widehat{b})^{k}}{k!}|n\rangle$$
$$= \frac{1}{Z}\sum_{n=0}^{\infty}\sqrt{n}\sum_{k=0}^{\infty}\frac{(-\beta\hbar\omega)^{k}}{k!}\langle n-1|(\widehat{b}^{+}\widehat{b})^{k}|n\rangle = 0.$$

c) Finally, the trace of the density operator is

$$Tr(\widehat{\rho}) = Tr\left(\frac{e^{-\beta\widehat{H}}}{Z}|n\rangle\langle n|\right) = Tr\left(\langle n|\frac{e^{-\beta\widehat{H}}}{Z}|n\rangle\right) = \sum_{n=0}^{\infty}\langle n|\frac{e^{-\beta\widehat{H}}}{Z}|n\rangle$$
$$= \frac{1}{Z}\sum_{n=0}^{\infty}\langle n|\sum_{k=0}^{\infty}\frac{(-\beta\hbar\omega\widehat{b}^{+}\widehat{b})^{k}}{k!}|n\rangle = \frac{1}{Z}\sum_{n=0}^{\infty}\sum_{k=0}^{\infty}\frac{(-\beta\hbar\omega)^{k}}{k!}\langle n|(\widehat{b}^{+}\widehat{b})^{k}|n\rangle$$
$$= \frac{1}{Z}\sum_{n=0}^{\infty}\sum_{k=0}^{\infty}\frac{(-\beta\hbar\omega)^{k}}{k!}n^{k} = \frac{1}{Z}\sum_{n=0}^{\infty}e^{-\beta\hbar\omega n} = \frac{Z}{Z} = 1.$$

Problem 4.27
Consider the Bose–Einstein density operator,

$$\widehat{\rho} = \frac{1}{\overline{N}+1}\sum_{n=0}^{\infty}\left(\frac{\overline{N}}{\overline{N}+1}\right)^{n}|n\rangle\langle n|,$$

where $|n\rangle = |0\rangle, |1\rangle, |2\rangle, \ldots$ are the number states and $\overline{N} = \frac{1}{e^{\beta\hbar\omega}-1}$. Calculate $Tr(\widehat{\rho})$.

By simple substitution, we find that

$$Tr(\widehat{\rho}) = Tr\left(\frac{1}{\overline{N}+1}\sum_{n=0}^{\infty}\left(\frac{\overline{N}}{\overline{N}+1}\right)^{n}|n\rangle\langle n|\right) = \frac{1}{\overline{N}+1}\sum_{n=0}^{\infty}\langle n|\left(\frac{\overline{N}}{\overline{N}+1}\right)^{n}|n\rangle$$

$$= \frac{1}{\overline{N}+1}\sum_{n=0}^{\infty}\left(\frac{\overline{N}}{\overline{N}+1}\right)^{n} = \frac{1}{\overline{N}+1}\frac{1}{1-\frac{\overline{N}}{\overline{N}+1}} = \frac{1}{\overline{N}+1}\frac{\overline{N}+1}{\overline{N}+1-\overline{N}} = 1.$$

In the preceding we have used that $\left(\frac{\overline{N}}{\overline{N}+1}\right)^{n}|n\rangle = |n\rangle\left(\frac{\overline{N}}{\overline{N}+1}\right)^{n}$.

This is an expected result, since the density matrix corresponds to the thermal density matrix for bosons. The result, $Tr(\widehat{\rho}) = 1$, is consistent with the fact that the density matrix corresponds to the thermal density matrix of bosons.

Problem 4.28

The Hamiltonian for a fermionic particle is given as $\widehat{H} = \hbar\omega\widehat{f}^{+}\widehat{f}$.

a) Calculate the partition function $Z = Tr\left(e^{-\beta\widehat{H}}\right)$, where β is a nonzero positive constant.
b) The density operator for this system is defined as $\widehat{\rho} = \frac{e^{-\beta\widehat{H}}}{Z}$. Calculate the ensemble average of the Hamiltonian $\langle\widehat{H}\rangle = Tr(\widehat{H}\widehat{\rho})$. Calculate $Tr(\widehat{\rho}\widehat{f})$ and $Tr(\widehat{\rho}\widehat{f}^{+})$.

---

a) There are only two eigenstates for the given Hamiltonian due to the Pauli exclusion principle. These are labeled as $|0\rangle$ (for spin "up") and $|1\rangle$ (for spin "down").

Then, we have $\widehat{f}\widehat{f}^{+}|0\rangle = \widehat{f}|1\rangle = |0\rangle$, and the partition function is

$$Z = Tr(e^{-\beta\widehat{H}}) = \langle 0|e^{-\beta\widehat{H}}|0\rangle + \langle 1|e^{-\beta\widehat{H}}|1\rangle = \langle 0|e^{-\beta\hbar\omega\widehat{f}^{+}\widehat{f}}|0\rangle + \langle 1|e^{-\beta\hbar\omega\widehat{f}^{+}\widehat{f}}|1\rangle,$$

$$\langle 0|e^{-\beta\hbar\omega\widehat{f}^{+}\widehat{f}}|0\rangle = \langle 0|\sum_{k=0}^{\infty}\frac{(-\beta\hbar\omega\widehat{f}^{+}\widehat{f})^{k}}{k!}|0\rangle$$

$$= \langle 0|0\rangle + \langle 0|\sum_{k=1}^{\infty}\frac{(-\beta\hbar\omega\widehat{f}^{+}\widehat{f})^{k-1}}{k!}(-\beta\hbar\omega\widehat{f}^{+}\widehat{f})|0\rangle = \langle 0|0\rangle + 0 = 1.$$

Because $\widehat{f}|0\rangle = 0$,

$$\langle 1|e^{-\beta\hbar\omega\widehat{f}^{+}\widehat{f}}|1\rangle = \langle 1|\sum_{k=0}^{\infty}\frac{(-\beta\hbar\omega\widehat{f}^{+}\widehat{f})^{k}}{k!}|1\rangle = \sum_{k=0}^{\infty}\frac{(-\beta\hbar\omega)^{k}}{k!}\langle 1|(\widehat{f}^{+}\widehat{f})^{k}|1\rangle,$$

$$\langle 1|(\widehat{f}^{+}\widehat{f})^{k}|1\rangle = \langle 1|(\widehat{f}^{+}\widehat{f})^{k-1}\widehat{f}^{+}\widehat{f}|1\rangle = \langle 1|(\widehat{f}^{+}\widehat{f})^{k-1}\widehat{f}^{+}|0\rangle = \langle 1|(\widehat{f}^{+}\widehat{f})^{k-1}|1\rangle.$$

Using $\langle 1|1\rangle = 1$, we obtain $\langle 1|(\widehat{f}^{+}\widehat{f})^{k}|1\rangle = 1$; for all integer $k \geq 0$ therefore:

$$\langle 1|e^{-\beta\hbar\omega\widehat{f}^{+}\widehat{f}}|1\rangle = \sum_{k=0}^{\infty}\frac{(-\beta\hbar\omega)^{k}}{k!} = e^{-\beta\hbar\omega}.$$

Finally, the partition function for the Fermi–Dirac distribution is

$$Z = Tr(e^{-\beta\widehat{H}}) = \langle 0|e^{-\beta\hbar\omega\widehat{f}^{+}\widehat{f}}|0\rangle + \langle 1|e^{-\beta\hbar\omega\widehat{f}^{+}\widehat{f}}|1\rangle = 1 + e^{-\beta\hbar\omega}.$$

b)

$$\langle \hat{H} \rangle = Tr(\hat{H}\hat{\rho}) = Tr\left(\hat{H}\frac{e^{-\beta\hat{H}}}{Z}|n\rangle\langle n|\right) = Tr\left(\langle n|\hat{H}\frac{e^{-\beta\hat{H}}}{Z}|n\rangle\right)$$
$$= \langle 0|\hat{H}\frac{e^{-\beta\hat{H}}}{Z}|0\rangle + \langle 1|\hat{H}\frac{e^{-\beta\hat{H}}}{Z}|1\rangle$$
$$= \frac{1}{Z}\langle 0|\hat{H}e^{-\beta\hat{H}}|0\rangle + \frac{1}{Z}\langle 1|\hat{H}e^{-\beta\hat{H}}|1\rangle,$$
$$\langle 0|\hat{H}e^{-\beta\hat{H}}|0\rangle = \langle 0|\hbar\omega\hat{f}^+\hat{f}\sum_{k=0}^{\infty}\frac{(-\beta\hbar\omega\hat{f}^+\hat{f})^k}{k!}|0\rangle = 0,$$
$$\langle 1|\hat{H}e^{-\beta\hat{H}}|1\rangle = \langle 1|\hbar\omega\hat{f}^+\hat{f}e^{-\beta\hat{H}}|1\rangle = \hbar\omega\langle 1|\hat{f}^+\hat{f}e^{-\beta\hat{H}}|1\rangle = \hbar\omega e^{-\beta\hbar\omega}.$$

Therefore, we obtain

$$\langle \hat{H} \rangle = 0 + \frac{\hbar\omega e^{-\beta\hbar\omega}}{Z} = \frac{\hbar\omega}{e^{\beta\hbar\omega}+1}.$$

Now we consider $Tr(\hat{f}\hat{\rho})$:

$$\langle \hat{f} \rangle = Tr(\hat{f}\hat{\rho}) = Tr\left(\hat{f}\frac{e^{-\beta\hat{H}}}{Z}|n\rangle\langle n|\right) = Tr\left(\langle n|\hat{f}\frac{e^{-\beta\hat{H}}}{Z}|n\rangle\right)$$
$$= \langle 0|\hat{f}\frac{e^{-\beta\hat{H}}}{Z}|0\rangle + \langle 1|\hat{f}\frac{e^{-\beta\hat{H}}}{Z}|1\rangle$$
$$= \frac{1}{Z}\langle 0|\hat{f}e^{-\beta\hat{H}}|0\rangle + \frac{1}{Z}\langle 1|\hat{f}e^{-\beta\hat{H}}|1\rangle,$$
$$\langle 0|\hat{f}e^{-\beta\hat{H}}|0\rangle = \langle 0|\hat{f}\sum_{k=0}^{\infty}\frac{(-\beta\hbar\omega\hat{f}^+\hat{f})^k}{k!}|0\rangle = 0,$$

where $\hat{f}|0\rangle = 0$ is taken into account. Further taking into account that $\hat{f}^+|0\rangle = |1\rangle$, $\hat{f}^+|1\rangle = 0$, $\langle 0|\hat{f} = \langle 1|$, and $\langle 1|\hat{f} = 0$,

$$\langle 1|\hat{f}e^{-\beta\hat{H}}|1\rangle = \langle 0|e^{-\beta\hat{H}}|1\rangle,$$
$$\langle 0|e^{-\beta\hbar\omega\hat{f}^+\hat{f}}|1\rangle = \langle 0|\sum_{k=0}^{\infty}\frac{(-\beta\hbar\omega\hat{f}^+\hat{f})^k}{k!}|1\rangle = \sum_{k=0}^{\infty}\frac{(-\beta\hbar\omega)^k}{k!}\langle 0|(\hat{f}^+\hat{f})^k|1\rangle,$$
$$\langle 0|(\hat{f}^+\hat{f})^k|1\rangle = \langle 0|(\hat{f}^+\hat{f})^{k-1}\hat{f}^+\hat{f}|1\rangle = \langle 0|(\hat{f}^+\hat{f})^{k-1}\hat{f}^+|0\rangle = \langle 0|(\hat{f}^+\hat{f})^{k-1}|1\rangle.$$

From $\langle 0|1\rangle = 0$, we obtain $\langle 0|(\hat{f}^+\hat{f})^k|1\rangle = 0$. Thus, $\langle 1|\hat{f}e^{-\beta\hat{H}}|1\rangle = 0$. We conclude that

$$\langle \hat{f} \rangle = Tr(\hat{f}\hat{\rho}) = 0.$$

Finally, we consider $Tr(\hat{f}^+\hat{\rho})$:

$$\langle \hat{f}^+ \rangle = Tr(\hat{f}^+\hat{\rho}) = \frac{1}{Z}\langle 0|\hat{f}^+e^{-\beta\hat{H}}|0\rangle + \frac{1}{Z}\langle 1|\hat{f}^+e^{-\beta\hat{H}}|1\rangle = 0.$$

Problem 4.29

One of the most popular ways to represent Bell's theorem is the *CHSH* inequality, which shows that correlations between measurements of two particles cannot be accounted for via a local hidden variables theory (Clauser et al., 1969). The CHSH inequality is given as

$$|S_{AB}| = \left| C\left[\widehat{A}_a\widehat{B}_b\right] + C\left[\widehat{A}_a\widehat{B}_{b'}\right] + C\left[\widehat{A}_{a'}\widehat{B}_b\right] - C\left[\widehat{A}_{a'}\widehat{B}_{b'}\right]\right| \leq 2,$$

where $C\left[\widehat{A}_a\widehat{B}_b\right] = \langle\widehat{A}_a\widehat{B}_b\rangle$ is the correlator between measuring a in the system A and b in system B. In the preceding, it is assumed that $\langle\widehat{A}_a\rangle^2 \leq 1$ for all measurements.

The CHSH inequality is fulfilled provided that (a) there are two measurements for each subsystem freely chosen by the different experimentalists, (b) the results for a, a' measurements are causally independent from the results for b, b' measurements (meaning the theory is local), and (c) the obtained results are assumed to exist at any moment in time (meaning the theory is real), not only upon the moment of measurement. If $S_{AB} \leq 2$, the studied system can be described by a local real theory characterized by local hidden variables. Otherwise, the system cannot be described by local hidden variables theory.

Here we would like to examine the CHSH inequality for specific examples.

a) Consider the Bell state $|\Psi\rangle = \frac{1}{\sqrt{2}}(|\uparrow_A,\uparrow_B\rangle + |\downarrow_A,\downarrow_B\rangle)$. Measure the spin of particle A in $\hat{z}$ and $\hat{x}$ directions and the spin of particle B in $\frac{1}{\sqrt{2}}(\hat{z}+\hat{x})$ and $\frac{1}{\sqrt{2}}(\hat{z}-\hat{x})$ directions and check that the CHSH inequality is not fulfilled; in particular, $S_{AB} = 2\sqrt{2}$.

b) Consider the density operator:

$$\rho_{AB} = \frac{1}{2}\begin{pmatrix} 1 & 0 & 0 & ae^{i\theta} \\ 0 & 0 & 0 & 0 \\ 0 & 0 & 0 & 0 \\ ae^{-i\theta} & 0 & 0 & 1 \end{pmatrix}.$$

Measure the spin of particle A in $\hat{z}$ and $\hat{x}$ directions and the spin of particle B in $\frac{1}{\sqrt{2}}(\hat{z}+\hat{x})$ and $\frac{1}{\sqrt{2}}(\hat{z}-\hat{x})$ directions and examine the validity of the CHSH inequality as a function of the parameter a.

c) Assume that the correlations of the measures in the two laboratories are given by $C[\widehat{A}_a\widehat{B}_b] = \langle\widehat{A}_a\widehat{B}_b\rangle = \int_\Lambda A_a(\lambda)B_b(\lambda)\rho(\lambda)d\lambda$, with $\rho(\lambda) \geq 0$, $\langle\widehat{A}\rangle^2 \leq 1$, and $\langle\widehat{B}\rangle^2 \leq 1$, where λ plays the role of a hidden local variable. Prove that, in this case, the CHSH inequality is fulfilled.

---

a) The two measurements for particle A correspond to measuring $a = \widehat{\sigma}_z$ and $a' = \widehat{\sigma}_x$, while the two measurements for particle B correspond to measuring $b = \frac{\widehat{\sigma}_z + \widehat{\sigma}_x}{\sqrt{2}}$ and $b' = \frac{\widehat{\sigma}_z - \widehat{\sigma}_x}{\sqrt{2}}$. We write

$$S_{AB} = C\left[\widehat{\sigma}_z \otimes \left(\frac{\widehat{\sigma}_z + \widehat{\sigma}_x}{\sqrt{2}}\right)\right] + C\left[\widehat{\sigma}_z \otimes \left(\frac{\widehat{\sigma}_z - \widehat{\sigma}_x}{\sqrt{2}}\right)\right] + C\left[\widehat{\sigma}_x \otimes \left(\frac{\widehat{\sigma}_z + \widehat{\sigma}_x}{\sqrt{2}}\right)\right]$$
$$- C\left[\widehat{\sigma}_x \otimes \left(\frac{\widehat{\sigma}_z - \widehat{\sigma}_x}{\sqrt{2}}\right)\right]$$

$$= \left\langle \hat{\sigma}_z \otimes \left(\frac{\hat{\sigma}_z + \hat{\sigma}_x}{\sqrt{2}} \right) \right\rangle + \left\langle \hat{\sigma}_z \otimes \left(\frac{\hat{\sigma}_z - \hat{\sigma}_x}{\sqrt{2}} \right) \right\rangle + \left\langle \hat{\sigma}_x \otimes \left(\frac{\hat{\sigma}_z + \hat{\sigma}_x}{\sqrt{2}} \right) \right\rangle - \left\langle \hat{\sigma}_x \otimes \left(\frac{\hat{\sigma}_z - \hat{\sigma}_x}{\sqrt{2}} \right) \right\rangle.$$

We further find that

$$\begin{aligned}\left\langle \hat{\sigma}_z \otimes \left(\frac{\hat{\sigma}_z + \hat{\sigma}_x}{\sqrt{2}} \right) \right\rangle &= \langle \Psi | \hat{\sigma}_z \otimes \left(\frac{\hat{\sigma}_z + \hat{\sigma}_x}{\sqrt{2}} \right) | \Psi \rangle \\ &= \frac{1}{2} (\langle \uparrow_A, \uparrow_B | + \langle \downarrow_A, \downarrow_B |) \left[\hat{\sigma}_z \otimes \left(\frac{\hat{\sigma}_z + \hat{\sigma}_x}{\sqrt{2}} \right) \right] (| \uparrow_A, \uparrow_B \rangle + | \downarrow_A, \downarrow_B \rangle) \\ &= \frac{1}{2} \left(\langle \uparrow_A | \hat{\sigma}_z | \uparrow_A \rangle \langle \uparrow_B | \frac{\hat{\sigma}_z + \hat{\sigma}_x}{\sqrt{2}} | \uparrow_B \rangle + \langle \downarrow_A | \hat{\sigma}_z | \uparrow_A \rangle \langle \downarrow_B | \frac{\hat{\sigma}_z + \hat{\sigma}_x}{\sqrt{2}} | \uparrow_B \rangle \right. \\ &\quad \left. + \langle \uparrow_A | \hat{\sigma}_z | \downarrow_A \rangle \langle \uparrow_B | \frac{\hat{\sigma}_z + \hat{\sigma}_x}{\sqrt{2}} | \downarrow_B \rangle + \langle \downarrow_A | \hat{\sigma}_z | \downarrow_A \rangle \langle \downarrow_B | \frac{\hat{\sigma}_z + \hat{\sigma}_x}{\sqrt{2}} | \downarrow_B \rangle \right) = \frac{1}{\sqrt{2}},\end{aligned}$$

where we have used that $\hat{\sigma}_z | \uparrow \rangle = | \uparrow \rangle, \hat{\sigma}_z | \downarrow \rangle = -| \downarrow \rangle, \hat{\sigma}_x | \uparrow \rangle = | \downarrow \rangle$, and $\hat{\sigma}_x | \downarrow \rangle = | \uparrow \rangle$. Similarly,

$$\begin{aligned}\left\langle \hat{\sigma}_z \otimes \left(\frac{\hat{\sigma}_z + \hat{\sigma}_x}{\sqrt{2}} \right) \right\rangle &= \left\langle \hat{\sigma}_z \otimes \left(\frac{\hat{\sigma}_z - \hat{\sigma}_x}{\sqrt{2}} \right) \right\rangle = \left\langle \hat{\sigma}_x \otimes \left(\frac{\hat{\sigma}_z + \hat{\sigma}_x}{\sqrt{2}} \right) \right\rangle \\ &= -\left\langle \hat{\sigma}_x \otimes \left(\frac{\hat{\sigma}_z - \hat{\sigma}_x}{\sqrt{2}} \right) \right\rangle = \frac{1}{\sqrt{2}}.\end{aligned}$$

Therefore, we obtain

$$S_{AB} = \frac{1}{\sqrt{2}} + \frac{1}{\sqrt{2}} + \frac{1}{\sqrt{2}} + \frac{1}{\sqrt{2}} = \frac{4}{\sqrt{2}} = 2\sqrt{2}.$$

b) We start again with the CHSH inequality for the same spin measurements for the two particles, which can be written as

$$\begin{aligned}S_{AB} &= \left\langle \hat{\sigma}_z \otimes \left(\frac{\hat{\sigma}_z + \hat{\sigma}_x}{\sqrt{2}} \right) \right\rangle + \left\langle \hat{\sigma}_z \otimes \left(\frac{\hat{\sigma}_z - \hat{\sigma}_x}{\sqrt{2}} \right) \right\rangle + \left\langle \hat{\sigma}_x \otimes \left(\frac{\hat{\sigma}_z + \hat{\sigma}_x}{\sqrt{2}} \right) \right\rangle - \left\langle \hat{\sigma}_x \otimes \left(\frac{\hat{\sigma}_z - \hat{\sigma}_x}{\sqrt{2}} \right) \right\rangle \\ &= Tr \left[\left(\hat{\sigma}_z \otimes \left(\frac{\hat{\sigma}_z + \hat{\sigma}_x}{\sqrt{2}} \right) + \hat{\sigma}_z \otimes \left(\frac{\hat{\sigma}_z - \hat{\sigma}_x}{\sqrt{2}} \right) + \hat{\sigma}_x \otimes \left(\frac{\hat{\sigma}_z + \hat{\sigma}_x}{\sqrt{2}} \right) - \hat{\sigma}_x \otimes \left(\frac{\hat{\sigma}_z - \hat{\sigma}_x}{\sqrt{2}} \right) \right) \cdot \hat{\rho}_{AB} \right],\end{aligned}$$

where

$$\begin{aligned}\hat{\sigma}_z \otimes \left(\frac{\hat{\sigma}_z + \hat{\sigma}_x}{\sqrt{2}} \right) &= \begin{pmatrix} +1 & 0 \\ 0 & -1 \end{pmatrix} \otimes \left(\frac{\hat{\sigma}_z + \hat{\sigma}_x}{\sqrt{2}} \right) \\ &= \begin{pmatrix} +\left(\frac{\hat{\sigma}_z + \hat{\sigma}_x}{\sqrt{2}} \right) & 0 \left(\frac{\hat{\sigma}_z + \hat{\sigma}_x}{\sqrt{2}} \right) \\ 0 \left(\frac{\hat{\sigma}_z + \hat{\sigma}_x}{\sqrt{2}} \right) & -\left(\frac{\hat{\sigma}_z + \hat{\sigma}_x}{\sqrt{2}} \right) \end{pmatrix} = \frac{1}{\sqrt{2}} \begin{pmatrix} 1 & 1 & 0 & 0 \\ 1 & -1 & 0 & 0 \\ 0 & 0 & -1 & -1 \\ 0 & 0 & -1 & 1 \end{pmatrix}.\end{aligned}$$

The rest of the matrices entering in S_{AB} are left as an exercise to the reader. The final result is

$$S_{AB} = Tr\left[\sqrt{2}\begin{pmatrix} 1 & 0 & 0 & 1 \\ 0 & -1 & 1 & 0 \\ 0 & 1 & -1 & 0 \\ 1 & 0 & 0 & 1 \end{pmatrix} \cdot \frac{1}{2}\begin{pmatrix} 1 & 0 & 0 & ae^{i\theta} \\ 0 & 0 & 0 & 0 \\ 0 & 0 & 0 & 0 \\ ae^{-i\theta} & 0 & 0 & 1 \end{pmatrix}\right] = \sqrt{2}(1+a).$$

To fulfill the CHSH inequality, we must have

$$S_{AB} = \sqrt{2}(1+a) \le 2 \to a \le (\sqrt{2}-1).$$

c) We start with the CHSH inequality:

$$S_{AB} = C[\widehat{A}_a\widehat{B}_b] + C[\widehat{A}_a\widehat{B}_{b'}] + C[\widehat{A}_{a'}\widehat{B}_b] - C[\widehat{A}_{a'}\widehat{B}_{b'}].$$

If we assume that the results of the measures in the two laboratories are local and causally independent, it means that the measures can be written as

$$C[\widehat{A}_a\widehat{B}_b] = \langle \widehat{A}_a\widehat{B}_b\rangle = \int_\Lambda A_a(\lambda)B_b(\lambda)\rho(\lambda)d\lambda,$$

where λ is a hidden local variable. Note that this description is real and local due to the fact that A measures a independently of the measure of b from B. Then, S_{AB} is

$$\begin{aligned} S_{AB} &= \langle \widehat{A}_a\widehat{B}_b\rangle + \langle \widehat{A}_a\widehat{B}_{b'}\rangle + \langle \widehat{A}_{a'}\widehat{B}_b\rangle - \langle \widehat{A}_{a'}\widehat{B}_{b'}\rangle \\ &= \int_\Lambda [A_a(\lambda)B_b(\lambda) + A_a(\lambda)B_{b'}(\lambda) + A_{a'}(\lambda)B_b(\lambda) - A_{a'}(\lambda)B_{b'}(\lambda)]\rho(\lambda)d\lambda. \end{aligned}$$

In addition to that, using the assumptions $\rho(\lambda) \ge 0$, $\langle \widehat{A}_a\rangle^2 \le 1$, and $\langle \widehat{B}_b\rangle^2 \le 1$, one sees immediately that

$$\begin{aligned} S_{AB} \le \|S_{AB}\| &= \left\| \int_\Lambda [A_a(\lambda)\left(B_b(\lambda) + B_{b'}(\lambda)\right) + A_{a'}(\lambda)\left(B_b(\lambda) + B_{b'}(\lambda)\right)]\rho(\lambda)d\lambda \right\| \\ &\le \int_\Lambda \|A_a(\lambda)\| \, \|B_b(\lambda)+B_{b'}(\lambda)\| \rho(\lambda)d\lambda + \int_\Lambda \|A_{a'}(\lambda)\| \, \|B_b(\lambda) - B_{b'}(\lambda)\| \rho(\lambda)d\lambda \\ &\le \int_\Lambda \|B_b(\lambda) + B_{b'}(\lambda)\| \rho(\lambda)d\lambda + \int_\Lambda \|B_b(\lambda) - B_{b'}(\lambda)\| \rho(\lambda)d\lambda, \end{aligned}$$

where we have used several well-known inequalities for integrals and summations. Now, without a loss of generality, we can assume that $\langle \widehat{B}_b\rangle \ge \langle \widehat{B}_{b'}\rangle \ge 0$; therefore

$$\begin{aligned} S_{AB} &\le \int_\Lambda \|B_b(\lambda) + B_{b'}(\lambda)\| \rho(\lambda)d\lambda + \int_\Lambda \|B_b(\lambda) - B_{b'}(\lambda)\| \rho(\lambda)d\lambda \\ &\le \int_\Lambda B_b(\lambda)\rho(\lambda)d\lambda + \int_\Lambda B_{b'}(\lambda)\rho(\lambda)d\lambda + \int_\Lambda B_b(\lambda)\rho(\lambda)d\lambda - \int_\Lambda B_{b'}(\lambda)\rho(\lambda)d\lambda \\ &\le 2\int_\Lambda B_b(\lambda)\rho(\lambda)d\lambda \le 2. \end{aligned}$$

This inequality is valid for real and local theories.

Food for thought: As we have seen, for some particular states in quantum mechanics, such as the Bell states, the inequality is not fulfilled. This brings us to the conclusion that at least one of the initial hypotheses (freedom of choice of the experiments, locality and reality) is not fulfilled in quantum mechanics.

5 Identical Particles and Elements of Second Quantization

Most problems we deal with in quantum mechanics involve systems that contain several or an infinite number of particles behaving in a collective manner. In addition to the probabilistic nature of processes of such multiparticle systems, quantum mechanical particles are truly indistinguishable when it comes to tracking their quantum numbers. While this is not a problem in classical mechanics where everything is tractable, two (or more) quantum particles that have the same charge, mass, or spin cannot be distinguished by any experiment regardless of the situation. Nature has given us two types of particles, *fermions* and *bosons*, such that the state function for fermions changes sign, while the state function for bosons does not change sign when two particles are interchanged. This situation has important consequences not only for a small, finite number of identical particles, but also for systems containing an extremely large or infinite number of identical particles.

Here we review basic concepts of composite fermion and boson systems followed by the exchange interaction and the different types of statistics each type of particle obeys. The problems we have selected here involve understanding of more traditional topics, such as addition of angular momenta and perturbation theory, for example. We give some explicit examples of three or more (but finite in number) identical particles and their composite states, which goes beyond the typical two-particle states found in most quantum mechanical books. Other problems illustrate important topics from chemistry, such as building the electronic shell structure in atoms and explicitly showing the application of Hund's rules. We also offer several problems with second quantization, which is needed when dealing with condensed matter systems, giving graphene as an example. The gradual buildup of complexity, especially in this chapter, illustrates the far-reaching consequences of the inherent statistics of quantum mechanical particles.

5.1 Wave Function for *N* Distinguishable Particles

$$\Phi(\boldsymbol{x}_1, \boldsymbol{x}_2, \ldots, \boldsymbol{x}_N) = \psi_{a_1}(\boldsymbol{x}_1)\,\psi_{a_2}(\boldsymbol{x}_2)\ldots\psi_{a_N}(\boldsymbol{x}_N)$$

The composite wave function is simply a product of the individual wave functions where the indices a_i collect all quantum numbers for a given state. For two distinguishable particles with principal quantum numbers n_i and spinors χ_{S_i}, for example, $\Phi(\boldsymbol{x}_1,\boldsymbol{x}_2)=\varphi_{n_1}(\boldsymbol{x}_1)\chi_{S_1}\varphi_{n_2}(\boldsymbol{x}_1)\chi_{S_2}$. This representation is appropriate when the particles can be considered as distinguishable, namely classical.

5.2 Wave Function for *N* Nondistinguishable Identical Bosons: *Slater Permanent*

$$\begin{aligned}\Phi(\boldsymbol{x}_1,\boldsymbol{x}_2,\ldots,\boldsymbol{x}_N) &= \frac{\sqrt{\prod_{r=1}^{R} n_r!}}{\sqrt{N!}}\sum_P P\left(\psi_{a_1}(\boldsymbol{x}_1)\psi_{a_2}(\boldsymbol{x}_2)\ldots\psi_{a_N}(\boldsymbol{x}_N)\right)\\ &= \frac{\sqrt{\prod_{r=1}^{R} n_r!}}{\sqrt{N!}}\left\|\begin{matrix}\psi_{a_1}(\boldsymbol{x}_1) & \psi_{a_1}(\boldsymbol{x}_2) & \cdots & \psi_{a_1}(\boldsymbol{x}_N)\\ \psi_{a_2}(\boldsymbol{x}_1) & \psi_{a_2}(\boldsymbol{x}_2) & \cdots & \psi_{a_2}(\boldsymbol{x}_N)\\ \vdots & \vdots & \ddots & \vdots\\ \psi_{a_N}(\boldsymbol{x}_1) & \psi_{a_N}(\boldsymbol{x}_2) & \cdots & \psi_{a_N}(\boldsymbol{x}_N)\end{matrix}\right\|\end{aligned}$$

The composite wave function is totally symmetric upon the number of total number of permutations P of quantum numbers a_i or positions $\boldsymbol{x}_i$. This composite function captures the possibility that there can be many bosons with the same quantum mechanical characteristics. Also, $\sqrt{\prod_{r=1}^{R} n_r!}$ takes into account the number of repetitions of the same one-particle quantum state. Conveniently, this can be expressed with a *Slater permanent* as shown above, and we will use $||A||$ to represent the permanent of the matrix A.

5.3 Wave Function for *N* Nondistinguishable Identical Fermions: *Slater Determinant*

$$\begin{aligned}\Phi(\boldsymbol{x}_1,\boldsymbol{x}_2,\ldots,\boldsymbol{x}_N) &= \frac{1}{\sqrt{N!}}\sum_P (-1)^P P\left(\psi_{a_1}(\boldsymbol{x}_1)\psi_{a_2}(\boldsymbol{x}_2)\ldots\psi_{a_N}(\boldsymbol{x}_N)\right)\\ &= \frac{1}{\sqrt{N!}}\left|\begin{matrix}\psi_{a_1}(\boldsymbol{x}_1) & \psi_{a_1}(\boldsymbol{x}_2) & \cdots & \psi_{a_1}(\boldsymbol{x}_N)\\ \psi_{a_2}(\boldsymbol{x}_1) & \psi_{a_2}(\boldsymbol{x}_2) & \cdots & \psi_{a_2}(\boldsymbol{x}_N)\\ \vdots & \vdots & \ddots & \vdots\\ \psi_{a_N}(\boldsymbol{x}_1) & \psi_{a_N}(\boldsymbol{x}_2) & \cdots & \psi_{a_N}(\boldsymbol{x}_N)\end{matrix}\right|\end{aligned}$$

The composite wave function is totally antisymmetric upon permutations (denoted as P) of quantum numbers a_i or positions $\boldsymbol{x}_i$. Conveniently, this can be expressed with a *Slater determinant* as shown above. The antisymmetric wave function is a consequence of the Pauli exclusion principle for fermions; that is, when there are two identical fermions, the composite wave function is zero.

5.4 Exchange Interaction for Two Particles

Exchange interaction for fermions with $\Phi(x_1,x_2)=\frac{1}{\sqrt{2}}(\psi_{a_1}(x_1)\psi_{a_2}(x_2)-\psi_{a_1}(x_2)\psi_{a_2}(x_1))$:

$$\left\langle(\widehat{r}_1-\widehat{r}_2)^2\right\rangle_f=\left\langle\widehat{r}^2\right\rangle_{a_1}+\left\langle\widehat{r}^2\right\rangle_{a_2}-2\left\langle\widehat{r}\right\rangle_{a_1}\left\langle\widehat{r}\right\rangle_{a_2}+2\left|\left\langle\widehat{r}\right\rangle_{a_1,a_2}\right|^2-2\mathbb{R}\mathrm{e}\left(\left\langle\widehat{r}^2\right\rangle_{a_1,a_2}\langle a_2\,|\,a_1\rangle\right),$$

Exchange interaction for bosons with $\Phi(x_1,x_2)=\frac{1}{\sqrt{2}}(\psi_{a_1}(x_1)\psi_{a_2}(x_2)+\psi_{a_1}(x_2)\psi_{a_2}(x_1))$:

$$\left\langle(\widehat{r}_1-\widehat{r}_2)^2\right\rangle_b=\left\langle\widehat{r}^2\right\rangle_{a_1}+\left\langle\widehat{r}^2\right\rangle_{a_2}-2\langle\widehat{r}\rangle_{a_1}\langle\widehat{r}\rangle_{a_2}-2\left|\langle\widehat{r}\rangle_{a_1,a_2}\right|^2+2\mathbb{R}\mathrm{e}\left(\left\langle\widehat{r}^2\right\rangle_{a_1,a_2}\langle a_2|a_1\rangle\right).$$

The fermionic nature acts as an effective repulsion by increasing the particle separation, while bosons effectively attract due to a decreased separation. Here $\left\langle\widehat{r}^n\right\rangle_{a_i}=\langle\psi_{a_i}|\widehat{r}^n|\psi_{a_i}\rangle$ and $\left\langle\widehat{r}^n\right\rangle_{a_1a_2}=\langle\psi_{a_1}|\widehat{r}^n|\psi_{a_2}\rangle$.

5.5 Fourier Analysis and Periodicity in Wave Functions

Fourier series of an N-dimensional periodic wave function $\langle x\,|\,f\rangle=f(x)=f(x+L)$, with $k_{n_i}=2\pi\frac{n_i}{L_i}$ and $V_N=\prod_{i=1}^{N}L_i$, such that $x\in\Omega_N=\bigcup_{i=1}^{N}\left[\frac{-L_i}{2},\frac{L_i}{2}\right]$ and $n\in\mathbb{Z}^N$.

$$f(x)=\langle x\,|\,f\rangle=\frac{1}{\sqrt{V_N}}\sum_{n\in\mathbb{Z}^N}e^{-ik_n\cdot x}\tilde{f}(k_n),$$

$$\tilde{f}(k_n)=\langle k_n\,|\,f\rangle=\frac{1}{\sqrt{V_N}}\int_{\Omega_N}dx e^{ik_n\cdot x}f(x),$$

$$\langle k_n\,|\,x\rangle=\frac{e^{ik_n\cdot x}}{\sqrt{V_N}};\ \langle k_n\,|\,k_m\rangle=\delta_{nm};\langle x\,|\,y\rangle=\delta(x-y).$$

Fourier series of an N-dimensional periodic wave function over a *Bravais* lattice $\{R_n\}_{n\in\mathbb{Z}^N}$ with $R_n=\sum_{i=1}^{N}n_ia_i$ defined in terms of the generating lattice vectors $\{a_i\}_{i=1}^N$: $\langle R_n\,|\,f\rangle=f(R_n)=f(R_n+R_m)$.

By using the definition of the reciprocal Bravais lattice as the set $\{b_i\}_{i=1}^N$: $a_i\cdot b_j=2\pi\delta_{ij}$, the volume of the Bravais zone as $V_{BZ}=\prod_{i=1}^N b_i$, and the Brillouin zone as $BZ=\bigcup_{i=1}^{N}\left[\frac{-b_i}{2},\frac{b_i}{2}\right]$. Thus, for $k\in BZ$ and $n\in\mathbb{Z}^N$,

$$f(R_n)=\langle R_n\,|\,f\rangle=\frac{1}{\sqrt{V_{BZ}}}\int_{BZ}dk e^{-ik\cdot R_n}\tilde{f}(k),$$

$$\tilde{f}(k)=\langle k\,|\,f\rangle=\frac{1}{\sqrt{V_{BZ}}}\sum_{n\in\mathbb{Z}^N}e^{ik\cdot R_n}f(R_n),$$

$$\langle k\,|\,R_n\rangle=\frac{e^{ik\cdot R_n}}{\sqrt{V_{BZ}}};\langle R_n\,|\,R_m\rangle=\delta_{nm};\langle k\,|\,q\rangle=\delta(k-q).$$

Fourier transform of an N-dimensional wave function $\langle x|f\rangle = f(x)$, with $x \in \mathbb{R}^N$ and $k \in \mathbb{R}^N$.

$$f(x) = \langle x \mid f\rangle = \frac{1}{(2\pi)^{N/2}} \int_{\mathbb{R}^N} dk e^{-ik\cdot x} \tilde{f}(k),$$

$$\tilde{f}(k) = \langle k \mid f\rangle = \frac{1}{(2\pi)^{N/2}} \int_{\mathbb{R}^N} dx e^{ik\cdot x} f(x),$$

$$\langle k \mid x\rangle = \frac{e^{ik\cdot x}}{(2\pi)^{N/2}}; \quad \langle k \mid q\rangle = \delta(k-q); \quad \langle x \mid y\rangle = \delta(x-y).$$

Discrete Fourier transform of an N-dimensional periodic wave function defined over the set $\{x_a\}_{a=\alpha}^{\alpha+M-1} \in \mathbb{R}^N$ with $x_{a_i} = a_i \Delta x_i$, then $\langle x_a|f\rangle = f(x_a) = f_a = f(x_a + L)$, with the condition $L_i = M_i \Delta x_i$. The discrete momentum belongs to the set $\{k_n\}_{n=\nu}^{\nu+M-1} \in \mathbb{R}^N$ with $k_{n_j} = 2\pi \frac{n_j}{L_n}$; then, defining $M^{(N)} = \prod_{i=1}^{N} M_i$:

$$f(x_a) = f_a = \langle x_a \mid f\rangle = \frac{1}{\sqrt{M^{(N)}}} \sum_{n=\nu}^{\nu+M-1} e^{-ik_n\cdot x_a} \tilde{f}(k_n),$$

$$\tilde{f}(k_n) = \tilde{f}_n = \langle k_n \mid f\rangle = \frac{1}{\sqrt{M^{(N)}}} \sum_{a=\alpha}^{\alpha+M-1} e^{ik_n\cdot x_a} f(x_a),$$

$$\langle k_n \mid x_a\rangle = \frac{e^{ik_n\cdot x_a}}{\sqrt{M^{(N)}}}; \langle k_n \mid k_m\rangle = \delta_{nm}; \langle x_a \mid x_b\rangle = \delta_{ab}.$$

5.6 Field Quantization Operators

Change of Basis and Creation, Annihilation Operators:

$$\widehat{c}_i = \sum_{m=1}^{N} \langle a_i \mid \alpha_m\rangle \widehat{d}_m; \quad \widehat{c}_m^+ = \sum_{i=1}^{N} \widehat{d}_i^+ \langle a_i \mid \alpha_m\rangle^*$$

The relations give the connection between the creation and annihilation operators corresponding to distinct discrete eigenbasis $|a_m\rangle$ and $|a_i\rangle$ with $|\alpha_m\rangle = \widehat{c}_m^+ |0\rangle$ and $|a_i\rangle = \widehat{d}_i^+ |0\rangle$ $(m, i = \{1, \ldots, N\})$. The change of discrete basis $|\alpha_m\rangle = \sum_{i=1}^{N} |a_i\rangle\langle a_i \mid \alpha_m\rangle$ relies on the identity $\sum_{i=1}^{N} |a_i\rangle\langle a_i| = \widehat{\mathbb{I}}_{NxN}$ operation and the wave functions $\langle a_i|\alpha_m\rangle$.

Change of Basis in Real Space and *Field* Operators:

$$\widehat{c}(r) = \sum_n \langle r \mid n\rangle \widehat{a}_n, \qquad \widehat{c}^+(r) = \sum_n \widehat{a}_n^+ \langle n \mid r\rangle,$$

$$\widehat{a}_n = \int d^3r \langle n \mid r\rangle \widehat{c}(r), \qquad \widehat{a}_n^+ = \int d^3r \widehat{c}^+(r) \langle r \mid n\rangle.$$

The relation between the creation and annihilation operators corresponding to the discrete eigenbasis $|n\rangle$ with creation (annihilation) operators $\widehat{a}_n^+$ $(\widehat{a}_n)$ and creation (annihilation) operators in real space $\widehat{c}^+(r)(\widehat{c}(r))$.

Field Operators in Momentum Space:

$$\widehat{C}(\boldsymbol{k}) = \int d^3\boldsymbol{r}\langle \boldsymbol{k} \mid \boldsymbol{r}\rangle \widehat{c}(\boldsymbol{r}) = \int d^3\boldsymbol{r}\frac{e^{i\boldsymbol{k}\cdot\boldsymbol{r}}}{(2\pi)^{3/2}}\widehat{c}(\boldsymbol{r}),$$

$$\widehat{C}^{+}(\boldsymbol{k}) = \int d^3\boldsymbol{r}\widehat{c}^{+}(\boldsymbol{r})\langle \boldsymbol{r} \mid \boldsymbol{k}\rangle = \int d^3\boldsymbol{r}\widehat{c}^{+}(\boldsymbol{r})\frac{e^{-i\boldsymbol{k}\cdot\boldsymbol{r}}}{(2\pi)^{3/2}}.$$

Using the Fourier transform relations, the operators in second quantization picture in momentum space can also be obtained.

Problem 5.1

Two particles (each with equal mass m) are moving in a 1D simple harmonic oscillator potential $V(\widehat{x}) = \frac{1}{2}k\widehat{x}^2$ and they also interact through an attractive force $F = -K|x_1 - x_2|$. Assuming that $K < \frac{3}{2}k$:

a) What are the energies of the lowest three states of this system of identical *distinguishable* particles?
b) What are the energies of the lowest three states of this system of *identical* bosonic particles?
c) What are the energies of the lowest three states of this system of identical fermionic particles?

We are dealing with two particles interacting with a force, and the problem asks about the states of distinguishable particles first.

Since this is a two-body problem in which the force only depends on the $|x_1 - x_2|$ (central force), we can represent the Hamiltonian by two independent composite particles. The Hamiltonian for this system is

$$\widehat{H} = \frac{\widehat{p}_1^2}{2m} + \frac{\widehat{p}_2^2}{2m} + \frac{1}{2}k\widehat{x}_1^2 + \frac{1}{2}k\widehat{x}_2^2 + \frac{1}{2}K\left(\widehat{x}_1 - \widehat{x}_2\right)^2.$$

After changing variables $\widehat{X} = \frac{1}{\sqrt{2}}\left(\widehat{x}_1 + \widehat{x}_2\right)$; $\widehat{Y} = \frac{1}{\sqrt{2}}\left(\widehat{x}_1 - \widehat{x}_2\right)$, the momentum becomes

$$\widehat{p}_{\widehat{X}} = -i\hbar\frac{\partial}{\partial \widehat{X}} = -i\hbar\left(\frac{\partial \widehat{x}_1}{\partial \widehat{X}}\frac{\partial}{\partial \widehat{x}_1} + \frac{\partial \widehat{x}_2}{\partial \widehat{X}}\frac{\partial}{\partial \widehat{x}_2}\right) = -i\hbar\left(\frac{1}{\sqrt{2}}\frac{\partial}{\partial \widehat{x}_1} + \frac{1}{\sqrt{2}}\frac{\partial}{\partial \widehat{x}_2}\right) = \frac{1}{\sqrt{2}}\left(\widehat{p}_1 + \widehat{p}_2\right),$$

and the Hamiltonian is found as

$$\widehat{H} = \frac{\widehat{p}_X^2}{2m} + \frac{1}{2}k\widehat{X}^2 + \frac{\widehat{p}_Y^2}{2m} + \frac{1}{2}(k + 2K)\widehat{Y}^2.$$

This $\widehat{H}$ represents two independent simple harmonic oscillators with frequencies $\omega_X = \sqrt{\frac{k}{m}}$; $\omega_Y = \sqrt{\frac{k+2K}{m}}$ (note that $\omega_X < \omega_Y < 2\omega_X$ as imposed by the condition $K < \frac{3}{2}k$). Therefore, the solutions of the eigenproblem $\widehat{H}\left|\psi_{n,m}(x_1, x_2)\right\rangle = E_{n,m}\left|\psi_{n,m}(x_1, x_2)\right\rangle$ are

$$\left|\psi_{n,m}(x_1, x_2)\right\rangle = \varphi_{n,X}(X)\varphi_{m,Y}(Y) = \varphi_{n,X}\left(\frac{x_1 + x_2}{\sqrt{2}}\right)\varphi_{m,Y}\left(\frac{x_1 - x_2}{\sqrt{2}}\right)$$

with energy

$$E_{n,m} = \hbar\omega_X\left(n+\frac{1}{2}\right) + \hbar\omega_Y\left(m+\frac{1}{2}\right),$$

where $\varphi_{n,\mu}(\xi) = \frac{1}{\sqrt{2^n n!}}\left(\frac{m\omega_\mu}{\pi\hbar}\right)^{\frac{1}{4}} H_n\left(\sqrt{\frac{m\omega_\mu}{\hbar}}\xi\right)e^{\frac{-m\omega_\mu}{2\hbar}\xi^2}$ are the eigenfunctions of each quantum oscillator with tabulated Hermite polynomials $H_n(u)$. The Hermite polynomials are of alternating even and odd symmetry, as captured in the property $H_n(-u) = (-1)^n H_n(u)$.

a) For identical *distinguishable* particles, the lowest three states and energies are the three eigenstates of the Hamiltonian with lowest energy:

$$\left|\Psi_0(x_1,x_2)\right\rangle = \left|\psi_{0,0}(x_1,x_2)\right\rangle, \text{ with } E_0 = E_{0,0} = \frac{\hbar}{2}(\omega_X+\omega_Y),$$

$$\left|\Psi_1(x_1,x_2)\right\rangle = \left|\psi_{1,0}(x_1,x_2)\right\rangle, \text{ with } E_1 = E_{1,0} = \frac{\hbar}{2}(3\omega_X+\omega_Y),$$

$$\left|\Psi_2(x_1,x_2)\right\rangle = \left|\psi_{0,1}(x_1,x_2)\right\rangle, \text{ with } E_2 = E_{0,1} = \frac{\hbar}{2}(\omega_X+3\omega_Y).$$

b) For identical *bosonic* particles, the states must be symmetric under the exchange of two bosons, $\left|\Psi_n(x_1,x_2)\right\rangle = +\left|\Psi_n(x_2,x_1)\right\rangle$. In general, a bosonic solution will be of the form

$$\begin{aligned}\left|\Psi(x_1,x_2)\right\rangle &= \frac{1}{\sqrt{2}}\left(\left|\psi_{n,m}(x_1,x_2)\right\rangle + \left|\psi_{n,m}(x_2,x_1)\right\rangle\right)\\ &= \frac{1}{\sqrt{2}}\left(\varphi_{n,X}\left(\frac{x_1+x_2}{\sqrt{2}}\right)\varphi_{m,Y}\left(\frac{x_1-x_2}{\sqrt{2}}\right) + \varphi_{n,X}\left(\frac{x_2+x_1}{\sqrt{2}}\right)\varphi_{m,Y}\left(\frac{x_2-x_1}{\sqrt{2}}\right)\right)\\ &= \frac{1}{\sqrt{2}}\varphi_{n,X}(X)\left(\varphi_{m,Y}(Y)+\varphi_{m,Y}(-Y)\right).\end{aligned}$$

Taking into account the symmetry of the Hermite polynomials, the lowest three states and energies for two *identical* bosonic particles are as follows:

$$\left|\Psi_0(x_1,x_2)\right\rangle = \frac{\left|\psi_{0,0}(x_1,x_2)\right\rangle + \left|\psi_{0,0}(x_2,x_1)\right\rangle}{\sqrt{2}} = \sqrt{2}\varphi_{0,X}(X)\varphi_{0,Y}(Y), \text{ with}$$

$$E_0 = E_{0,0} = \frac{\hbar}{2}(\omega_X+\omega_Y),$$

$$\left|\Psi_1(x_1,x_2)\right\rangle = \frac{\left|\psi_{1,0}(x_1,x_2)\right\rangle + \left|\psi_{1,0}(x_2,x_1)\right\rangle}{\sqrt{2}} = \sqrt{2}\varphi_{1,X}(X)\varphi_{0,Y}(Y), \text{ with}$$

$$E_1 = E_{1,0} = \frac{\hbar}{2}(3\omega_X+\omega_Y),$$

$$\left|\Psi_2(x_1,x_2)\right\rangle = \frac{\left|\psi_{2,0}(x_1,x_2)\right\rangle + \left|\psi_{2,0}(x_2,x_1)\right\rangle}{\sqrt{2}} = \sqrt{2}\varphi_{2,X}(X)\varphi_{0,Y}(Y), \text{ with}$$

$$E_2 = E_{2,0} = \frac{\hbar}{2}(5\omega_X+\omega_Y).$$

Note that $E_{2,0} < E_{0,2}$ because $\omega_X < \omega_Y$.

c) For identical *fermionic* particles, the states must be antisymmetric under the exchange of two fermionic particles, that is to say, $|\Psi_n(x_1,x_2)\rangle = -|\Psi_n(x_2,x_1)\rangle$. Therefore, the general fermionic state is of the following form:

$$\begin{aligned}|\Psi(x_1,x_2)\rangle &= \frac{1}{\sqrt{2}}\left(|\psi_{n,m}(x_1,x_2)\rangle - |\psi_{n,m}(x_2,x_1)\rangle\right)\\ &= \frac{1}{\sqrt{2}}\left(\varphi_{n,X}\left(\frac{x_1+x_2}{\sqrt{2}}\right)\varphi_{m,Y}\left(\frac{x_1-x_2}{\sqrt{2}}\right) - \varphi_{n,X}\left(\frac{x_2+x_1}{\sqrt{2}}\right)\varphi_{m,Y}\left(\frac{x_2-x_1}{\sqrt{2}}\right)\right)\\ &= \frac{1}{\sqrt{2}}\left(\varphi_{n,X}(X)\varphi_{m,Y}(Y) - \varphi_{n,X}(X)\varphi_{m,Y}(-Y)\right)\\ &= \frac{1}{\sqrt{2}}\varphi_{n,X}(X)\left(\varphi_{m,Y}(Y) - \varphi_{m,Y}(-Y)\right).\end{aligned}$$

Taking into account the symmetry of the Hermite polynomials, we find

$$|\Psi_0(x_1,x_2)\rangle = \frac{|\psi_{0,1}(x_1,x_2)\rangle - |\psi_{0,1}(x_2,x_1)\rangle}{\sqrt{2}} = \sqrt{2}\varphi_{0,X}(X)\varphi_{1,Y}(Y), \text{ with}$$

$$E_0 = E_{0,1} = \frac{\hbar}{2}(\omega_X + 3\omega_Y),$$

$$|\Psi_1(x_1,x_2)\rangle = \frac{|\psi_{1,1}(x_1,x_2)\rangle - |\psi_{1,1}(x_2,x_1)\rangle}{\sqrt{2}} = \sqrt{2}\varphi_{1,X}(X)\varphi_{1,Y}(Y), \text{ with}$$

$$E_1 = E_{1,1} = \frac{\hbar}{2}(3\omega_X + 3\omega_Y),$$

$$|\Psi_2(x_1,x_2)\rangle = \frac{|\psi_{2,1}(x_1,x_2)\rangle - |\psi_{2,1}(x_2,x_1)\rangle}{\sqrt{2}} = \sqrt{2}\varphi_{2,X}(X)\varphi_{1,Y}(Y), \text{ with}$$

$$E_2 = E_{2,1} = \frac{\hbar}{2}(5\omega_X + 3\omega_Y).$$

Note that $E_{2,1} < E_{0,3}$ because $\omega_X < \omega_Y$.

Problem 5.2

Consider the case of three identical fermionic particles with spin $s = 1/2$ whose states can be represented by $|\uparrow\rangle = \begin{pmatrix}1\\0\end{pmatrix}$ and $|\downarrow\rangle = \begin{pmatrix}0\\1\end{pmatrix}$. Express the composite states $|S = \frac{3}{2}, m\rangle$ as a linear combination of the states $|m_1\, m_2\, m_3\rangle$, where $m_{1,2,3}$ are s_z projections of the individual spins.

When adding angular moment for two particles, one typically uses the Clebsch–Gordan coefficients table, which gives the coefficients connecting the basis of the composite state and the states of the individual particles for two particles. In the case of three particles, one cannot readily use this table. Here we illustrate how this can be done for three $s = \frac{1}{2}$ particles using a tensor product representation of their spin states $\frac{1}{2}\otimes\frac{1}{2}\otimes\frac{1}{2}$:

$$\frac{1}{2}\otimes\frac{1}{2}\otimes\frac{1}{2} = \left(\frac{1}{2}\otimes\frac{1}{2}\right)\otimes\frac{1}{2} = (1\oplus 0)\otimes\frac{1}{2} = \left(1\otimes\frac{1}{2}\right)\oplus\left(0\otimes\frac{1}{2}\right) = \frac{3}{2}\oplus\frac{1}{2}\oplus\frac{1}{2}.$$

The preceding relation shows that we will have a representation of total spin $S=\frac{3}{2}$, and a two-dimensional representation of spin $S=\frac{1}{2}$, one coming from the coupling of the singlet of two fermions with the remaining fermion, and the other from the coupling of the triplet of two fermions with the third fermion. In fact, the two $S=\frac{1}{2}$ representations cannot be distinguished a priori. Note that while the sign $\otimes$ represents the tensor product between different spin spaces, the sign $\oplus$ represents the direct sum of spin spaces.

When $S=\frac{3}{2}$, then $m=\left\{-\frac{3}{2},-\frac{1}{2},\frac{1}{2},\frac{3}{2}\right\}$, and when $S=\frac{1}{2}$, then $m=\left\{-\frac{1}{2},\frac{1}{2}\right\}$. Starting at the top of the ladder for the case of $S=\frac{3}{2}$,

$$\left|\frac{3}{2},\frac{3}{2}\right\rangle=|\uparrow\uparrow\uparrow\rangle,$$

where $s_z|\uparrow\rangle=\frac{1}{2}|\uparrow\rangle$ and $s_z|\downarrow\rangle=-\frac{1}{2}|\downarrow\rangle$ of each $s=1/2$ particle.

Applying the lowering operator $\widehat{J}_-=\widehat{J}_{1-}\otimes\widehat{\mathbb{I}}_2\otimes\widehat{\mathbb{I}}_3+\widehat{\mathbb{I}}_1\otimes\widehat{J}_{2-}\otimes\widehat{\mathbb{I}}_3+\widehat{\mathbb{I}}_1\otimes\widehat{\mathbb{I}}_2\otimes\widehat{J}_{3-}$ on both sides by using $\widehat{J}_\pm|J,M\rangle=\hbar\sqrt{J(J+1)-M(M\pm1)}\,|J,M\pm1\rangle$, we find

$$\widehat{J}_-\left|\frac{3}{2},\frac{3}{2}\right\rangle=\left(\widehat{J}_{1-}+\widehat{J}_{2-}+\widehat{J}_{3-}\right)|\uparrow\uparrow\uparrow\rangle,$$

$$\hbar\sqrt{\frac{3}{2}\left(\frac{3}{2}+1\right)-\frac{3}{2}\left(\frac{3}{2}-1\right)}\left|\frac{3}{2},\frac{1}{2}\right\rangle=\hbar\sqrt{\frac{1}{2}\left(\frac{1}{2}+1\right)-\frac{1}{2}\left(\frac{1}{2}-1\right)}\,(|\downarrow\uparrow\uparrow\rangle+|\uparrow\downarrow\uparrow\rangle+|\uparrow\uparrow\downarrow\rangle),$$

$$\left|\frac{3}{2},\frac{1}{2}\right\rangle=\frac{1}{\sqrt{3}}(|\downarrow\uparrow\uparrow\rangle+|\uparrow\downarrow\uparrow\rangle+|\uparrow\uparrow\downarrow\rangle).$$

In a similar successive application of the lowering ladder operator, we obtain

$$\left|\frac{3}{2},-\frac{1}{2}\right\rangle=\frac{1}{\sqrt{3}}(|\downarrow\downarrow\uparrow\rangle+|\downarrow\uparrow\downarrow\rangle+|\uparrow\downarrow\downarrow\rangle),$$

$$\left|\frac{3}{2},-\frac{3}{2}\right\rangle=|\downarrow\downarrow\downarrow\rangle.$$

The cases shown here can easily be related to the appropriate combinations of the triplet $|1,m\rangle$ and the third spin $|\uparrow\rangle$ or $|\downarrow\rangle$,

$$\begin{array}{ll}|\uparrow\uparrow\rangle\otimes|\uparrow\rangle; & |\uparrow\uparrow\rangle\otimes|\downarrow\rangle\\ (|\uparrow\downarrow\rangle+|\downarrow\uparrow\rangle)\otimes|\uparrow\rangle; & (|\uparrow\downarrow\rangle+|\downarrow\uparrow\rangle)\otimes|\downarrow\rangle\\ |\downarrow\downarrow\rangle\otimes|\uparrow\rangle; & |\downarrow\downarrow\rangle\otimes|\downarrow\rangle,\end{array}$$

which after normalization yield the results for $\left|\frac{3}{2},\pm\frac{3}{2}\right\rangle$, $\left|\frac{3}{2},\pm\frac{1}{2}\right\rangle$.

Note that all states of $S=\frac{3}{2}$ are ***symmetric*** under the exchange of the spin of any two particles.

Next, we consider the case of $S=\frac{1}{2}$ with $m=\pm\frac{1}{2}$. Here, we need to combine the singlet state of two spins with the third to obtain the states $\left|\frac{1}{2},\pm\frac{1}{2}\right\rangle$. Let us focus on the $\left|\frac{1}{2},\frac{1}{2}\right\rangle$ first. Three different states can be constructed combining the fermion singlet $|0,0\rangle$ and the third spin $|\uparrow\rangle$:

$$|\varphi_{+,1}\rangle = \frac{1}{\sqrt{2}}(|\uparrow\uparrow\downarrow\rangle - |\uparrow\downarrow\uparrow\rangle),$$
$$|\varphi_{+,2}\rangle = \frac{1}{\sqrt{2}}(|\uparrow\uparrow\downarrow\rangle - |\downarrow\uparrow\uparrow\rangle),$$
$$|\varphi_{+,3}\rangle = \frac{1}{\sqrt{2}}(|\uparrow\downarrow\uparrow\rangle - |\downarrow\uparrow\uparrow\rangle).$$

It is easy to verify that $\widehat{J}^2|\varphi_{+,\mu}\rangle = \frac{1}{2}\left(\frac{1}{2}+1\right)|\varphi_{+,\mu}\rangle$ and $\widehat{J}_z|\varphi_{+,\mu}\rangle = \frac{1}{2}|\varphi_{+,\mu}\rangle$. These states are not linearly independent, as is evident from the $|\varphi_{+,1}\rangle + |\varphi_{+,3}\rangle = |\varphi_{+,2}\rangle$ relation. To construct linearly independent states for the $S = \frac{1}{2}$, we ensure that $\langle\varphi_{+,1}\,|\,\psi_{+,1}\rangle = 0$ and $\langle\psi_{+,1}\,|\,\psi_{+,1}\rangle = 1$, where $|\psi_{+,1}\rangle = \alpha|\varphi_{+,2}\rangle + \beta|\varphi_{+,3}\rangle$. We find that

$$|\psi_{+,1}\rangle = \frac{1}{\sqrt{6}}(|\uparrow\uparrow\downarrow\rangle - 2|\downarrow\uparrow\uparrow\rangle + |\uparrow\downarrow\uparrow\rangle).$$

Now applying the $\widehat{J}_-$ operator, one obtains

$$|\varphi_{-,1}\rangle = \frac{1}{\sqrt{2}}(|\downarrow\uparrow\downarrow\rangle - |\downarrow\downarrow\uparrow\rangle),$$
$$|\psi_{-,1}\rangle = \frac{-1}{\sqrt{6}}(|\downarrow\downarrow\uparrow\rangle - 2|\uparrow\downarrow\downarrow\rangle + |\downarrow\uparrow\downarrow\rangle).$$

Summarizing the composite states for the three identical spin $s = 1/2$ particles are

$$\left|\frac{3}{2},\frac{3}{2}\right\rangle = |\uparrow\uparrow\uparrow\rangle,$$
$$\left|\frac{3}{2},\frac{1}{2}\right\rangle = \frac{1}{\sqrt{3}}(|\downarrow\uparrow\uparrow\rangle + |\uparrow\downarrow\uparrow\rangle + |\uparrow\uparrow\downarrow\rangle),$$
$$\left|\frac{3}{2},\frac{-1}{2}\right\rangle = \frac{1}{\sqrt{3}}(|\downarrow\downarrow\uparrow\rangle + |\downarrow\uparrow\downarrow\rangle + |\uparrow\downarrow\downarrow\rangle),$$
$$\left|\frac{3}{2},\frac{-3}{2}\right\rangle = |\downarrow\downarrow\downarrow\rangle,$$
$$\left|\frac{1}{2},\frac{1}{2}\right\rangle^{(1)} = \frac{1}{\sqrt{2}}(|\uparrow\uparrow\downarrow\rangle - |\uparrow\downarrow\uparrow\rangle),$$
$$\left|\frac{1}{2},\frac{-1}{2}\right\rangle^{(1)} = \frac{1}{\sqrt{2}}(|\downarrow\uparrow\downarrow\rangle - |\downarrow\downarrow\uparrow\rangle),$$
$$\left|\frac{1}{2},\frac{1}{2}\right\rangle^{(2)} = \frac{1}{\sqrt{6}}(|\uparrow\uparrow\downarrow\rangle - 2|\downarrow\uparrow\uparrow\rangle + |\uparrow\downarrow\uparrow\rangle),$$
$$\left|\frac{1}{2},\frac{-1}{2}\right\rangle^{(2)} = \frac{-1}{\sqrt{6}}(|\downarrow\downarrow\uparrow\rangle - 2|\uparrow\downarrow\downarrow\rangle + |\downarrow\uparrow\downarrow\rangle).$$

Food for thought: Any orthonormal basis of the two-dimensional space formed by states $|\varphi_{+,\mu}\rangle$ is a valid basis for the representation of spin 1/2 states. In particular, Chakrabarti (1964) suggests the use of

$$\left|\frac{1}{2},\frac{1}{2}\right\rangle^{(1)} = \frac{1}{\sqrt{3}}\left(|\uparrow\uparrow\downarrow\rangle + e^{-2\pi i/3}|\uparrow\downarrow\uparrow\rangle + e^{2\pi i/3}|\downarrow\uparrow\uparrow\rangle\right),$$

$$\left|\frac{1}{2},\frac{1}{2}\right\rangle^{(2)} = \frac{1}{\sqrt{3}}\left(|\uparrow\uparrow\downarrow\rangle + e^{2\pi i/3}|\uparrow\downarrow\uparrow\rangle + e^{-2\pi i/3}|\downarrow\uparrow\uparrow\rangle\right),$$

because of the high symmetry properties of those states under permutations.

Problem 5.3

Consider the case of three identical bosonic particles with spin $s = 1$. Express the composite states $|S = 3, m\rangle$ as a linear combination of the states $|m_1\, m_2\, m_3\rangle$, where $m_{1,2,3}$ are s_z projections of the individual spins.

This is another example of adding the angular momenta of more than two identical particles, an operation that cannot be done with the help of the Clebsch–Gordan coefficients table right away. Now we add three $s = 1$ identical particles in the $1 \otimes 1 \otimes 1$ representation with tensor product operations, highlighting again the general strategy for such problems:

$$\begin{aligned} 1\otimes 1\otimes 1 &= (1\otimes 1)\otimes 1 = (2\oplus 1\oplus 0)\otimes 1 = (2\otimes 1)\oplus(1\otimes 1)\oplus(0\otimes 1) \\ &= (3\oplus 2\oplus 1)\oplus(2\oplus 1\oplus 0)\oplus(1) = 3\oplus 2\oplus 2\oplus 1\oplus 1\oplus 1\oplus 0. \end{aligned}$$

We have a representation of total spin $J = 3$, a two-dimensional representation of spin $J = 2$, one coming from the coupling of the singlet of two bosons with the third one, and the other one coming from the coupling of the triplet of two bosons with the third one. Additionally, there is a three-dimensional representation of spin $J = 1$, and a one-dimensional representation of spin $J = 0$. The tensor product sign $\otimes$ and direct spin space summation sign $\oplus$ have the same meaning as in the preceding problem.

When the total spin is $J = 3$, then $m = \{-3,-2,-1,0,+1,+2,+3\}$, and we build those states as combinations of $S = 1$ particles, with $\widehat{S_z}|m\rangle = m|m\rangle$ where $m = \{-1,0,1\} = \{-,0,+\}$ and $|+\rangle = \begin{pmatrix}1\\0\\0\end{pmatrix}, |0\rangle = \begin{pmatrix}0\\1\\0\end{pmatrix}, |-\rangle = \begin{pmatrix}0\\0\\1\end{pmatrix}$. Starting at the top of the ladder

$$|3,3\rangle = |+++\rangle,$$

and applying the lowering operator $\widehat{J_-} = \widehat{J_{1-}}\otimes\widehat{\mathbb{I}_2}\otimes\widehat{\mathbb{I}_3} + \widehat{\mathbb{I}_1}\otimes\widehat{J_{2-}}\otimes\widehat{\mathbb{I}_3} + \widehat{\mathbb{I}_1}\otimes\widehat{\mathbb{I}_2}\otimes\widehat{J_{3-}}$ on both sides with $\widehat{J_\pm}|J,M\rangle = \hbar\sqrt{J(J+1)-M(M\pm 1)}|J,M\pm 1\rangle$, we find

$$|3,2\rangle = \frac{1}{\sqrt{3}}\left(|0++\rangle + |+0+\rangle + |++0\rangle\right),$$

$$|3,1\rangle = \frac{1}{\sqrt{15}}(2|00+\rangle + 2|0+0\rangle + 2|+00\rangle + |-++\rangle + |+-+\rangle + |++-\rangle),$$

$$|3,0\rangle = \frac{1}{\sqrt{10}}(|0-+\rangle + |-0+\rangle + |-+0\rangle + 2|000\rangle + |0+-\rangle + |+0-\rangle + |+-0\rangle),$$

$$|3,-1\rangle = \frac{1}{\sqrt{15}}(2|00-\rangle + 2|0-0\rangle + 2|-00\rangle + |+--\rangle + |-+-\rangle + |--+\rangle),$$

$$|3,-2\rangle = \frac{1}{\sqrt{3}}(|0--\rangle + |-0-\rangle + |--0\rangle),$$

$$|3,-3\rangle = |---\rangle.$$

Note that all states of $S = 3$ are ***symmetric*** under the exchange of the spin of any two particles.

Next, we obtain the composite states corresponding to $S = 2$. We start with $|2,2\rangle^{(1)}$ which is a combination of the triplet state of two spins $|1,1\rangle$ combined with the third one $|+\rangle$,

$$|2,2\rangle^{(1)} = |1,1\rangle \otimes |+\rangle = \frac{1}{\sqrt{2}}(|0++\rangle - |+0+\rangle).$$

The additional, linearly independent state with $S = 2$, $m = 2$ can be found by constructing $|2,2\rangle^{(2)} = \alpha|0++\rangle + \beta|+0+\rangle + \gamma|++0\rangle$, which must be normalized and orthogonal to $|2,2\rangle^{(1)}$ and $|3,2\rangle$. This state also must fulfil $\hat{J}^2|2,2\rangle^{(2)} = 2\left(2+\frac{1}{2}\right)|2,2\rangle^{(2)}$ As a result,

$$|2,2\rangle^{(2)} = \frac{1}{\sqrt{6}}(|0++\rangle + |+0+\rangle - 2|++0\rangle).$$

Applying the lowering operator $\hat{J}_-$ on both sides of $|2,2\rangle^{(1)}$ and $|2,2\rangle^{(2)}$ gives the following composite states:

$$|2,1\rangle^{(1)} = \frac{1}{2}(|-++\rangle - |+-+\rangle + |0+0\rangle - |+00\rangle),$$

$$|2,0\rangle^{(1)} = \frac{1}{\sqrt{12}}(|-0+\rangle + 2|-+0\rangle - |0-+\rangle + |0+-\rangle - 2|+-0\rangle - |+0-\rangle),$$

$$|2,-1\rangle^{(1)} = \frac{1}{2}(|-00\rangle - |0-0\rangle + |-+-\rangle - |+--\rangle),$$

$$|2,-2\rangle^{(1)} = \frac{1}{\sqrt{2}}(|-0-\rangle - |0--\rangle),$$

$$|2,1\rangle^{(2)} = \frac{1}{\sqrt{12}}(|-++\rangle + 2|00+\rangle - |0+0\rangle + |+-+\rangle - |+00\rangle - 2|++-\rangle),$$

$$|2,0\rangle^{(2)} = \frac{1}{2}(|-0+\rangle + |0-+\rangle - |0+-\rangle - |+0-\rangle),$$

$$|2,-1\rangle^{(2)} = \frac{1}{\sqrt{12}}(|-00\rangle + |0-0\rangle - |-+-\rangle - |+--\rangle + 2|--+\rangle - 2|00-\rangle),$$

$$|2,-2\rangle^{(2)} = \frac{1}{\sqrt{6}}(2|--0\rangle - |-0-\rangle - |0--\rangle).$$

Now we focus on the $S = 1$ composite state with $m = 1$. We form the following possible states by coupling the singlet of any two spins with the third one:

$$|\varphi_1\rangle = \frac{1}{\sqrt{3}}(|++-\rangle - |+00\rangle + |+-+\rangle),$$
$$|\varphi_2\rangle = \frac{1}{\sqrt{3}}(|-++\rangle - |0+0\rangle + |++-\rangle),$$
$$|\varphi_3\rangle = \frac{1}{\sqrt{3}}(|-++\rangle - |00+\rangle + |+-+\rangle).$$

These are eigenstates of $\widehat{J}^2$, since $\widehat{J}^2|\varphi_k\rangle = 1(1+1)|\varphi_k\rangle$, and $\widehat{J}_z|\varphi_k\rangle = 1|\varphi_k\rangle$. However, $\langle\varphi_\mu\,|\,\varphi_\nu\rangle = \frac{1}{3} + \frac{2}{3}\delta_{\mu\nu}$, meaning that $|\varphi_{1,2,3}\rangle$ are not ortho-normal. Diagonalizing the matrix representation of $\widehat{J}^2$ in the basis $|\varphi_{1,2,3}\rangle$ yields

$$|1,1\rangle^{(1)} = \frac{1}{\sqrt{15}}(2|-++\rangle + 2|+-+\rangle + 2|++-\rangle - |00+\rangle - |0+0\rangle - |+00\rangle),$$
$$|1,1\rangle^{(2)} = \frac{1}{2}(|0+0\rangle - |00+\rangle + |+-+\rangle - |++-\rangle),$$
$$|1,1\rangle^{(3)} = \frac{1}{\sqrt{12}}(2|-++\rangle - |+-+\rangle - |++-\rangle + 2|+00\rangle - |00+\rangle - |0+0\rangle).$$

Again, after applying $\widehat{J}_-$ on both sides of the above relations, we find

$$|1,0\rangle^{(1)} = \frac{1}{\sqrt{15}}(|-0+\rangle + |-+0\rangle + |0-+\rangle - 3|000\rangle + |0+-\rangle + |+-0\rangle + |+0-\rangle),$$
$$|1,-1\rangle^{(1)} = \frac{1}{\sqrt{15}}(2|+--\rangle + 2|-+-\rangle + 2|--+\rangle - |-00\rangle - |0-0\rangle - |00-\rangle),$$

$$|1,0\rangle^{(2)} = \frac{1}{2}(-|-0+\rangle + |-+0\rangle + |+-0\rangle - |+0-\rangle),$$
$$|1,-1\rangle^{(2)} = \frac{1}{2}(|-+-\rangle - |--+\rangle + |0-0\rangle - |00-\rangle),$$

$$|1,0\rangle^{(3)} = \frac{1}{\sqrt{12}}(|-0+\rangle + |-+0\rangle - 2|0-+\rangle - 2|0+-\rangle + |+-0\rangle + |+0-\rangle),$$
$$|1,-1\rangle^{(3)} = \frac{1}{\sqrt{12}}(2|+--\rangle - |-+-\rangle - |--+\rangle + 2|-00\rangle - |0-0\rangle - |00-\rangle).$$

Finally, there is only one state corresponding to $|0,0\rangle = \alpha_1|-0+\rangle + \alpha_2|0-+\rangle + \alpha_3|+-0\rangle - \alpha_4|-+0\rangle + \alpha_5|0+-\rangle + \alpha_6|+0-\rangle + \alpha_7|000\rangle$. Making sure that $|0,0\rangle$ is normalized and orthogonal to $|3,0\rangle$, $|2,0\rangle^{(1)}, |2,0\rangle^{(2)}$, $|1,0\rangle^{(1)}$, $|1,0\rangle^{(2)}$, and $|1,0\rangle^{(3)}$ yields

$$|0,0\rangle = \frac{1}{\sqrt{6}}(|-0+\rangle - |0-+\rangle + |+-0\rangle - |-+0\rangle + |0+-\rangle - |+0-\rangle).$$

We verify that $\widehat{J}^2|0,0\rangle = 0$ and $\widehat{J}_z = |0,0\rangle = 0$, as expected.

Food for thought: The main difficulty in the preceding hands-on procedure is constructing a representative for each subgroup through which the application of the lowering (or raising) operator can generate the complete subspace. For example, after constructing $|2,2\rangle^{(1)}$ and $|2,2\rangle^{(2)}$ we can easily find the rest of the states corresponding to this subspace by using the $\widehat{J}_-$ operator.

Another starting point for each subspace can be obtained in a different (but related) way. For the case of the three $\widehat{S}=1$ particles, let us consider all possible combinations with $\widehat{S}_z = 0$:

$$\{|-+0\rangle, |-0+\rangle, |0+-\rangle, |0-+\rangle, |+0-\rangle, |+-0\rangle, |000\rangle\}.$$

The matrix representation of $\widehat{S^2}$ in the preceding basis is

$$\widehat{S^2} = \hbar^2 \begin{pmatrix} 4 & 2 & 2 & 0 & 0 & 0 & 2 \\ 2 & 4 & 0 & 2 & 0 & 0 & 2 \\ 2 & 0 & 4 & 0 & 2 & 0 & 2 \\ 0 & 2 & 0 & 4 & 0 & 2 & 2 \\ 0 & 0 & 2 & 0 & 4 & 2 & 2 \\ 0 & 0 & 0 & 2 & 2 & 4 & 2 \\ 2 & 2 & 2 & 2 & 2 & 2 & 6 \end{pmatrix}.$$

Its eigenvalues are $\{12,6,6,2,2,2,0\}\hbar^2$, which correspond to $S(S+1)\hbar^2$ with $S=\{3,2,1,0\}$. The eigenstates are

$$\{|3,0\rangle, |2,0\rangle^{(1)}, |2,0\rangle^{(2)}, |1,0\rangle^{(1)}, |1,0\rangle^{(2)}, |1,0\rangle^{(3)}, |0,0\rangle\},$$

as given earlier. Now by applying $\widehat{S}_+$ and $\widehat{S}_-$ (whenever appropriate), one can obtain all composite states forming an ortho-normalized basis, as already found earlier. (Try it!)

Problem 5.4

a) Two identical noninteracting fermions with $s = 1/2$ are in a 1D infinite square well potential $V(x) = \begin{cases} 0, & 0 < x < a \\ \infty, & otherwise \end{cases}$. Suppose $s_z = 1/2$ for one particle and $s_z = -1/2$ for the other particle. Construct the ground, first, and second excited states.

b) Two identical noninteracting fermions with $s = 1/2$ are in a 1D infinite square well potential $V(x) = \begin{cases} 0, & 0 < x < a \\ \infty, & otherwise \end{cases}$. Suppose $s_z = 1/2$ for one particle and $s_z = +1/2$ for the other particle. Construct the ground, first, and second excited states.

c) Suppose now that a third identical fermion with $s = 1/2$ and $s_z = +1/2$ is added to the two fermions from parts (a) and (b). Construct the ground states for these cases.

This is a problem to practice basic rules for constructing composite states of identical fermions.

We recall that the eigenstates and eigenenergies for the given Hamiltonian are $\psi_n = \sqrt{\frac{2}{a}}\, sin\left(\frac{n\pi x}{a}\right)$; $E_n = \frac{n^2\pi^2\hbar^2}{2ma^2}$

We also know that in all situations in this problem, the total wave function for the identical particles must be antisymmetric since the particles are fermions.

a) The ground state for this case is the product of a symmetric radial part and the antisymmetric singlet spin state for a pair of electrons

$$\Psi_0(x_1,x_2) = \frac{1}{\sqrt{2!}}\begin{vmatrix} \psi_1(x_1)|\uparrow\rangle & \psi_1(x_2)|\uparrow\rangle \\ \psi_1(x_1)|\downarrow\rangle & \psi_1(x_2)|\downarrow\rangle \end{vmatrix} = \psi_1(x_1)\psi_1(x_2)\chi_S;$$

$$\chi_S = \frac{1}{\sqrt{2}}[|\uparrow\downarrow\rangle - |\downarrow\uparrow\rangle],$$

$$E_0 = \frac{\hbar^2\pi^2}{2ma^2}(1^2+1^2) = \frac{\hbar^2\pi^2}{ma^2}.$$

The first excited state has two possibilities: the total wave function, $\Psi_1^{(I)}$, is a product of a symmetric radial part and the antisymmetric singlet spin state, or the total wave function, $\Psi_1^{(II)}$, is a product of an antisymmetric radial part and the symmetric triplet spin state. The energy is the same for both cases.

$$\Psi_1^{(I)}(x_1,x_2) = \frac{1}{\sqrt{2}}\left(\frac{1}{\sqrt{2!}}\begin{vmatrix} \psi_1(x_1)|\uparrow\rangle & \psi_1(x_2)|\uparrow\rangle \\ \psi_2(x_1)|\downarrow\rangle & \psi_2(x_2)|\downarrow\rangle \end{vmatrix} - \frac{1}{\sqrt{2!}}\begin{vmatrix} \psi_1(x_1)|\downarrow\rangle & \psi_1(x_2)|\downarrow\rangle \\ \psi_2(x_1)|\uparrow\rangle & \psi_2(x_2)|\uparrow\rangle \end{vmatrix}\right)$$

$$= \frac{1}{\sqrt{2!}}[\psi_1(x_1)\psi_2(x_2) + \psi_2(x_1)\psi_1(x_2)]\chi_S = \frac{1}{\sqrt{2!}}\left\|\begin{matrix} \psi_1(x_1) & \psi_1(x_2) \\ \psi_2(x_1) & \psi_2(x_2) \end{matrix}\right\|\chi_S,$$

$$\Psi_1^{(II)}(x_1,x_2) = \frac{1}{\sqrt{2}}\left(\frac{1}{\sqrt{2!}}\begin{vmatrix} \psi_1(x_1)|\uparrow\rangle & \psi_1(x_2)|\uparrow\rangle \\ \psi_2(x_1)|\downarrow\rangle & \psi_2(x_2)|\downarrow\rangle \end{vmatrix} + \frac{1}{\sqrt{2!}}\begin{vmatrix} \psi_1(x_1)|\downarrow\rangle & \psi_1(x_2)|\downarrow\rangle \\ \psi_2(x_1)|\uparrow\rangle & \psi_2(x_2)|\uparrow\rangle \end{vmatrix}\right)$$

$$= \frac{1}{\sqrt{2!}}\begin{vmatrix} \psi_1(x_1) & \psi_1(x_2) \\ \psi_2(x_1) & \psi_2(x_2) \end{vmatrix}\chi_T,$$

$$\chi_T = \frac{1}{\sqrt{2}}[|\uparrow\downarrow\rangle + |\downarrow\uparrow\rangle],$$

$$E_1 = \frac{\hbar^2\pi^2}{2ma^2}(2^2+1^2) = \frac{5\hbar^2\pi^2}{2ma^2}.$$

The second excited state has only one possibility of a symmetric radial part multiplied by the antisymmetric singlet state:

$$\Psi_2(x_1,x_2) = \frac{1}{\sqrt{2!}}\begin{vmatrix} \psi_2(x_1)|\uparrow\rangle & \psi_2(x_2)|\uparrow\rangle \\ \psi_2(x_1)|\downarrow\rangle & \psi_2(x_2)|\downarrow\rangle \end{vmatrix} = \psi_2(x_1)\psi_2(x_2)\chi_S,$$

$$E_2 = \frac{\hbar^2\pi^2}{2ma^2}(2^2+2^2) = \frac{4\hbar^2\pi^2}{ma^2}.$$

b) In this case, the two spins can only be in the triplet state $\chi_T = |+\,+\rangle$, which is symmetric. Therefore, the total wave function must be a product of an antisymmetric radial part and the symmetric triplet state. This means that the radial functions for the two particles must have different n-quantum numbers:

$$\Psi_0(x_1,x_2) = \frac{1}{\sqrt{2!}} \begin{vmatrix} \psi_1(x_1)|\uparrow\rangle & \psi_1(x_2)|\uparrow\rangle \\ \psi_2(x_1)|\uparrow\rangle & \psi_2(x_2)|\uparrow\rangle \end{vmatrix} = \frac{1}{\sqrt{2}} \begin{vmatrix} \psi_1(x_1) & \psi_1(x_2) \\ \psi_2(x_1) & \psi_2(x_2) \end{vmatrix} |\uparrow\uparrow\rangle,$$

$$E_0 = \frac{\hbar^2\pi^2}{2ma^2}\left(2^2+1^2\right) = \frac{5\hbar^2\pi^2}{2ma^2}.$$

The first excited state is

$$\Psi_1(x_1,x_2) = \frac{1}{\sqrt{2!}} \begin{vmatrix} \psi_1(x_1)|\uparrow\rangle & \psi_1(x_2)|\uparrow\rangle \\ \psi_3(x_1)|\uparrow\rangle & \psi_3(x_2)|\uparrow\rangle \end{vmatrix} = \frac{1}{\sqrt{2}} \begin{vmatrix} \psi_1(x_1) & \psi_1(x_2) \\ \psi_3(x_1) & \psi_3(x_2) \end{vmatrix} |\uparrow\uparrow\rangle,$$

$$E_1 = \frac{\hbar^2\pi^2}{2ma^2}\left(3^2+1^2\right) = \frac{5\hbar^2\pi^2}{ma^2}.$$

The second excited state is

$$\Psi_2(x_1,x_2) = \frac{1}{\sqrt{2!}} \begin{vmatrix} \psi_2(x_1)|\uparrow\rangle & \psi_2(x_2)|\uparrow\rangle \\ \psi_3(x_1)|\uparrow\rangle & \psi_3(x_2)|\uparrow\rangle \end{vmatrix} = \frac{1}{\sqrt{2}} \begin{vmatrix} \psi_2(x_1) & \psi_2(x_2) \\ \psi_3(x_1) & \psi_3(x_2) \end{vmatrix} |\uparrow\uparrow\rangle,$$

$$E_2 = \frac{\hbar^2\pi^2}{2ma^2}\left(3^2+2^2\right) = \frac{13\hbar^2\pi^2}{2ma^2}.$$

c) When a third noninteracting $s = 1/2$ particle is added to the case of the two fermions from part (a), we recognize that the composite state wave function must be antisymmetric. Applying the Slater determinant for the ground state, we have

$$|\Psi_0\rangle = \frac{1}{\sqrt{3!}} \begin{vmatrix} \psi_1(x_1)|\uparrow\rangle & \psi_1(x_2)|\uparrow\rangle & \psi_1(x_3)|\uparrow\rangle \\ \psi_1(x_1)|\downarrow\rangle & \psi_1(x_2)|\downarrow\rangle & \psi_1(x_3)|\downarrow\rangle \\ \psi_2(x_1)|\uparrow\rangle & \psi_2(x_2)|\uparrow\rangle & \psi_2(x_3)|\uparrow\rangle \end{vmatrix},$$

$$E_0 = \frac{\hbar^2\pi^2}{2ma^2}\left(1^2+1^2+2^2\right) = \frac{3\hbar^2\pi^2}{ma^2}.$$

From here we find

$$|\Psi_0\rangle = \frac{1}{\sqrt{6}}\left[\psi_1(x_2)\left(\psi_1(x_1)\psi_2(x_3)-\psi_1(x_3)\psi_2(x_1)\right)|\uparrow\downarrow\uparrow\rangle + \psi_1(x_1)\left(\psi_2(x_2)\psi_1(x_3)\right.\right.$$
$$\left.\left.-\psi_2(x_3)\psi_1(x_2)\right)|\downarrow\uparrow\uparrow\rangle + \psi_1(x_3)\left(\psi_2(x_1)\psi_1(x_2)-\psi_2(x_2)\psi_1(x_1)\right)|\uparrow\uparrow\downarrow\rangle\right].$$

From Problem 5.2, $|\uparrow\downarrow\uparrow\rangle$, $|\downarrow\uparrow\uparrow\rangle$, and $|\uparrow\uparrow\downarrow\rangle$ can be expressed in terms of the $\widehat{S}^2$, $\widehat{S}_z$ eigenstates of the added angular momenta, such that

$$|\uparrow\downarrow\uparrow\rangle = \frac{1}{\sqrt{3}}\left|\frac{3}{2},\frac{1}{2}\right\rangle + \frac{1}{\sqrt{6}}\left|\frac{1}{2},\frac{1}{2}\right\rangle^{(2)} - \frac{1}{\sqrt{2}}\left|\frac{1}{2},\frac{1}{2}\right\rangle^{(1)},$$

$$|\downarrow\uparrow\uparrow\rangle = \frac{1}{\sqrt{3}}\left|\frac{3}{2},\frac{1}{2}\right\rangle - \frac{\sqrt{2}}{\sqrt{3}}\left|\frac{1}{2},\frac{1}{2}\right\rangle^{(2)},$$

$$|\uparrow\uparrow\downarrow\rangle = \frac{1}{\sqrt{3}}\left|\frac{3}{2},\frac{1}{2}\right\rangle + \frac{1}{\sqrt{6}}\left|\frac{1}{2},\frac{1}{2}\right\rangle^{(2)} + \frac{1}{\sqrt{2}}\left|\frac{1}{2},\frac{1}{2}\right\rangle^{(1)}.$$

Clearly, the composite wave function cannot be simply decomposed into symmetric/antisymmetric terms for the space and spin degrees of freedom, as it is the case of two identical fermions from part (a).

When a third noninteracting $s = 1/2$ particle is added to the case of the two fermions from part (b), the Slater determinant and corresponding ground state energy for the three particles becomes

$$|\Psi_0\rangle = \frac{1}{\sqrt{3!}} \begin{vmatrix} \psi_1(x_1)|\uparrow\rangle & \psi_1(x_2)|\uparrow\rangle & \psi_1(x_3)|\uparrow\rangle \\ \psi_2(x_1)|\uparrow\rangle & \psi_2(x_2)|\uparrow\rangle & \psi_2(x_3)|\uparrow\rangle \\ \psi_3(x_1)|\uparrow\rangle & \psi_3(x_2)|\uparrow\rangle & \psi_3(x_3)|\uparrow\rangle \end{vmatrix}$$

$$= \frac{1}{\sqrt{6}} \begin{vmatrix} \psi_1(x_1) & \psi_1(x_2) & \psi_1(x_3) \\ \psi_2(x_1) & \psi_2(x_2) & \psi_2(x_3) \\ \psi_3(x_1) & \psi_3(x_2) & \psi_3(x_3) \end{vmatrix} \left| \frac{3}{2}, \frac{3}{2} \right\rangle,$$

$$E_0 = \frac{\hbar^2 \pi^2}{2ma^2}\left(1^2 + 2^2 + 3^2\right) = \frac{7\hbar^2\pi^2}{ma^2},$$

where we have taken into account that $|\uparrow\uparrow\uparrow\rangle = \left|\frac{3}{2}, \frac{3}{2}\right\rangle$ from Problem 5.2. In this case, the wave function is a product of an antisymmetric space-dependent term and a symmetric angular momentum term.

Food for thought: For extra practice, construct the first excited states for part (c).

Problem 5.5

a) Two identical noninteracting bosons with $s = 1$ are in a 1D infinite square well potential $V(x) = \begin{cases} 0, & 0 < x < a \\ \infty, & otherwise \end{cases}$. Suppose $s_z = 0$ for one particle and $s_z = 1$ for the other particle. Construct the ground, first, second, and third excited states.

b) Suppose that a third identical boson with $s = 1$ and $s_z = -1$ is added to the two bosons from (a). Construct the ground and first excited states.

This is a basic problem to practice the construction of composite states of noninteracting identical bosons by following the established symmetrization procedure for their wave function.

The eigenstates and eigenenergies for the given Hamiltonian for each particle are $\psi_n = \sqrt{\frac{2}{a}} \sin\left(\frac{n\pi x}{a}\right)$; $E_n = \frac{n^2\pi^2\hbar^2}{2ma^2}$.

We also note that in all situations in this problem, the total wave function for the identical particles must be symmetric since the particles are bosons.

a) Let's remind ourselves of the composite spin states for the two $s = 1$ bosons as read from the Clebsch–Gordan coefficients table using the notation of $|S,m\rangle$ for the composite spin and $|m_1\,m_2\rangle$ for the spins of the individual particles. From the Clebsch–Gordan coefficients table, we find the following possible states:

$$|S=2, m=2\rangle = |+\,+\rangle,$$
$$|2,1\rangle = \frac{1}{\sqrt{2}}[|+\,0\rangle + |0+\rangle],$$
$$|2,0\rangle = \frac{1}{\sqrt{6}}|+\,-\rangle + \sqrt{\frac{2}{3}}|0\,0\rangle + \frac{1}{\sqrt{6}}|-\,+\rangle,$$
$$|2,-1\rangle = \frac{1}{\sqrt{2}}[|0-\rangle + |-\,0\rangle],$$
$$|2,-2\rangle = |-\,-\rangle$$
$$|1,1\rangle = \frac{1}{\sqrt{2}}[|+\,0\rangle - |0+\rangle],$$
$$|1,0\rangle = \frac{1}{\sqrt{2}}[|+\,-\rangle - |-\,+\rangle],$$
$$|1,-1\rangle = \frac{1}{\sqrt{2}}[|0-\rangle - |-\,0\rangle],$$
$$|0,0\rangle = \frac{1}{\sqrt{3}}[|+\,-\rangle - |0\,0\rangle + |-\,+\rangle].$$

For the case in this problem one particle has $s_z = 0$, while the other has $s_z = +1$. Therefore, we have the following possibilities of the spin states:

$$|2,1\rangle = \frac{1}{\sqrt{2}}[|+\,0\rangle + |0+\rangle],$$
$$|1,1\rangle = \frac{1}{\sqrt{2}}[|+\,0\rangle - |0+\rangle].$$

Note that $|2,1\rangle$ is symmetric under exchange of spins, while $|1,1\rangle$ is antisymmetric.

Given that the total wave function is always symmetric, for the ground state of the two identical bosons we find

$$\Psi_0 = \psi_1(x_1)\,\psi_1(x_2)\,|2,1\rangle,$$
$$E_0 = \frac{\hbar^2\pi^2}{2ma^2}\left(1^2 + 1^2\right) = \frac{\hbar^2\pi^2}{ma^2}.$$

For the first excited state we have two options,

$$\Psi_1^{(I)} = \frac{1}{\sqrt{2}}\left[\psi_1(x_1)\,\psi_2(x_2) + \psi_1(x_2)\,\psi_2(x_1)\right]|2,1\rangle, \text{or}$$
$$\Psi_1^{(II)} = \frac{1}{\sqrt{2}}\left[\psi_1(x_1)\,\psi_2(x_2) - \psi_1(x_2)\,\psi_2(x_1)\right]|1,1\rangle$$
$$E_1 = \frac{\hbar^2\pi^2}{2ma^2}\left(2^2 + 1^2\right) = \frac{5\hbar^2\pi^2}{2ma^2}.$$

For the second excited state,

$$\Psi_2 = \psi_2(x_1)\psi_2(x_2)|2,1\rangle,$$
$$E_2 = \frac{\hbar^2\pi^2}{2ma^2}\left(2^2+2^2\right) = \frac{4\hbar^2\pi^2}{2ma^2}.$$

For the third excited state, we have two options again,

$$\Psi_3^{(I)} = \frac{1}{\sqrt{2}}\left[\psi_2(x_1)\psi_3(x_2)+\psi_3(x_2)\psi_2(x_1)\right]|2,1\rangle \text{ or}$$
$$\Psi_3^{(II)} = \frac{1}{\sqrt{2}}\left[\psi_2(x_1)\psi_3(x_2)-\psi_3(x_2)\psi_2(x_1)\right]|1,1\rangle,$$
$$E_3 = \frac{\hbar^2\pi^2}{2ma^2}\left(2^2+3^2\right) = \frac{13\hbar^2\pi^2}{2ma^2}.$$

b) Adding a third boson with $s=1$, $s_z=-1$ can be done using the symmetrized Slater permanent. For the ground state,

$$|\Psi_0\rangle = \frac{1}{\sqrt{3!}}\left\|\begin{matrix} \psi_1(x_1)|+\rangle & \psi_1(x_2)|+\rangle & \psi_1(x_3)|+\rangle \\ \psi_1(x_1)|0\rangle & \psi_1(x_2)|0\rangle & \psi_1(x_3)|0\rangle \\ \psi_1(x_1)|-\rangle & \psi_1(x_2)|-\rangle & \psi_1(x_3)|-\rangle \end{matrix}\right\|$$
$$= \frac{1}{\sqrt{6}}\psi_1(x_1)\psi_1(x_2)\psi_1(x_3)\left(|+0-\rangle+|-+0\rangle+|0-+\rangle+|-0+\rangle+|+-0\rangle+|0+-\rangle\right).$$

Using the results from Problem 5.3, the spinor states can be given in terms of eigenstates to the total S^2, S_z operators.

$$|+0-\rangle+|-+0\rangle+|0-+\rangle+|-0+\rangle+|+-0\rangle+|0+-\rangle = \sqrt{\frac{3}{5}}|3,0\rangle+\sqrt{\frac{2}{5}}|1,0\rangle^{(1)}.$$

In summary, the composite ground state and energy are

$$|\Psi_0\rangle = \psi_1(x_1)\psi_1(x_2)\psi_1(x_3)\left(\sqrt{\frac{3}{5}}|3,0\rangle+\sqrt{\frac{2}{5}}|1,0\rangle^{(1)}\right),$$
$$E_0 = \frac{\hbar^2\pi^2}{2ma^2}\left(1^2+1^2+1^2\right) = \frac{3\hbar^2\pi^2}{2ma^2}.$$

Similarly, for the first excited state,

$$|\Psi_1\rangle^{(I)} = \frac{1}{\sqrt{3!}}\left\|\begin{matrix} \psi_1(x_1)|+\rangle & \psi_1(x_2)|+\rangle & \psi_1(x_3)|+\rangle \\ \psi_1(x_1)|0\rangle & \psi_1(x_2)|0\rangle & \psi_1(x_3)|0\rangle \\ \psi_2(x_1)|-\rangle & \psi_2(x_2)|-\rangle & \psi_2(x_3)|-\rangle \end{matrix}\right\|$$
$$= \frac{1}{\sqrt{6}}\left(\psi_2(x_1)\psi_1(x_2)\psi_1(x_3)\left[|-+0\rangle+|-0+\rangle\right]\right.$$
$$+\psi_1(x_1)\psi_2(x_2)\psi_1(x_3)\left[|0-+\rangle+|0+-\rangle\right]$$
$$\left.+\psi_1(x_1)\psi_1(x_2)\psi_2(x_3)\left[|+0-\rangle+|+-0\rangle\right]\right).$$

The preceding result can also be given in terms of the composite states for the angular momenta, as derived in Problem 5.3:

$$
\begin{aligned}
|\Psi_1\rangle^{(I)} = \frac{1}{\sqrt{24}} \Big(& [\psi_2(x_1)\psi_1(x_2)\psi_1(x_3) + \psi_1(x_1)\psi_2(x_2)\psi_1(x_3) + \psi_1(x_1)\psi_1(x_2)\psi_2(x_3)] \\
& \times \left[\frac{\sqrt{8}}{\sqrt{5}}|3,0\rangle + \frac{4}{\sqrt{15}}|1,0\rangle^{(1)} - \frac{1}{\sqrt{3}}|1,0\rangle^{(3)} \right] \\
& + \sqrt{3}|2,0\rangle^{(1)} \left([\psi_2(x_1)\psi_1(x_2) - \psi_1(x_1)\psi_2(x_2)]\psi_1(x_3) \right) \\
& + |2,0\rangle^{(2)} \left([\psi_2(x_1)\psi_1(x_2) + \psi_1(x_1)\psi_2(x_2)]\psi_1(x_3) - 2\psi_1(x_1)\psi_1(x_2)\psi_2(x_3) \right) \\
& + |1,0\rangle^{(2)} \left(\psi_1(x_1)[\psi_2(x_2)\psi_1(x_3) - \psi_1(x_2)\psi_2(x_3)] \right) \\
& + \sqrt{3}\psi_2(x_1)\psi_1(x_2)\psi_1(x_3)|1,0\rangle^{(3)} \Big).
\end{aligned}
$$

Two additional options for the first excited state are also possible, which we only give with the Slater permanents:

$$
|\Psi_1\rangle^{(II)} = \frac{1}{\sqrt{3!}} \left\| \begin{matrix} \psi_1(x_1)|+\rangle & \psi_1(x_2)|+\rangle & \psi_1(x_3)|+\rangle \\ \psi_2(x_1)|0\rangle & \psi_2(x_2)|0\rangle & \psi_2(x_3)|0\rangle \\ \psi_1(x_1)|-\rangle & \psi_1(x_2)|-\rangle & \psi_1(x_3)|-\rangle \end{matrix} \right\|,
$$

$$
|\Psi_1\rangle^{(III)} = \frac{1}{\sqrt{3!}} \left\| \begin{matrix} \psi_2(x_1)|+\rangle & \psi_2(x_2)|+\rangle & \psi_2(x_3)|+\rangle \\ \psi_1(x_1)|0\rangle & \psi_1(x_2)|0\rangle & \psi_1(x_3)|0\rangle \\ \psi_1(x_1)|-\rangle & \psi_1(x_2)|-\rangle & \psi_1(x_3)|-\rangle \end{matrix} \right\|,
$$

$$
E_1 = \frac{\hbar^2\pi^2}{2ma^2}\left(1^2 + 1^2 + 2^2\right) = \frac{3\hbar^2\pi^2}{ma^2}.
$$

Problem 5.6

Two noninteracting identical *electrons* are in a simple 1D harmonic oscillator potential $V(\hat{x}) = \frac{1}{2}m\omega^2\hat{x}^2$. The total energy for the composite state of the two electrons is found to be $E = 4\hbar\omega$ and the electrons are in a singlet spin state.

What is the wave function of the two noninteracting identical electrons? Also, calculate $\langle(\hat{x}_1 - \hat{x}_2)^2\rangle$.

This problem requires basic knowledge of eigenenergies and eigenstates of the simple harmonic oscillator. It also probes our understanding of exchange interaction for fermionic systems whose wave functions must always be antisymmetric.

The total energy of the composite electronic state is $E_{n_1n_2} = \hbar\omega(n_1 + n_2 + 1)$, where n_1, n_2 are integer numbers. The given $E = 4\hbar\omega$ can be obtained by two possibilities,

$$
E_{12} = E_{21} = \hbar\omega\left(1 + \frac{1}{2}\right) + \hbar\omega\left(2 + \frac{1}{2}\right) = \hbar\omega\left(2 + \frac{1}{2}\right) + \hbar\omega\left(1 + \frac{1}{2}\right) = 4\hbar\omega,
$$

$$
E_{30} = E_{03} = \hbar\omega\left(3 + \frac{1}{2}\right) + \hbar\omega\left(0 + \frac{1}{2}\right) = \hbar\omega\left(0 + \frac{1}{2}\right) + \hbar\omega\left(3 + \frac{1}{2}\right) = 4\hbar\omega.
$$

The electrons are in a singlet state; therefore, the spin state of the composite system can only be the antisymmetric state $\chi_S = |0,0\rangle = \frac{1}{\sqrt{2}}[|\uparrow\downarrow\rangle - |\downarrow\uparrow\rangle]$. Therefore, the full wave function only can be antisymmetric if the spatial part is symmetric, such that

$$
|\Psi_{1,2}\rangle = \frac{1}{\sqrt{2}}[\Phi_1(x_1)\Phi_2(x_2) + \Phi_2(x_1)\Phi_1(x_2)]\chi_S,
$$

$$|\Psi_{3,0}\rangle = \frac{1}{\sqrt{2}}\left[\Phi_3(x_1)\Phi_0(x_2) + \Phi_0(x_1)\Phi_3(x_2)\right]\chi_S,$$

where $\Phi_n(x) = \left(\frac{m\omega}{\pi\hbar}\right)^{\frac{1}{4}} \frac{1}{\sqrt{2^n n!}} H_n\left(\sqrt{\frac{m\omega}{\hbar}}x\right) e^{-\frac{m\omega x^2}{2\hbar}}$ are the eigenfunctions for the harmonic oscillator and $H_n\left(\sqrt{\frac{m\omega}{\hbar}}x\right)$ are the Hermite polynomials (see, for example, Wikipedia, n.d.).

Let us now calculate

$$\langle(\hat{x}_1 - \hat{x}_2)^2\rangle = \langle\Psi|(\hat{x}_1^2 + \hat{x}_2^2 - 2\hat{x}_1\hat{x}_2)|\Psi\rangle,$$

using the total wave functions found earlier. For this purpose, we represent $\Phi_n \to |n\rangle$ and $\hat{x}_{1,2} = \sqrt{\frac{\hbar}{2m\omega}}\left(\hat{a}_{1,2} + \hat{a}_{1,2}^{+}\right)$, where $\hat{a}_{1,2}, \hat{a}_{1,2}^{+}$ are the raising and lowering operators for both particles. Thus,

$$\left\langle\Psi_{1,2}\left|\hat{x}_1^2\right|\Psi_{1,2}\right\rangle = \frac{1}{2}\frac{\hbar}{2m\omega}\left[\langle 1,2| + \langle 2,1|\right]\left(\hat{a}_1^2 + (\hat{a}_1^{+})^2 + \hat{a}_1\hat{a}_1^{+} + \hat{a}_1^{+}\hat{a}_1\right)\left[|1,2\rangle + |2,1\rangle\right].$$

By using $\hat{a}|n\rangle = \sqrt{n}|n-1\rangle$, $\hat{a}^{+}|n\rangle = \sqrt{n+1}|n+1\rangle$, we easily find for the first wave function

$$\left\langle\Psi_{1,2}\left|\hat{x}_1^2\right|\Psi_{1,2}\right\rangle = \left\langle\Psi_{1,2}\left|\hat{x}_2^2\right|\Psi_{1,2}\right\rangle = \frac{2\hbar}{m\omega},$$

$$\left\langle\Psi_{1,2}\left|2\hat{x}_1\hat{x}_2\right|\Psi_{1,2}\right\rangle = \frac{2}{2}\frac{\hbar}{2m\omega}\left[\langle 1,2| + \langle 2,1|\right](\hat{a}_1 + \hat{a}_1^{+})(\hat{a}_2 + \hat{a}_2^{+})\left[|1,2\rangle + |2,1\rangle\right],$$

$$\left\langle\Psi_{1,2}\left|2\hat{x}_1\hat{x}_2\right|\Psi_{1,2}\right\rangle = \frac{\hbar}{m\omega},$$

$$\langle(\hat{x}_1 - \hat{x}_2)^2\rangle = \left\langle\Psi_{1,2}\left|\left(\hat{x}_1^2 + \hat{x}_2^2 - 2\hat{x}_1\hat{x}_2\right)\right|\Psi_{1,2}\right\rangle$$

$$\langle(\hat{x}_1 - \hat{x}_2)^2\rangle = \left\langle\Psi_{1,2}\left|\hat{x}_1^2\right|\Psi_{1,2}\right\rangle + \left\langle\Psi_{1,2}\left|\hat{x}_2^2\right|\Psi_{1,2}\right\rangle - 2\left\langle\Psi_{1,2}\left|\hat{x}_1\hat{x}_2\right|\Psi_{1,2}\right\rangle = \frac{2\hbar}{m\omega}.$$

By using the second wave function, we find in a similar way,

$$\langle(\hat{x}_1 - \hat{x}_2)^2\rangle = \left\langle\Psi_{3,0}\left|\left(\hat{x}_1^2 + \hat{x}_2^2 - 2\hat{x}_1\hat{x}_2\right)\right|\Psi_{3,0}\right\rangle = \frac{4\hbar}{m\omega}.$$

Food for thought: Due to the fermionic statistics of the particles, the exchange term coming from $\langle\Psi|2\hat{x}_1\hat{x}_2|\Psi\rangle$ is negative. As a result, there is an effective repulsion between the two particles.

Problem 5.7

a) What is the exchange splitting energy of the energy levels of a system of two identical electrons when the Coulomb interaction between them is considered as a perturbation?

b) Calculate the effect of exchange splitting for two helium electrons in their ground state.

a) The wave function of the two noninteracting electrons can be written as

$$\Psi_S = \varphi_S(x_1,x_2)\chi_S = \frac{1}{\sqrt{2}}\left[\psi_{a_1}(x_1)\psi_{a_2}(x_2) + \psi_{a_1}(x_2)\psi_{a_2}(x_1)\right]\chi_S,$$

$$\Psi_T = \varphi_T(x_1,x_2)\chi_T = \frac{1}{\sqrt{2}}\left[\psi_{a_1}(x_1)\psi_{a_2}(x_2) - \psi_{a_1}(x_2)\psi_{a_2}(x_1)\right]\chi_T,$$

where χ_S, χ_T are the singlet and triplet states, separated from the composite quantum numbers $a_{1,2}$.

The perturbative Coulomb interaction between the electrons is of the form $U(|\boldsymbol{r}_1 - \boldsymbol{r}_2|)$. Within perturbation theory, the first-order correction to the energy of the noninteracting particles E_0 is:

$$\begin{aligned}
\delta E_n^{(1)} &= \langle \Psi_n(\boldsymbol{x}_1,\boldsymbol{x}_2)\left|U(|\hat{\boldsymbol{r}}_1 - \hat{\boldsymbol{r}}_2|)\right|\Psi_n(\boldsymbol{x}_1,\boldsymbol{x}_2)\rangle \\
&= \int d\boldsymbol{x}_1 d\boldsymbol{x}_2 \varphi_S^*(\boldsymbol{x}_1,\boldsymbol{x}_2)U(|\boldsymbol{x}_1 - \boldsymbol{x}_2|)\varphi_S(\boldsymbol{x}_1,\boldsymbol{x}_2)\langle\chi_S|\chi_S\rangle \\
&= \frac{1}{2}\int d\boldsymbol{x}_1 d\boldsymbol{x}_2 \left[\psi_{a_1}^*(\boldsymbol{x}_1)\psi_{a_2}^*(\boldsymbol{x}_2) + \psi_{a_1}^*(\boldsymbol{x}_2)\psi_{a_2}^*(\boldsymbol{x}_1)\right] \\
&\quad \times U(|\boldsymbol{x}_1 - \boldsymbol{x}_2|)\left[\psi_{a_1}(\boldsymbol{x}_1)\psi_{a_2}(\boldsymbol{x}_2) + \psi_{a_1}(\boldsymbol{x}_2)\psi_{a_2}(\boldsymbol{x}_1)\right] \\
&= \frac{1}{2}\int d\boldsymbol{x}_1 d\boldsymbol{x}_2 \left|\psi_{a_1}(\boldsymbol{x}_1)\right|^2\left|\psi_{a_2}(\boldsymbol{x}_2)\right|^2 U(|\boldsymbol{x}_1 - \boldsymbol{x}_2|) \\
&\quad + \frac{1}{2}\int d\boldsymbol{x}_1 d\boldsymbol{x}_2 \psi_{a_1}^*(\boldsymbol{x}_1)\psi_{a_2}^*(\boldsymbol{x}_2)U(|\boldsymbol{x}_1 - \boldsymbol{x}_2|)\psi_{a_1}(\boldsymbol{x}_2)\psi_{a_2}(\boldsymbol{x}_1) \\
&\quad + \frac{1}{2}\int d\boldsymbol{x}_1 d\boldsymbol{x}_2 \psi_{a_1}^*(\boldsymbol{x}_2)\psi_{a_2}^*(\boldsymbol{x}_1)U(|\boldsymbol{x}_1 - \boldsymbol{x}_2|)\psi_{a_1}(\boldsymbol{x}_1)\psi_{a_2}(\boldsymbol{x}_2) \\
&\quad + \frac{1}{2}\int d\boldsymbol{x}_1 d\boldsymbol{x}_2 \left|\psi_{a_1}(\boldsymbol{x}_2)\right|^2\left|\psi_{a_2}(\boldsymbol{x}_1)\right|^2 U(|\boldsymbol{x}_1 - \boldsymbol{x}_2|).
\end{aligned}$$

Using that $U(|\boldsymbol{x}_1 - \boldsymbol{x}_2|) = U(|\boldsymbol{x}_2 - \boldsymbol{x}_1|)$, we get

$$\begin{aligned}
\delta E_S^{(1)} &= \int d\boldsymbol{x}_1 d\boldsymbol{x}_2 \left|\psi_{a_1}(\boldsymbol{x}_1)\right|^2\left|\psi_{a_2}(\boldsymbol{x}_2)\right|^2 U(|\boldsymbol{x}_1 - \boldsymbol{x}_2|) \\
&\quad + \int d\boldsymbol{x}_1 d\boldsymbol{x}_2 \psi_{a_1}^*(\boldsymbol{x}_1)\psi_{a_2}^*(\boldsymbol{x}_2)U(|\boldsymbol{x}_1 - \boldsymbol{x}_2|)\psi_{a_1}(\boldsymbol{x}_2)\psi_{a_2}(\boldsymbol{x}_1).
\end{aligned}$$

For the second type of wave function, we find

$$\begin{aligned}
\delta E_T^{(1)} &= \int d\boldsymbol{x}_1 d\boldsymbol{x}_2 \left|\psi_{a_1}(\boldsymbol{x}_1)\right|^2\left|\psi_{a_2}(\boldsymbol{x}_2)\right|^2 U(|\boldsymbol{x}_1 - \boldsymbol{x}_2|) \\
&\quad - \int d\boldsymbol{x}_1 d\boldsymbol{x}_2 \psi_{a_1}^*(\boldsymbol{x}_1)\psi_{a_2}^*(\boldsymbol{x}_2)U(|\boldsymbol{x}_1 - \boldsymbol{x}_2|)\psi_{a_1}(\boldsymbol{x}_2)\psi_{a_2}(\boldsymbol{x}_1).
\end{aligned}$$

The exchange interaction is represented by

$$J = \int d\boldsymbol{x}_1 d\boldsymbol{x}_2 \psi_{a_1}^*(\boldsymbol{x}_1)\psi_{a_2}^*(\boldsymbol{x}_2)U(|\boldsymbol{x}_1 - \boldsymbol{x}_2|)\psi_{a_1}(\boldsymbol{x}_2)\psi_{a_2}(\boldsymbol{x}_1),$$

thus, the perturbative exchange splitting is $\pm J$ between the singlet and triplet spin states.

b) The composite wave function in this case is constructed from $\psi_0(\boldsymbol{x}_{1,2}) = \sqrt{\frac{8}{\pi a_0^3}} e^{-2r_{1,2}/a_0}$, where a_0 is the Bohr radius for the He atom and $r_n = \sqrt{x_n^2 + y_n^2 + z_n^2}$ is the radial component of the position vector:

$$J = \int d\boldsymbol{x}_1 d\boldsymbol{x}_2 \psi_0^*(\boldsymbol{x}_1)\psi_0^*(\boldsymbol{x}_2) U(|\boldsymbol{x}_1 - \boldsymbol{x}_2|)\psi_0(\boldsymbol{x}_2)\psi_0(\boldsymbol{x}_1)$$

$$= \int d\boldsymbol{x}_1 d\boldsymbol{x}_2 \psi_0^2(r_1)\psi_0^2(r_2)\frac{e^2}{|\boldsymbol{x}_1 - \boldsymbol{x}_2|} = \frac{64e^2}{\pi^2 a_0^6}\int d\boldsymbol{x}_1 d\boldsymbol{x}_2 \frac{e^{-\frac{4(r_1+r_2)}{a_0}}}{|\boldsymbol{x}_1 - \boldsymbol{x}_2|}.$$

Let's perform the integral over $\boldsymbol{x}_2$ first by writing $|\boldsymbol{x}_1 - \boldsymbol{x}_2| = \sqrt{r_1^2 + r_2^2 - 2r_1 r_2 \cos(\theta_2)}$, which yields

$$\int d\boldsymbol{x}_2 \frac{e^{-\frac{4r_2}{a_0}}}{\sqrt{r_1^2 + r_2^2 - 2r_1 r_2 \cos(\theta_2)}}$$

$$= \int_0^{2\pi} d\varphi_2 \int_0^{\pi} d\theta_2 \sin(\theta_2) \int_0^{\infty} dr_2 r_2^2 \frac{e^{-\frac{4r_2}{a_0}}}{\sqrt{r_1^2 + r_2^2 - 2r_1 r_2 \cos(\theta_2)}}$$

$$= 4\pi \int_0^{\infty} dr_2 r_2 e^{-\frac{4r_2}{a_0}} \left[\frac{r_2}{r_1}\Theta(r_1 - r_2) + \Theta(r_2 - r_1)\right]$$

$$= 4\pi \left[\frac{1}{r_1}\int_0^{r_1} e^{-\frac{4r_2}{a_0}} r_2^2 dr_2 + \int_{r_1}^{\infty} e^{-\frac{4r_2}{a_0}} r_2 dr_2\right] = \frac{\pi a_0^3}{8r_1}\left[1 - \left(1 + \frac{2r_1}{a_0}\right)e^{-\frac{4r_1}{a_0}}\right].$$

Next, we continue with the integration over the $\boldsymbol{r}_1$ variable:

$$J = \frac{64e^2\pi a_0^3}{8\pi^2 a_0^6}\int \left[1 - \left(1 + \frac{2r_1}{a_0}\right)e^{-\frac{4r_1}{a_0}}\right] e^{-\frac{4r_1}{a_0}} r_1 \sin(\theta_1) dr_1 d\theta_1 d\varphi_1 = \frac{8e^2}{\pi a_0^3} 4\pi \frac{5a_0^2}{128} = \frac{5e^2}{4a_0}.$$

> *Food for thought:* Can you provide numerical estimates for the energy of the two identical He electrons and their Coulomb perturbative interaction?

Problem 5.8

a) Consider two identical noninteracting spinless bosons with mass m as suggested in Tamvakis (2019). The particles are moving in 1D and their wave functions are $\psi_1(x) = \left(\frac{\alpha}{\pi}\right)^{\frac{1}{4}} e^{-\frac{\alpha(x-a)^2}{2}}; \psi_2(x) = \left(\frac{\alpha}{\pi}\right)^{\frac{1}{4}} e^{-\frac{\alpha(x+a)^2}{2}}$. What is the composite wave function for the two particles? What is the expectation value of $\langle(\widehat{x}_1 - \widehat{x}_2)^2\rangle$? Calculate the expectation energy of the composite state of the two particles.

b) Consider the same problem, but now take the case of two identical spinless fermions.

c) When can the two identical particles be considered as classical particles?

---

a) The wave function for the identical noninteracting bosons is

$$\Psi(x_1, x_2) = C[\psi_1(x_1)\psi_2(x_2) + \psi_2(x_1)\psi_1(x_2)].$$

The normalization constant C can be found from

$$
\begin{aligned}
1 = \langle \Psi \mid \Psi \rangle &= \int_{-\infty}^{+\infty} dx_1 \int_{-\infty}^{+\infty} dx_2 \left|\Psi(x_1,x_2)\right|^2 \\
&= |C|^2 \left[\int_{-\infty}^{+\infty} dx_1 \psi_1^2(x_1) \int_{-\infty}^{+\infty} dx_2 \psi_2^2(x_2) \right. \\
&\quad + \int_{-\infty}^{+\infty} dx_1 \psi_2^2(x_1) \int_{-\infty}^{+\infty} dx_2 \psi_1^2(x_2) \\
&\quad \left. + 2 \int_{-\infty}^{+\infty} dx_1 \psi_1(x_1)\psi_2(x_1) \int_{-\infty}^{+\infty} dx_2 \psi_1(x_2)\psi_2(x_2) \right] \\
&= |C|^2 \left[2 + 2A^2\right],
\end{aligned}
$$

where $A = e^{-\alpha a^2}$. Thus, the wave function becomes

$$
\Psi(x_1,x_2) = \frac{1}{2\sqrt{1+A^2}} \left[\psi_1(x_1)\psi_2(x_2) + \psi_2(x_1)\psi_1(x_2)\right].
$$

To find the expectation value of the square of the relative distance, we need to calculate

$$
\begin{aligned}
\left\langle (\widehat{x}_1 - \widehat{x}_2)^2 \right\rangle &= \int_{-\infty}^{+\infty} dx_1 \int_{-\infty}^{+\infty} dx_2 \left(x_1^2 + x_2^2 - 2x_1x_2\right) \left|\Psi(x_1,x_2)\right|^2 \\
&= \frac{1}{\alpha} + 2a^2 \left[1 + \tanh\left(\alpha a^2\right)\right].
\end{aligned}
$$

The expectation energy is $E = \int_{-\infty}^{+\infty} dx_1 \int_{-\infty}^{+\infty} dx_2 \left\langle \Psi(x_1,x_2) | \widehat{H} | \Psi(x_1,x_2) \right\rangle$, where $\widehat{H} = \frac{\widehat{p}_1^2}{2m} + \frac{\widehat{p}_2^2}{2m}$. Therefore,

$$
E = \frac{\alpha \hbar^2}{2m} \left(1 + \alpha a^2 \left[\tanh\left(\alpha a^2\right) - 1\right]\right).
$$

b) For the case of two identical noninteracting fermions, the wave function must be taken as $\Psi(x_1,x_2) = \frac{1}{2\sqrt{1-A^2}} \left[\psi_1(x_1)\psi_2(x_2) - \psi_2(x_1)\psi_1(x_2)\right]$. Then, $\langle (\widehat{x}_1 - \widehat{x}_2)^2 \rangle$ becomes

$$
\begin{aligned}
\langle (\widehat{x}_1 - \widehat{x}_2)^2 \rangle &= \int_{-\infty}^{+\infty} dx_1 \int_{-\infty}^{+\infty} dx_2 \left(x_1^2 + x_2^2 - 2x_1x_2\right) \left|\Psi(x_1,x_2)\right|^2 \\
&= \frac{1}{\alpha} + 2a^2 \left[1 + \frac{1}{\tanh\left(\alpha a^2\right)}\right].
\end{aligned}
$$

The energy of the composite fermion state for the $\widehat{H} = \frac{\widehat{p}_1^2}{2m} + \frac{\widehat{p}_2^2}{2m}$ Hamiltonian is

$$
E = \frac{\alpha \hbar^2}{2m} \left(1 + \alpha a^2 \left[\frac{1}{\tanh\left(\alpha a^2\right)} - 1\right]\right).
$$

c) When the particles are classical, there is no overlap between the two wave functions. In fact, as the particles move in space, their energy changes depending on the overlap of the two wave functions, which is controlled by $A^2 = e^{-2\alpha a^2}$. The limit of $2\alpha a^2 \gg 1$ gives two particles with no overlap, meaning they can be considered as classical distinguishable particles. In this case, the wave function and corresponding energy are

$$\Psi(x_1,x_2) = \psi_1(x_1)\psi_2(x_2), \quad E = \frac{\alpha\hbar^2}{2m},$$

since in the limit of $\alpha a^2 \to \infty, \tanh(\alpha a^2) \to 1$.

To find the expectation value of the square of the relative distance, we calculate

$$\left\langle (\hat{x}_1 - \hat{x}_2)^2 \right\rangle = \int_{-\infty}^{+\infty} dx_1 \int_{-\infty}^{+\infty} dx_2 \left(x_1^2 + x_2^2 - 2x_1x_2\right) \left|\Psi(x_1,x_2)\right|^2 = \frac{1}{\alpha} + 4a^2.$$

Problem 5.9

Consider two identical $s = 1/2$ fermions in an infinite square well given as $V(x) = \begin{cases} 0, & 0 < x < a \\ \infty, & otherwise \end{cases}$. The following perturbation is then introduced, $\hat{V}_1(\hat{x}_1,\hat{x}_2,t) = \beta\delta(\hat{x}_1 - \hat{x}_2)\hat{S}_{1z}\hat{S}_{2z}e^{-t/\tau}$, where x_1, x_2 are the positions for the two particles and $\hat{S}_{1z}, \hat{S}_{2z}$ are the z-projections of the spin operators for the two particles.

Within the first order of perturbation theory, find the probability for the identical particles to transition from their ground state to their first and second excited states.

This problem combines identical particles and time-dependent perturbation theory because the perturbation interaction is time-dependent. We remember that the eigenstates and eigenenergies for a particle in this infinite square well are $\varphi_n(x) = \sqrt{\frac{2}{a}}\, sin\left(\frac{n\pi x}{a}\right)$; $E_n = \frac{n^2\pi^2\hbar^2}{2ma^2}$. We also recall that the first-order time-dependent transition probability is given by the first coefficient in the Dyson series, given in what follows [Sakurai].

The total wave function for the identical $s = 1/2$ fermions must be antisymmetric. Therefore, the ground state wave function and energy are

$$|\Psi_0\rangle = \varphi_1(x_1)\varphi_1(x_2)\chi_S; \text{ where } \chi_S = \frac{1}{\sqrt{2}}[|\uparrow\downarrow\rangle - |\downarrow\uparrow\rangle],$$

$$E_0 = \frac{\pi^2\hbar^2}{2ma^2}\left(1^2 + 1^2\right) = \frac{\pi^2\hbar^2}{ma^2}.$$

The wave function and energy for the first excited state are

$$|\Psi_{1,S}\rangle = \frac{1}{\sqrt{2}}\left[\varphi_1(x_1)\varphi_2(x_2) + \varphi_2(x_1)\varphi_1(x_2)\right]\chi_S$$

$$\text{or} \quad |\Psi_{1,T}\rangle = \frac{1}{\sqrt{2}}\left[\varphi_1(x_1)\varphi_2(x_2) - \varphi_2(x_1)\varphi_1(x_2)\right]\chi_T,$$

$$\text{where } \chi_T = \left\{ \begin{array}{c} |\uparrow\uparrow\rangle \\ \frac{1}{\sqrt{2}}[|\uparrow\downarrow\rangle + |\downarrow\uparrow\rangle] \\ |\downarrow\downarrow\rangle \end{array} \right\},$$

$$E_1 = \frac{\pi^2\hbar^2}{2ma^2}\left(2^2 + 1^2\right) = \frac{\pi^2\hbar^2}{2ma^2}\left(1^2 + 2^2\right) = \frac{5\pi^2\hbar^2}{2ma^2}.$$

The wave function and energy for the second excited state are

$$|\Psi_2\rangle = \varphi_2(x_1)\varphi_2(x_2)\chi_S,$$
$$E_2 = \frac{\pi^2\hbar^2}{2ma^2}\left(2^2+2^2\right) = \frac{4\pi^2\hbar^2}{ma^2}.$$

To find the time-dependent transition probability, we use the first term in the Dyson's series for the evolution operator $\widehat{U}(t,t_0) = \widehat{\mathfrak{T}}e^{-\frac{i}{\hbar}\int_{t_0}^{t}\widehat{V}_I(\tau)d\tau}$, with $\widehat{V}_I(t) = e^{it\frac{\widehat{H}_0}{\hbar}}\widehat{V}_1(t)e^{-it\frac{\widehat{H}_0}{\hbar}}$ the perturbation in the interaction picture (see, for example, Sakurai & Napolitano, 2017) since the potential is time-dependent,

$$\widehat{U}(t,t_0) = \widehat{\mathbb{1}} - \frac{i}{\hbar}\int_{t_0}^{t}\widehat{V}_I(t_1)dt_1 + \left(\frac{-i}{\hbar}\right)^2\int_{t_0}^{t}dt_1\int_{t_0}^{t_1}dt_2\widehat{V}_I(t_1)\widehat{V}_I(t_2)+\ldots,$$
$$\begin{aligned}
\left|\Psi_0(t)\right\rangle &= \widehat{U}(t,t_0)\left|\Psi_0(t_0)\right\rangle \\
&= \left|\Psi_0(t_0)\right\rangle - \frac{i}{\hbar}\int_{t_0}^{t}\widehat{V}_I(t_1)\left|\Psi_0(t_0)\right\rangle dt_1 \\
&\quad + \left(\frac{-i}{\hbar}\right)^2\int_{t_0}^{t}dt_1\int_{t_0}^{t_1}dt_2\widehat{V}_I(t_1)\widehat{V}_I(t_2)\left|\Psi_0(t_0)\right\rangle+\ldots, \\
\left\langle\Psi_1(t)\mid\Psi_0(t)\right\rangle &= \left\langle\Psi_1(t)\middle|\widehat{U}(t,t_0)\middle|\Psi_0(t_0)\right\rangle \\
&= \left\langle\Psi_1(t)\mid\Psi_0(t_0)\right\rangle - \frac{i}{\hbar}\int_{t_0}^{t}\left\langle\Psi_1(t)\middle|\widehat{V}_I(t_1)\middle|\Psi_0(t_0)\right\rangle dt_1 \\
&\quad + \left(\frac{-i}{\hbar}\right)^2\int_{t_0}^{t}dt_1\int_{t_0}^{t_1}dt_2\left\langle\Psi_1(t)\middle|\widehat{V}_I(t_1)\widehat{V}_I(t_2)\middle|\Psi_0(t_0)\right\rangle+\ldots
\end{aligned}$$

Then, using that $|\Psi_n(t)\rangle = |\Psi_n\rangle$ in the interaction picture when $\widehat{H}\neq f(t)$, the time-dependent coefficients for ground-first excited state transitions are

$$\begin{aligned}
C^{(1)}_{0\to 1,s}(t) &= -\frac{i}{\hbar}\int_0^t e^{i\frac{E_1-E_0}{\hbar}t_1}\left\langle\Psi_{1,S}\middle|\widehat{V}_1(x_1,x_2,t_1)\middle|\Psi_0\right\rangle dt_1 \\
&= -\frac{i\beta}{\hbar}\int_0^t e^{i\frac{E_1-E_0}{\hbar}t_1}e^{-t_1/\tau}dt_1\int_0^a dx_1\int_0^a dx_2\frac{1}{\sqrt{2}} \\
&\quad\times\left[\varphi_1^*(x_1)\varphi_2^*(x_2)+\varphi_2^*(x_1)\varphi_1^*(x_2)\right]\delta(x_1-x_2)\varphi_1(x_1)\varphi_1(x_2)\left\langle\chi_S\middle|\widehat{S}_{1z}\widehat{S}_{2z}\middle|\chi_S\right\rangle
\end{aligned}$$

and

$$\begin{aligned}
C^{(1)}_{0\to 1,T}(t) &= -\frac{i}{\hbar}\int_0^t e^{i\frac{E_1-E_0}{\hbar}t_1}\left\langle\Psi_{1,T}\middle|\widehat{V}_1(x_1,x_2,t_1)\middle|\Psi_0\right\rangle dt_1 \\
&= -\frac{i\beta}{\hbar}\int_0^t e^{i\frac{E_1-E_0}{\hbar}t_1}e^{-t_1/\tau}dt_1\int_0^a dx_1\int_0^a dx_2\frac{1}{\sqrt{2}} \\
&\quad\times\left[\varphi_1^*(x_1)\varphi_2^*(x_2)-\varphi_2^*(x_1)\varphi_1^*(x_2)\right]\delta(x_1-x_2)\varphi_1(x_1)\varphi_1(x_2)\left\langle\chi_T\middle|\widehat{S}_{1z}\widehat{S}_{2z}\middle|\chi_S\right\rangle.
\end{aligned}$$

Since $\left\langle\chi_S\middle|\widehat{S}_{1z}\widehat{S}_{2z}\middle|\chi_S\right\rangle = \frac{-1}{4}$ and $\left\langle\chi_T\middle|\widehat{S}_{1z}\widehat{S}_{2z}\middle|\chi_S\right\rangle = 0$, only the first integral survives, giving the following result:

$$C^{(1)}_{0\to 1,S}(t) = \frac{i\beta}{2\sqrt{2}\hbar}\int_0^t e^{i\frac{E_1-E_0}{\hbar}t_1}e^{-t_1/\tau}dt_1\int_0^a dx_1\varphi_1^*(x_1)\varphi_2^*(x_1)\varphi_1(x_1)\varphi_1(x_1),$$

where

$$\int_0^a dx_1\varphi_1^*(x_1)\varphi_2^*(x_1)\varphi_1(x_1)\varphi_1(x_1)=\frac{4}{a^2}\int_0^a dx_1\left[\sin\left(\frac{\pi x_1}{a}\right)\right]^3\sin\left(\frac{2\pi x_1}{a}\right)=0.$$

Thus, we conclude that the transition $|\Psi_0\rangle\to|\Psi_1\rangle$ is forbidden in the first order of time-dependent perturbation theory.

For the second excited state $|\Psi_2\rangle$, we have

$$C^{(1)}_{0\to 2}(t)=-\frac{i}{\hbar}\int_0^t e^{i\frac{E_2-E_0}{\hbar}t_1}\left\langle\Psi_2\left|\widehat{V}_1(x_1,x_2,t_1)\right|\Psi_0\right\rangle dt_1=-\frac{i\beta}{\hbar}\int_0^t e^{i\frac{E_2-E_0}{\hbar}t_1}e^{-t_1/\tau}dt_1\int_0^a dx_1$$
$$\times\int_0^a dx_2\varphi_2^*(x_1)\varphi_2^*(x_2)\delta(x_1-x_2)\varphi_1(x_1)\varphi_1(x_2)\left\langle\chi_S\left|\widehat{S}_{1z}\widehat{S}_{2z}\right|\chi_S\right\rangle$$
$$=\frac{i\beta}{4\hbar}\int_0^t e^{i\frac{E_2-E_0}{\hbar}t_1}e^{-t_1/\tau}dt_1\int_0^a dx_1\varphi_2^*(x_1)\varphi_2^*(x_1)\varphi_1(x_1)\varphi_1(x_1),$$

$$\int_0^a dx_1\varphi_2^*(x_1)\varphi_2^*(x_1)\varphi_1(x_1)\varphi_1(x_1)=\frac{4}{a^2}\int_0^a dx_1\left[\sin\left(\frac{\pi x_1}{a}\right)\right]^2\left[\sin\left(\frac{2\pi x_1}{a}\right)\right]^2=\frac{1}{a},$$

$$C^{(1)}_{0\to 2}(t)=\frac{i\beta}{4a\hbar}\int_0^t e^{i\frac{E_2-E_0}{\hbar}t_1}e^{-t_1/\tau}dt_1=\frac{i\beta\tau}{4a}\frac{1-e^{-t\left(\frac{1}{\tau}+i\frac{E_0-E_2}{\hbar}\right)}}{\hbar+i\tau(E_0-E_2)}.$$

Therefore, the probability becomes

$$P_{0\to 2}(t)=\left|C^{(1)}_{0\to 2}(t)\right|^2=\frac{\tau^2\beta^2}{16a^2}\frac{1+e^{-2t/\tau}-e^{-t/\tau}\cos\left(t\frac{E_0-E_2}{\hbar}\right)}{\hbar^2+\tau^2(E_0-E_2)^2},$$

$$\frac{1}{T}=\frac{E_2-E_0}{\hbar}=\frac{\pi^2\hbar}{ma^2},$$

$$P_{0\to 2}(t)=\left|C^{(1)}_{0\to 2}(t)\right|^2=\frac{\tau^2\beta^2T^2}{16a^2\hbar^2}\frac{1+e^{-2t/\tau}-e^{-t/\tau}\cos\left(\frac{t}{T}\right)}{T^2+\tau^2}.$$

> *Food for thought:* Can you work out the probability within second order perturbation theory for the transition from the ground state to the first excited state?

Problem 5.10

Two identical spinless bosons are in their ground state. Each particle has the same electric charge, and it interacts with a charged center, but the interaction between the two bosons is neglected.

Calculate the expectation values of (a) electric dipole moment $\widehat{\boldsymbol{d}}$, (b) the magnetic dipole moment $\widehat{\boldsymbol{\mu}}$, (c) and the electrical quadrupole moment $\widehat{\boldsymbol{Q}}_{ij}$.

This problem requires the definitions of electric and magnetic dipoles, and electric quadrupoles. Since the particles interact via the Coulomb interaction, it is convenient to represent the eigenstates of each particle as hydrogenic orbitals. The geometrical representation with characteristic distances is given schematically in Figure 5.1.

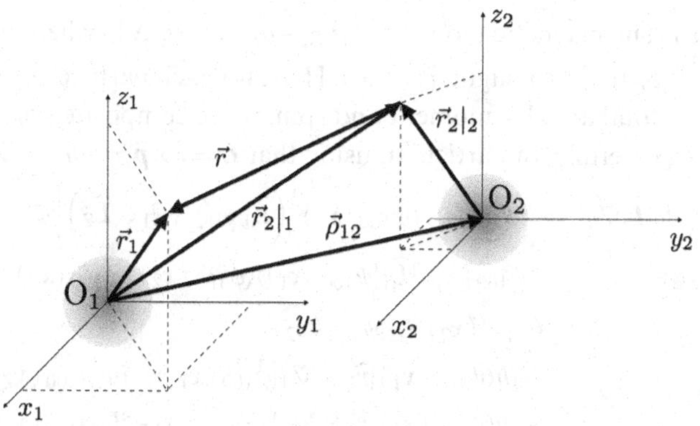

Figure 5.1 Geometrical schematic of the two dipoles located in O_1 and O_2 with characteristic distances.

For bosons, the composite wave function is symmetric, thus for the ground state of the two spinless particles, we have

$$|\Psi_0\rangle = \psi_{100}(\boldsymbol{x}_1)\psi_{100}(\boldsymbol{x}_2)\rangle,$$

where $\psi_{nlm}(\boldsymbol{x})$ are the eigenstates of the hydrogenic atom.

An important point here is the reference frame for evaluating all quantities. The electric dipole moment operator is $\widehat{\boldsymbol{d}} = q\widehat{\boldsymbol{\rho}}_{12}$, where q is the charge and $\widehat{\boldsymbol{\rho}}_{12}$ is the distance vector operator between the atoms represented as circular formations in Figure 5.1). To find the expectation value of the electric dipole moment, then

$$\begin{aligned}\langle\Psi_0|\widehat{\boldsymbol{d}}|\Psi_0\rangle &= q\langle\psi_{100}(\boldsymbol{x}_1)\psi_{100}(\boldsymbol{x}_2)\,|\widehat{\boldsymbol{\rho}}_{12}|\,\psi_{100}(\boldsymbol{x}_1)\psi_{100}(\boldsymbol{x}_2)\rangle \\ &= q\langle\psi_{100}(\boldsymbol{x}_1)\psi_{100}(\boldsymbol{x}_2)\,|\widehat{\boldsymbol{r}}_1 - \widehat{\boldsymbol{r}}_2|_1|\psi_{100}(\boldsymbol{x}_1)\psi_{100}(\boldsymbol{x}_2)\rangle,\end{aligned}$$

where the subscript 1 indicates that the reference frame for the calculations is one associated with particle (1), as shown in Figure 5.1).

The relationship between the system of reference centered at particle 1 and the one centered at the position of particle 2 is

$$\widehat{\boldsymbol{r}}_2\big|_1 = \widehat{\boldsymbol{\rho}}_{12} + \widehat{\boldsymbol{r}}_2\big|_2.$$

a) From here, we find

$$\begin{aligned}\langle\Psi_0|\widehat{\boldsymbol{d}}|\Psi_0\rangle &= q\big(\langle\psi_{100}(\boldsymbol{x}_1)\,|\widehat{\boldsymbol{r}}_1|\,\psi_{100}(\boldsymbol{x}_1)\rangle\langle\psi_{100}(\boldsymbol{x}_2)\,|\,\psi_{100}(\boldsymbol{x}_2)\rangle \\ &\quad - \langle\psi_{100}(\boldsymbol{x}_2)\,|\widehat{\boldsymbol{r}}_2\big|_1\,|\,\psi_{100}(\boldsymbol{x}_2)\rangle\langle\psi_{100}(\boldsymbol{x}_1)\,|\,\psi_{100}(\boldsymbol{x}_1)\rangle\big) \\ &= q\big(\langle\psi_{100}(\boldsymbol{x}_1)\,|\widehat{\boldsymbol{r}}_1|\,\psi_{100}(\boldsymbol{x}_1)\rangle - \langle\psi_{100}(\boldsymbol{x}_2)\,|\widehat{\boldsymbol{r}}_2|_1|\psi_{100}(\boldsymbol{x}_2)\rangle\big) \\ &= q\big(\langle\psi_{100}(\boldsymbol{x}_1)\,|\widehat{\boldsymbol{r}}_1|\,\psi_{100}(\boldsymbol{x}_1)\rangle - \langle\psi_{100}(\boldsymbol{x}_2)\,|\widehat{\boldsymbol{\rho}}_{12} + \widehat{\boldsymbol{r}}_2|_2|\psi_{100}(\boldsymbol{x}_2)\rangle\big) \\ &= \big(\langle\psi_{100}(\boldsymbol{x}_1)\,|\widehat{\boldsymbol{r}}_1|\psi_{100}(\boldsymbol{x}_1)\rangle - \boldsymbol{\rho}_{12}\langle\psi_{100}(\boldsymbol{x}_2)\,|\psi_{100}(\boldsymbol{x}_2)\rangle \\ &\quad - \langle\psi_{100}(\boldsymbol{x}_2)\,|\widehat{\boldsymbol{r}}_2|_2|\psi_{100}(\boldsymbol{x}_2)\rangle\big) = q\,(\boldsymbol{0} - \boldsymbol{\rho}_{12} - \boldsymbol{0}) = q\boldsymbol{\rho}_{12},\end{aligned}$$

where we have used that $\langle\psi_{100}(\boldsymbol{x}_1)\,|\widehat{\boldsymbol{r}}_1|\psi_{100}(\boldsymbol{x}_1)\rangle = \boldsymbol{0}$ because of parity.

b) The magnetic moment is $\widehat{\boldsymbol{\mu}} = -\mu_B\left(\widehat{\boldsymbol{L}} + g_e\widehat{\boldsymbol{S}}\right)$ (where μ_B is the Bohr magneton, and g_e the gyromagnetic ratio). Here, for each particle, we have $\boldsymbol{L} = \boldsymbol{0}$ and $\boldsymbol{S} = \boldsymbol{0}$, but the total angular moment and spin of the composite system has to be calculated more carefully. In particular, using that $\widehat{\boldsymbol{L}} = \widehat{\boldsymbol{r}}\times\widehat{\boldsymbol{p}} = -i\hbar\widehat{\boldsymbol{r}}\times\boldsymbol{\nabla}$, we obtain

$$\begin{aligned}
\langle\Psi_0|\widehat{\boldsymbol{L}}|\Psi_0\rangle &= \left\langle\psi_{100}(\boldsymbol{x}_1)\psi_{100}(\boldsymbol{x}_2)\middle|\left(\widehat{\boldsymbol{L}}_1\otimes\widehat{\mathbb{1}}_2 + \widehat{\mathbb{1}}_1\otimes\widehat{\boldsymbol{L}}_2\right)\middle|\psi_{100}(\boldsymbol{x}_1)\psi_{100}(\boldsymbol{x}_2)\right\rangle \\
&= \langle\psi_{100}(\boldsymbol{x}_1)|\widehat{\boldsymbol{L}}_1|\psi_{100}(\boldsymbol{x}_1)\rangle\langle\psi_{100}(\boldsymbol{x}_2)\mid\psi_{100}(\boldsymbol{x}_2)\rangle + \langle\psi_{100}(\boldsymbol{x}_1)\mid\psi_{100}(\boldsymbol{x}_1)\rangle \\
&\quad \langle\psi_{100}(\boldsymbol{x}_2)|\widehat{\boldsymbol{L}}_2|\psi_{100}(\boldsymbol{x}_2)\rangle \\
&= -i\hbar\langle\psi_{100}(\boldsymbol{x}_1)|\widehat{\boldsymbol{r}}_1\times\boldsymbol{\nabla}_1|\psi_{100}(\boldsymbol{x}_1)\rangle - i\hbar\langle\psi_{100}(\boldsymbol{x}_2)|\widehat{\boldsymbol{r}}_2|_1\times\boldsymbol{\nabla}_2\mid\psi_{100}(\boldsymbol{x}_2)\rangle \\
&= -i\hbar\langle\psi_{100}(\boldsymbol{x}_1)|\widehat{\boldsymbol{r}}_1\times\boldsymbol{\nabla}_1|\psi_{100}(\boldsymbol{x}_1)\rangle - i\hbar\langle\psi_{100}(\boldsymbol{x}_2)|\left(\widehat{\boldsymbol{\rho}}_{12} + \widehat{\boldsymbol{r}}_2\big|_2\right)\times\boldsymbol{\nabla}_2|\psi_{100}(\boldsymbol{x}_2)\rangle \\
&= -i\hbar\langle\psi_{100}(\boldsymbol{x}_1)|\widehat{\boldsymbol{r}}_1\times\boldsymbol{\nabla}_1|\psi_{100}(\boldsymbol{x}_1)\rangle - i\hbar\langle\psi_{100}(\boldsymbol{x}_2)|\widehat{\boldsymbol{\rho}}_{12}\times\boldsymbol{\nabla}_2|\psi_{100}(\boldsymbol{x}_2)\rangle \\
&\quad - i\hbar\langle\psi_{100}(\boldsymbol{x}_2)|\widehat{\boldsymbol{r}}_2|_2\times\boldsymbol{\nabla}_2\mid\psi_{100}(\boldsymbol{x}_2)\rangle \\
&= \langle\psi_{100}(\boldsymbol{x}_1)|\widehat{\boldsymbol{L}}_1|\psi_{100}(\boldsymbol{x}_1)\rangle + \boldsymbol{\rho}_{12}\times\langle\psi_{100}(\boldsymbol{x}_2)|\widehat{\boldsymbol{p}}_2|\psi_{100}(\boldsymbol{x}_2)\rangle \\
&\quad + \langle\psi_{100}(\boldsymbol{x}_2)|\widehat{\boldsymbol{L}}_2|\psi_{100}(\boldsymbol{x}_2)\rangle.
\end{aligned}$$

The above result indicates that the expectation value of the total angular momentum is the sum of the angular momentum of each particle (in its own reference frame) plus the "classical" contribution due to the particular geometry of the system. Therefore, since the angular momentum is odd under parity transformation while $\psi_{100}(\boldsymbol{x})$ is even, we have $\langle\psi_{100}(\boldsymbol{x})|\widehat{\boldsymbol{L}}|\psi_{100}(\boldsymbol{x})\rangle = \boldsymbol{0}$. Due to the same argument, we have that $\langle\psi_{100}(\boldsymbol{x})|\widehat{\boldsymbol{p}}|\psi_{100}(\boldsymbol{x})\rangle = \boldsymbol{0}$ as well; therefore, we have

$$\langle\Psi_0|\widehat{\boldsymbol{L}}|\Psi_0\rangle = \boldsymbol{0}.$$

The total spin of the system is obtained as

$$\begin{aligned}
\langle\Psi_0|\widehat{\boldsymbol{S}}|\Psi_0\rangle &= \langle\psi_{100}(\boldsymbol{x}_1)\psi_{100}(\boldsymbol{x}_2)|(\widehat{\boldsymbol{S}}_1\otimes\widehat{\mathbb{1}}_2 + \widehat{\mathbb{1}}_1\otimes\widehat{\boldsymbol{S}}_2)|\psi_{100}(\boldsymbol{x}_1)\psi_{100}(\boldsymbol{x}_2)\rangle \\
&= \langle\psi_{100}(\boldsymbol{x}_1)|\widehat{\boldsymbol{S}}_1|\psi_{100}(\boldsymbol{x}_1)\rangle\langle\psi_{100}(\boldsymbol{x}_2)\mid\psi_{100}(\boldsymbol{x}_2)\rangle \\
&\quad + \langle\psi_{100}(\boldsymbol{x}_1)\mid\psi_{100}(\boldsymbol{x}_1)\rangle\langle\psi_{100}(\boldsymbol{x}_2)|\widehat{\boldsymbol{S}}_2|\psi_{100}(\boldsymbol{x}_2)\rangle \\
&= \langle\psi_{100}(\boldsymbol{x}_1)|\widehat{\boldsymbol{S}}_1|\psi_{100}(\boldsymbol{x}_1)\rangle + \langle\psi_{100}(\boldsymbol{x}_2)|\widehat{\boldsymbol{S}}_2|\psi_{100}(\boldsymbol{x}_2)\rangle.
\end{aligned}$$

From here, using that the two particles have spin 0, $\langle\psi_{100}(\boldsymbol{x}_1)|\widehat{\boldsymbol{S}}_1|\psi_{100}(\boldsymbol{x}_1)\rangle = \langle\psi_{100}(\boldsymbol{x}_2)|\widehat{\boldsymbol{S}}_2|\psi_{100}(\boldsymbol{x}_2)\rangle = \boldsymbol{0}$, we obtain that $\langle\Psi_0|\widehat{\boldsymbol{S}}|\Psi_0\rangle = \boldsymbol{0}$.

Then, we conclude that

$$\langle\Psi_0|\widehat{\boldsymbol{\mu}}|\Psi_0\rangle = \boldsymbol{0}.$$

c) The quadruple moment is a tensor with components $\widehat{Q}_{ij} = q\left(3\widehat{r}_i\widehat{r}_j - \delta_{ij}\widehat{\boldsymbol{r}}^2\right)$, where $\widehat{r}_i$ are the x, y, z components of $\widehat{\boldsymbol{r}} = \widehat{\boldsymbol{r}}_1 - \widehat{\boldsymbol{r}}_2\big|_1 = \widehat{\boldsymbol{r}}_1 - \left(\widehat{\boldsymbol{\rho}}_{12} + \widehat{\boldsymbol{r}}_2\big|_2\right) = \widehat{\boldsymbol{r}}_1 - \widehat{\boldsymbol{r}}_2\big|_2 - \widehat{\boldsymbol{\rho}}_{12}$, where the subscript 2 indicates the reference frame for particle 2. From here, using that the wave function normalizations and the parity arguments are as discussed above, we find that

$$\begin{aligned}
\langle\Psi_0|\widehat{r}_i\widehat{r}_j|\Psi_0\rangle &= \langle\psi_{100}(\boldsymbol{x}_1)\psi_{100}(\boldsymbol{x}_2)|\left(\widehat{r}_{1,i} - \widehat{r}_{2,i}\big|_2 - \widehat{\rho}_{12,i}\right)\left(\widehat{r}_{1,j} - \widehat{r}_{2,j}\big|_2 - \widehat{\rho}_{12,j}\right)|\psi_{100}(\boldsymbol{x}_1)\psi_{100}(\boldsymbol{x}_2)\rangle \\
&= \langle\psi_{100}(\boldsymbol{x}_1)|\widehat{r}_{1,i}\widehat{r}_{1,j}|\psi_{100}(\boldsymbol{x}_1)\rangle + \langle\psi_{100}(\boldsymbol{x}_2)|\widehat{r}_{2,i}\big|_2\widehat{r}_{2,j}\big|_2|\psi_{100}(\boldsymbol{x}_2)\rangle + \rho_{12,i}\rho_{12,j},
\end{aligned}$$

where $\langle\psi_{100}(\boldsymbol{x}_1)|\psi_{100}(\boldsymbol{x}_1)\rangle = 1$ and the parity property $\langle\psi_{100}(\boldsymbol{x}_1)|\widehat{r}_{1,i}|\psi_{100}(\boldsymbol{x}_1)\rangle = \langle\psi_{100}(\boldsymbol{x}_2)\left|\widehat{r}_{2,i|2}\right|\psi_{100}(\boldsymbol{x}_2)\rangle = 0$ are taken into account.

Using that $\langle\psi_{100}(\boldsymbol{x})\left|\widehat{r}_i\widehat{r}_j\right|\psi_{100}(\boldsymbol{x})\rangle = a_B^2\delta_{ij}$ and the spherical symmetry of $\psi_{100}(\boldsymbol{x})$, for all components $i = \{1,2,3\}$, we further find

$$\begin{aligned}\langle\Psi_0|\widehat{r}_i\widehat{r}_j|\Psi_0\rangle &= 2\langle\psi_{100}(\boldsymbol{x}_1)|\widehat{r}_{1,i}\widehat{r}_{1,j}|\psi_{100}(\boldsymbol{x}_1)\rangle + \rho_{12,i}\rho_{12,j}\\ &= 2\delta_{ij}\langle\psi_{100}(\boldsymbol{x})|\widehat{r}_i^2|\psi_{100}(\boldsymbol{x})\rangle + \rho_{12,i}\rho_{12,j} = 2\delta_{ij}a_B^2 + \rho_{12,i}\rho_{12,j}.\end{aligned}$$

Therefore,

$$\begin{aligned}\langle\Psi_0\left|\widehat{Q}_{ij}\right|\Psi_0\rangle &= \langle\Psi_0\left|q\left(3\widehat{r}_i\widehat{r}_j - \delta_{ij}\widehat{\boldsymbol{r}}^2\right)\right|\Psi_0\rangle = q\left(3\langle\Psi_0\left|\widehat{r}_i\widehat{r}_j\right|\Psi_0\rangle - \delta_{ij}\langle\Psi_0\left|\widehat{\boldsymbol{r}}^2\right|\Psi_0\rangle\right)\\ &= q\left(3\left[2\delta_{ij}a_B^2 + \rho_{12,i}\rho_{12,j}\right] - \delta_{ij}\left[6a_B^2 + \rho_{12}^2\right]\right) = q\left(3\rho_{12,i}\rho_{12,j} - \delta_{ij}\rho_{12}^2\right).\end{aligned}$$

Thus, we conclude that the quadrupole moment of two identical spinless bosons in their ground state only depends on their relative distance.

Problem 5.11

The purpose of this problem is to illustrate the application of Hund's rules in the construction of the electronic shell structure of atoms from the periodic table. The atomic shell structure is constructed by considering that the electrons are identical *fermionic* particles placed in the Coulomb potential from the nucleus and the electron–electron interactions are neglected.

Often in chemistry and physics, electronic shell configurations $1s; 2s; 2p; \ldots$. are convenient, in other times the following notation ${}^{2S+1}L_J$ is more suitable, where S is the total spin, J is the grand total angular momentum, and L is the letter equivalent of the orbital angular momentum.

First Hund's rule – state with highest S has the lowest energy.

Second Hund's rule – for a given S, the state with highest L has the lowest energy.

Third Hund's rule – for a less than half-filled subshell (n,ℓ) one has $J = |L - S|$; otherwise, $J = L + S$.

Conveniently, these rules are put together in a working scheme (shown to the right in Figure 5.2) according to Madelung (Levine, 2008; Szabo & Ostlund, 1989). Simply by following the arrows starting with the first shell for $\ell = 0$, it is quite easy to construct the electronic shell configurations of all atoms in the periodic table. For practice,

a) Construct the electronic shell configuration for nitrogen and give the appropriate ${}^{2S+1}L_J$ notation;
b) Construct the electronic shell configuration for iron and give the appropriate ${}^{2S+1}L_J$ notation.

---

a) Nitrogen has seven identical electrons, thus using the above scheme the electron spin distribution is given schematically as follows:

$(1s)^2$	↑↓		
$(2s)^2$	↑↓		
$(2p)^3$	↑	↑	↑

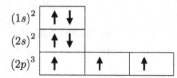

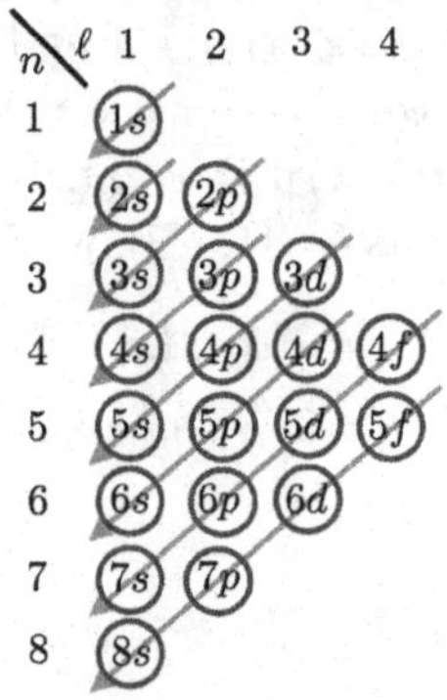

Figure 5.2 Schematic of the atomic shell structure.

In the above construction, we have used the following:

First Hund's rule – the unfilled *p*-shell will have to accommodate first the same-spin orientation, so $S = \frac{3}{2}$ and $S_z = \pm\frac{3}{2}$.

Second Hund's rule – for this $S = \frac{3}{2}$ the state with the highest angular momentum is $L = 0$, because only in this case can the spatial wave function be antisymmetric. We can check this by comparing the quantum numbers of these three *p*-electrons (labeled as $1, 2, 3$), as we have done in the following table below:

	1	**2**	**3**
n	2	2	2
s	1/2	1/2	1/2
s_z	1/2	1/2	1/2
l	1	1	1
m	1	0	−1

By the Pauli exclusion principle, we cannot have two electrons with the same quantum numbers; therefore, the only possible quantum number *m* must be different for the three electrons. As a consequence, the projection over the *z*-axis of the composite angular momentum is always zero, and this fact limits the composite angular momentum as zero as well, thus $L = 0$.

Third Hund's rule – $J = |L+S|$ since the $(2p)$ shell is half-filled, thus we take $J = |L+S| = \left|0+\frac{3}{2}\right| = \frac{3}{2}$.

So, the electronic shell configuration is

$$(1s)^2(2s)^2(2p)^3 \text{ and } {}^4S_{3/2}.$$

Note that when the shells are half or completely filled, for example $(3p)^3$, $(3p)^6$, $(3d)^5$, $(3d)^{10}$, the second Hund's rule always yields $L = 0$ because of the necessity to antisymmetrize the spatial part of the wave function since all electrons are up ($m_s = \frac{1}{2}$).

b) Construct the electron shell structure and give the $^{2S+1}L_J$ notation for iron.

Iron has 26 identical electrons; thus, using the preceding scheme, we write

$(1s)^2$	↑↓				
$(2s)^2$	↑↓				
$(2p)^6$	↑↓	↑↓	↑↓		
$(3s)^2$	↑↓				
$(3p)^6$	↑↓	↑↓	↑↓		
$(4s)^2$	↑↓				
$(3d)^6$	↑↓	↑	↑	↑	↑

First Hund's rule – the unfilled d-shell will have to accommodate first the same-spin orientation, so $S = 2$ and $S_z = \pm 2$.

Second Hund's rule – for this $S = 2$ the state with the highest angular momentum is $L = 2$. We can check this by comparing the quantum numbers of the four p-electrons (labeled as $1, 2, 3, 4$), remembering that all quantum numbers have to be different because of the Pauli exclusion principle:

	1	2	3	4
n	3	3	3	3
s	1/2	1/2	1/2	1/2
s_z	1/2	1/2	1/2	1/2
l	2	2	2	2
m	2	1	0	−1

Third Hund's rule – as the $(3d)$ shell is more than half-filled, we have $J = L + S = 4$.

Following Hund's rules as just shown, we get

$$(1s)^{2}(2s)^{2}(2p)^{6}(3s)^{2}(3p)^{6}(4s)^{2}(3d)^{6} \text{ and } {}^{5}D_4.$$

Problem 5.12

We have N identical noninteracting particles whose total Hamiltonian is a sum of the single-particle Hamiltonians.

a) What is the energy for the ground state if the particles are spinless bosons?
b) What is the energy for the ground state if the particles are electrons?
c) Write down explicitly the wave functions and energies for three identical spinless bosons and three identical electrons.

The Hamiltonian for the N particles is $\widehat{H} = \sum_{n=1}^{N} \widehat{H}_n$ and the eigenstates and eigenenergies for each individual single-particle Hamiltonian $\widehat{H}_n$ are $\widehat{H}_n \left|\psi_k^{(n)}\right\rangle = E_k^{(n)} \left|\psi_k^{(n)}\right\rangle$.

a) If the particles are spinless bosons, then all particles are found in the same single-particle ground state with energy

$$E = \sum_{n=1}^{N} E_1 = NE_1,$$

where E_1 is the single-particle ground energy.

b) If the particles are spin = 1/2 fermions, then we have to take into account Pauli's exclusion principle that two identical fermions cannot occupy the same state. This means every energy level $\left|\psi_k^{(n)}\right\rangle$ can only host two electrons with spin up and down. Thus, the total energy is $E = 2\sum_{i=1}^{N} E_i$,

$$E = 2 \sum_{n=1}^{M=N/2} E_n$$

for an even number $N = 2M$ of fermions, or

$$E = 2 \sum_{n=1}^{M=(N-1)/2} E_n + E_{M+1}$$

for an odd number $N = (2M+1)$ of fermions. The prefactor 2 comes from the fact that $s = \frac{1}{2}$.

c) For three spinless bosons, the wave function is $\Psi = \psi_1(x_1)\psi_1(x_2)\psi_1(x_3)$ with $E = 3E_1$, as shown in Figure 5.3.

For 3 $s = \frac{1}{2}$ fermions, the wave function is any linear combination of

$$\left|\Psi_\uparrow(x_1, x_2, x_3)\right\rangle = \frac{1}{\sqrt{3!}} \begin{vmatrix} \psi_{1\uparrow}(x_1) & \psi_{1\uparrow}(x_2) & \psi_{1\uparrow}(x_3) \\ \psi_{1\downarrow}(x_1) & \psi_{1\downarrow}(x_2) & \psi_{1\downarrow}(x_3) \\ \psi_{2\uparrow}(x_1) & \psi_{2\uparrow}(x_2) & \psi_{2\uparrow}(x_3) \end{vmatrix}$$

and

$$\left|\Psi_\downarrow(x_1, x_2, x_3)\right\rangle = \frac{1}{\sqrt{3!}} \begin{vmatrix} \psi_{1\uparrow}(x_1) & \psi_{1\uparrow}(x_2) & \psi_{1\uparrow}(x_3) \\ \psi_{1\downarrow}(x_1) & \psi_{1\downarrow}(x_2) & \psi_{1\downarrow}(x_3) \\ \psi_{2\downarrow}(x_1) & \psi_{2\downarrow}(x_2) & \psi_{2\downarrow}(x_3) \end{vmatrix}$$

with energy $E = 2E_1 + E_2$, as shown in Figure 5.3.

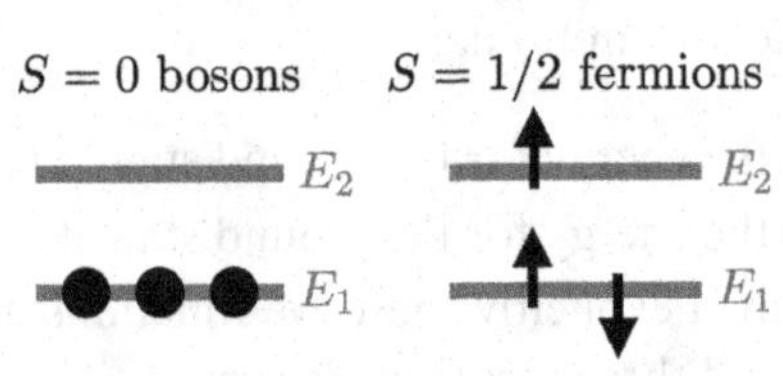

Figure 5.3 Ground-state atomic-level populations for spinless bosons and $s = \frac{1}{2}$ fermions.

Problem 5.13

A porphyrin ring is a type of a biological molecule, which can be represented as having 18 electrons constrained in a one-dimensional circular path of a constant radius.

a) What are the one-particle eigenfunctions of the systems when the mutual electron–electron interaction is neglected?
b) What is the distribution of electrons for the ground state of this molecule?

---

a) To continue with this problem, we recognize that the circular periodic boundary conditions must be imposed on the Schrödinger equation, which yields the eigenenergies and eigenfunctions for the identical $s = \frac{1}{2}$ fermions.

The Schrödinger equation can be written in 2D polar coordinates, but since $r = const$, the problem is dependent on the angular variable θ only. Thus, we have

$$\widehat{H}\psi(\theta) = -\frac{\hbar^2}{2mr^2}\frac{d^2}{d\theta^2}\psi(\theta) = E\psi(\theta),$$

$$\psi(\theta) = \frac{e^{ik\theta}}{\sqrt{2\pi}};\ E_k = \frac{\hbar^2 k^2}{2mr^2}.$$

From the periodic boundary conditions $\psi(\theta) = \psi(\theta + 2\pi)$, we find that $k = \{0, \pm 1, \pm 2, \ldots\}$. Which makes the energy E_k quantized. In summary, we have

$$\psi_k(\theta) = \frac{e^{ik\theta}}{\sqrt{2\pi}};\ E_k \frac{\hbar^2 k^2}{2mr^2};\ k = 0, \pm 1, \pm 2, \ldots$$

b) The electronic distribution is such that E_0 can accommodate two electrons with spin up and down, while every other E_i will accommodate four electrons since $k = \{\pm 1,\ \pm 2, \ldots\}$, as see Figure 5.4.

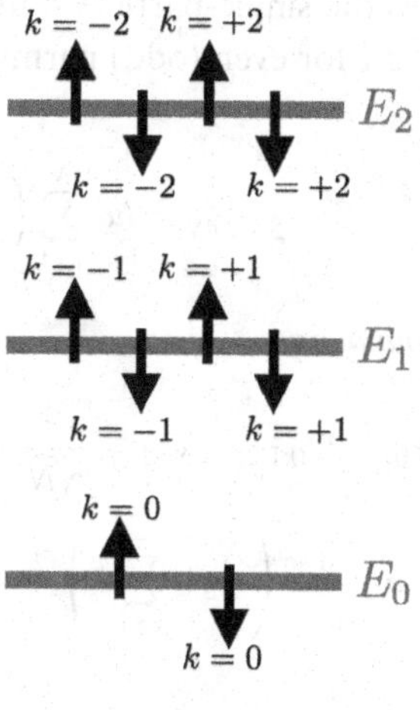

Figure 5.4 Ground-state atomic-level populations for electrons in the porphyrin ring.

The ground state of these 18 electrons is then:

$$(k=0)^2(k=1)^4(k=2)^4(k=3)^4(k=4)^4.$$

Problem 5.14

Consider N noninteracting spinless (aka ignoring the spin degree of freedom) fermions of mass m placed in a one-dimensional harmonic oscillator $V(\widehat{x}) = \frac{1}{2}m\omega^2\widehat{x}^2$.

a) Obtain the normalized wave functions for the composite state with the corresponding energies. Indicate the degeneracies for the first three levels and calculate their energies explicitly.

b) Calculate $\left\langle \sum_{i=1}^{N} \widehat{x}_i^2 \right\rangle$ for the first three levels of the composite state.

---

a) The Hamiltonian for the system is a sum of all the individual Hamiltonians for each particle,

$$\widehat{H} = \sum_{i=1}^{N}\left(\frac{\widehat{p}_i^2}{2m} + \frac{1}{2}m\omega^2\widehat{x}_i^2\right).$$

The wave function for the composite fermionic state can be found from the Slater determinant,

$$\Psi_{n_1n_2n_3\cdots n_N}(\boldsymbol{x}_1,\boldsymbol{x}_2,\ldots,\boldsymbol{x}_N) = \frac{1}{\sqrt{N!}}\begin{vmatrix} \psi_{n_1}(\boldsymbol{x}_1) & \psi_{n_1}(\boldsymbol{x}_2) & \cdots & \psi_{n_1}(\boldsymbol{x}_N) \\ \psi_{n_2}(\boldsymbol{x}_1) & \psi_{n_2}(\boldsymbol{x}_1) & \cdots & \psi_{n_2}(\boldsymbol{x}_N) \\ \vdots & \vdots & \ddots & \vdots \\ \psi_{n_N}(\boldsymbol{x}_1) & \psi_{n_N}(\boldsymbol{x}_2) & \cdots & \psi_{n_N}(\boldsymbol{x}_N) \end{vmatrix}$$

$$= \frac{1}{\sqrt{N!}}\sum_P(-1)^P P\left[\psi_{n_1}(\boldsymbol{x}_1)\psi_{n_2}(\boldsymbol{x}_2)\cdots\psi_{n_N}(\boldsymbol{x}_N)\right],$$

where n_i labels the single-particle states and P denotes the permutation operation with $(-1)^P = \pm 1$ for even (odd) permutations. The energy for this composite state is

$$E_{n_1,n_2,\ldots,n_N} = \hbar\omega\sum_{i=1}^{N}\left(\frac{1}{2}+n_i\right) = \hbar\omega\left(\frac{N}{2}+\sum_{i=1}^{N}n_i\right).$$

For the *ground state*,

$$|\Psi_0\rangle \equiv \Psi_{0,1,2,\ldots,N-1} = \frac{1}{\sqrt{N!}}\sum_P(-1)^P P\left[\psi_0(\boldsymbol{x}_1)\psi_1(\boldsymbol{x}_2)\cdots\psi_{N-1}(\boldsymbol{x}_N)\right],$$

$$E_{0,1,\ldots,N-1} = \hbar\omega\left(\frac{N}{2}+\sum_{k=1}^{N-1}k\right) = \hbar\omega\left(\frac{N}{2}+\frac{N(N-1)}{2}\right) = \frac{\hbar\omega}{2}N^2.$$

For the *first excited state,*

$$\left|\Psi_N^{N-1}\right\rangle \equiv \Psi_{0,1,2,\ldots,N-2,N}$$
$$= \frac{1}{\sqrt{N!}}\sum_P (-1)^P P\left[\psi_0(\boldsymbol{x}_1)\psi_1(\boldsymbol{x}_2)\cdots\psi_{N-2}(\boldsymbol{x}_{N-1})\psi_N(\boldsymbol{x}_N)\right],$$
$$E_{0,1,\ldots,N-2,N} = \hbar\omega\left(\frac{N}{2}+\sum_{k=1}^{N-2}k+N\right) = \frac{\hbar\omega}{2}\left(N^2+2\right).$$

Let us note here that the first excited state can equivalently be represented using the creation, annihilation fermionic operators $\widehat{a}_i^+$, $\widehat{a}_i (i = 1,\ldots N)$, such that $\left|\Psi_N^{N-1}\right\rangle = \widehat{a}_N^+\widehat{a}_{N-1}\left|\Psi_0\right\rangle$.

The *second excited state* is doubly degenerated; we have

$$\left|\Psi_{N+1}^{N-1}\right\rangle \equiv \Psi_{0,1,\ldots,N-2,N+1} = \frac{1}{\sqrt{N!}}\sum_P (-1)^P P\left[\psi_0(\boldsymbol{x}_1)\psi_1(\boldsymbol{x}_2)\cdots\psi_{N-2}(\boldsymbol{x}_{N-1})\psi_{N+1}(\boldsymbol{x}_N)\right],$$

and

$$\left|\Psi_N^{N-2}\right\rangle \equiv \Psi_{0,1,\ldots,N-3,N-1,N} = \frac{1}{\sqrt{N!}}\sum_P (-1)^P P\left[\psi_0(\boldsymbol{x}_1)\cdots\psi_{N-3}(\boldsymbol{x}_{N-2})\psi_{N-1}(\boldsymbol{x}_{N-1})\psi_N(\boldsymbol{x}_N)\right],$$

with

$$E_{0,1,\ldots,N-2,N+1} = E_{0,1,\ldots,N-3,N-1,N} = \hbar\omega\left(\frac{N}{2}+\sum_{k=1}^{N-2}k+(N+1)\right) = \frac{\hbar\omega}{2}\left(N^2+4\right).$$

Equivalently, $\left|\Psi_{N+1}^{N-1}\right\rangle = \widehat{a}_{N+1}^+\widehat{a}_{N-1}\left|\Psi_0\right\rangle$ and $\left|\Psi_N^{N-2}\right\rangle = \widehat{a}_N^+\widehat{a}_{N-2}\left|\Psi_0\right\rangle$. It is now easy to see that the nth excited state has degeneracy n. *Can you explain why?*

b) To calculate $\left\langle \sum_{i=1}^{N} \widehat{x}_i^2 \right\rangle$, we can use the virial theorem for the harmonic oscillator

$$\langle T\rangle = \langle V\rangle = \frac{E}{2},$$

where the averaging is done with respect to the corresponding state with energy E. Basic derivations and properties of the virial theorem can be found in Griffiths (2004), for example. From the Hamiltonian $\widehat{H} = \sum_{i=1}^{N}\left(\frac{\widehat{p}_i^2}{2m}+\frac{1}{2}m\omega^2\widehat{x}_i^2\right)$, we obtain $\langle V\rangle = \left\langle \sum_{i=1}^{N}\left(\frac{1}{2}m\omega^2\widehat{x}_i^2\right)\right\rangle = \frac{1}{2}m\omega^2\left\langle \sum_{i=1}^{N}\widehat{x}_i^2\right\rangle = \frac{E}{2}$, thus $\left\langle \sum_{i=1}^{N}\widehat{x}_i^2\right\rangle = \frac{E}{m\omega^2}$.

Therefore, for the *ground state*, $\left\langle \sum_{i=1}^{N}\widehat{x}_i^2\right\rangle = \frac{\hbar}{2m\omega}N^2$.

For the *first excited state*, $\left\langle \sum_{i=1}^{N}\widehat{x}_i^2\right\rangle = \frac{\hbar}{2m\omega}(N^2+2)$.

For the *second excited state*, $\left\langle \sum_{i=1}^{N}\widehat{x}_i^2\right\rangle = \frac{\hbar}{2m\omega}(N^2+4)$.

Problem 5.15

A many-particle state can be written as $\left|n_1 n_2\cdots n_k\cdots\right\rangle = \left[\prod_{k=1}\frac{(\widehat{a}_k^+)^{n_k}}{\sqrt{n_k!}}\right]\left|0\right\rangle$, where $\widehat{a}_k^+$ are the creation operators for the kth particle with occupation n_i and $\left|0\right\rangle$ is the vacuum. In the case of fermions, n_k is either 0 or 1 according to Pauli's exclusion principle, while in the case of bosons, n_k can be any positive integer number.

Prove that for fermions, we have $\{\widehat{a}_m, \widehat{a}_\ell^+\} = \delta_{m\ell}$, while for bosons we have $[\widehat{a}_m, \widehat{a}_\ell^+] = \delta_{m\ell}$. For this purpose, you can assume that $\{\widehat{a}_m^+, \widehat{a}_\ell^+\} = 0 = \{\widehat{a}_m, \widehat{a}_\ell\}$ for fermions and $[\widehat{a}_m^+, \widehat{a}_\ell^+] = 0 = [\widehat{a}_m, \widehat{a}_\ell]$ for bosons.

We begin with the case $\ell \neq m$ by considering the different possibilities for action of the anticommutator $\{\widehat{a}_m, \widehat{a}_\ell^+\}$, as determined by the many-particle state occupations,

$$
\begin{aligned}
&\{\widehat{a}_m, \widehat{a}_\ell^+\} |n_1, \ldots, n_m = 0, \ldots, n_\ell = 0, \ldots, n_R\rangle \\
&\quad = \widehat{a}_m \widehat{a}_\ell^+ |n_1, \ldots, 0_m, \ldots, 0_\ell \ldots, n_R\rangle + \widehat{a}_\ell^+ \widehat{a}_m |n_1, \ldots, 0_m, \ldots, 0_\ell \ldots, n_R\rangle \\
&\quad = \widehat{a}_m (-1)^{\sum_{k=1}^{\ell-1} n_k} |n_1, \ldots, 0_m, \ldots, 1_\ell \ldots, n_R\rangle + 0 = 0, \\
&\{\widehat{a}_m, \widehat{a}_\ell^+\} |n_1, \ldots, n_m = 1, \ldots, n_\ell = 0, \ldots, n_R\rangle \\
&\quad = \widehat{a}_m \widehat{a}_\ell^+ |n_1, \ldots, 1_m, \ldots, 0_\ell \ldots, n_R\rangle + \widehat{a}_\ell^+ \widehat{a}_m |n_1, \ldots, 1_m, \ldots, 0_\ell \ldots, n_R\rangle \\
&\quad = (-1)^{\sum_{k=1}^{\ell-1} n_k} \widehat{a}_m |n_1, \ldots, 1_m, \ldots, 1_\ell, \ldots, n_R\rangle + (-1)^{\sum_{k=1}^{m-1} n_k} \widehat{a}_\ell^+ |n_1, \ldots, 0_m, \ldots, 0_\ell \ldots, n_R\rangle \\
&\quad = (-1)^{\sum_{k=1}^{\ell-1} n_k} (-1)^{\sum_{k=1}^{m-1} n_k} |n_1, \ldots, 0_m, \ldots, 1_\ell, \ldots, n_R\rangle \\
&\qquad + (-1)^{\sum_{k=1}^{m-1} n_k} (-1)^{-1+\sum_{k=1}^{\ell-1} n_k} |n_1, \ldots, 0_m, \ldots, 1_\ell, \ldots, n_R\rangle \\
&\quad = (-1)^{\sum_{k=m-1}^{\ell-1} n_k} |n_1, \ldots, 0_m, \ldots, 1_\ell, \ldots, n_R\rangle - (-1)^{\sum_{k=m-1}^{\ell-1} n_k} |n_1, \ldots, 0_m, \ldots, 1_\ell, \ldots, n_R\rangle = 0.
\end{aligned}
$$

Note that, in the second term, we have applied one less exchange because we have already destroyed the particle $|1_m\rangle$.

$$
\begin{aligned}
&\{\widehat{a}_m, \widehat{a}_\ell^+\} |n_1, \ldots, n_m = 0, \ldots, n_\ell = 1, \ldots, n_R\rangle \\
&\quad = \widehat{a}_m \widehat{a}_\ell^+ |n_1, \ldots, 0_m, \ldots, 1_\ell \ldots, n_R\rangle + \widehat{a}_\ell^+ \widehat{a}_m |n_1, \ldots, 0_m, \ldots, 1_\ell, \ldots, n_R\rangle = 0 + 0 = 0, \\
&\{\widehat{a}_m, \widehat{a}_\ell^+\} |n_1, \ldots, n_m = 1, \ldots, n_\ell = 1, \ldots, n_R\rangle \\
&\quad = \widehat{a}_m \widehat{a}_\ell^+ |n_1, \ldots, 1_m, \ldots, 1_\ell \ldots, n_R\rangle + \widehat{a}_\ell^+ \widehat{a}_m |n_1, \ldots, 1_m, \ldots, 1_\ell, \ldots, n_R\rangle \\
&\quad = 0 + \widehat{a}_\ell^+ (-1)^{\sum_{k=1}^{m-1} n_k} |n_1, \ldots, 0_m, \ldots, 1_\ell, \ldots, n_R\rangle = 0.
\end{aligned}
$$

Then, if $\ell \neq m$, we have $\{\widehat{a}_m, \widehat{a}_\ell^+\} |n_1, \ldots, n_R\rangle = 0$.

Now we consider the different cases for $m = \ell$, as dictated by the many-particle state occupations:

$$
\begin{aligned}
&\{\widehat{a}_\ell, \widehat{a}_\ell^+\} |n_1, \ldots, n_\ell = 0, \ldots, n_R\rangle \\
&\quad = \widehat{a}_\ell \widehat{a}_\ell^+ |n_1, \ldots, 0_\ell \ldots, n_R\rangle + \widehat{a}_\ell^+ \widehat{a}_\ell |n_1, \ldots, 0_\ell \ldots, n_R\rangle = \widehat{a}_\ell (-1)^{\sum_{k=1}^{\ell-1} n_k} |n_1, \ldots, 1_\ell, \ldots, n_R\rangle + 0 \\
&\quad = (-1)^{2\sum_{k=1}^{\ell-1} n_k} |n_1, \ldots, 0_\ell, \ldots, n_R\rangle = |n_1, \ldots, 0_\ell, \ldots, n_R\rangle, \\
&\{\widehat{a}_\ell, \widehat{a}_\ell^+\} |n_1, \ldots, n_\ell = 1, \ldots, n_R\rangle \\
&\quad = \widehat{a}_\ell \widehat{a}_\ell^+ |n_1, \ldots, 1_\ell \ldots, n_R\rangle + \widehat{a}_\ell^+ \widehat{a}_\ell |n_1, \ldots, 1_\ell \ldots, n_R\rangle = 0 + \widehat{a}_\ell^+ (-1)^{\sum_{k=1}^{\ell-1} n_k} |n_1, \ldots, 0_\ell, \ldots, n_R\rangle
\end{aligned}
$$

$$= (-1)^{2\sum_{k=1}^{\ell-1} n_k} |n_1, \ldots, 1_\ell \ldots, n_R\rangle = |n_1, \ldots, 1_\ell \ldots, n_R\rangle.$$

Then, if $\ell = m$, we have $\{\widehat{a}_\ell, \widehat{a}_\ell^+\}|n_1, \ldots, n_R\rangle = |n_1, \ldots, n_R\rangle$. In summary, $\{\widehat{a}_m, \widehat{a}_\ell^+\} = \delta_{m\ell}$.

The commutation relation for bosons $\left[\widehat{a}_m, \widehat{a}_\ell^+\right] = \delta_{m\ell}$ is left as an exercise.

Problem 5.16

Consider the number operator $\widehat{n}_\alpha = \widehat{c}_\alpha^+ \widehat{c}_\alpha$ for the case of fermionic particles. Show that $\left[\widehat{n}_\alpha, \widehat{n}_\beta\right] = 0$.

Using $\left\{\widehat{c}_\alpha, \widehat{c}_\beta^+\right\} = \delta_{\alpha\beta}$ and $\left\{\widehat{c}_\alpha, \widehat{c}_\beta\right\} = \left\{\widehat{c}_\alpha^+, \widehat{c}_\beta^+\right\} = 0$, we obtain

$$\begin{aligned}
\left[\widehat{n}_\alpha, \widehat{n}_\beta\right] &= \widehat{c}_\alpha^+ \widehat{c}_\alpha \widehat{c}_\beta^+ \widehat{c}_\beta - \widehat{c}_\beta^+ \widehat{c}_\beta \widehat{c}_\alpha^+ \widehat{c}_\alpha = \widehat{n}_\alpha \widehat{n}_\beta - \widehat{c}_\beta^+ \left(\delta_{\alpha\beta} - \widehat{c}_\alpha^+ \widehat{c}_\beta\right) \widehat{c}_\alpha \\
&= \widehat{n}_\alpha \widehat{n}_\beta - \delta_{\alpha\beta} \widehat{c}_\beta^+ \widehat{c}_\alpha + \widehat{c}_\beta^+ \widehat{c}_\alpha^+ \widehat{c}_\beta \widehat{c}_\alpha = \widehat{n}_\alpha \widehat{n}_\beta - \delta_{\alpha\beta} \widehat{c}_\beta^+ \widehat{c}_\alpha + \left(-\widehat{c}_\alpha^+ \widehat{c}_\beta^+\right)\left(-\widehat{c}_\alpha \widehat{c}_\beta\right) \\
&= \widehat{n}_\alpha \widehat{n}_\beta + \widehat{c}_\alpha^+ \widehat{c}_\beta^+ \widehat{c}_\alpha \widehat{c}_\beta - \delta_{\alpha\beta} \widehat{c}_\beta^+ \widehat{c}_\alpha = \widehat{n}_\alpha \widehat{n}_\beta + \widehat{c}_\alpha^+ \left(\delta_{\alpha\beta} - \widehat{c}_\alpha \widehat{c}_\beta^+\right) \widehat{c}_\beta - \delta_{\alpha\beta} \widehat{c}_\beta^+ \widehat{c}_\alpha \\
&= \widehat{n}_\alpha \widehat{n}_\beta + \delta_{\alpha\beta} \widehat{c}_\alpha^+ \widehat{c}_\beta - \widehat{c}_\alpha^+ \widehat{c}_\alpha \widehat{c}_\beta^+ \widehat{c}_\beta - \delta_{\alpha\beta} \widehat{c}_\beta^+ \widehat{c}_\alpha \\
&= \widehat{n}_\alpha \widehat{n}_\beta - \widehat{n}_\alpha \widehat{n}_\beta + \delta_{\alpha\beta} \left(\widehat{c}_\alpha^+ \widehat{c}_\beta - \widehat{c}_\beta^+ \widehat{c}_\alpha\right) = \delta_{\alpha\beta} \left(\widehat{c}_\alpha^+ \widehat{c}_\alpha - \widehat{c}_\alpha^+ \widehat{c}_\alpha\right) = 0.
\end{aligned}$$

Problem 5.17

The Hamiltonian of a chain of even N particles separated by a distance a can be given as

$$\widehat{H} = \frac{1}{2} \sum_{\mu=-N/2}^{N/2} \left[\widehat{p}_\mu^2 + \left(\widehat{q}_\mu - \widehat{q}_{\mu-1}\right)^2\right],$$

where $\widehat{p}_\mu$ and $\widehat{q}_\mu$ are the momentum and position operators for each particle $\mu = \{1, 2, 3, \ldots, N\}$. The operators are Hermitian and satisfy the commutation relations $\left[\widehat{q}_\mu, \widehat{q}_\nu\right] = \left[\widehat{p}_\mu, \widehat{p}_\nu\right] = 0$; $\left[\widehat{q}_\mu, \widehat{p}_\nu\right] = i\delta_{\mu\nu}$. There are also periodic boundary conditions, such that $\widehat{q}_{\mu+N} = \widehat{q}_\mu$.

Let us define new operators via the relations:

$$\widehat{q}_\mu = \frac{1}{N} \sum_{n=-N/2}^{N/2} \widehat{Q}_{k_n} e^{-ik_n \mu a}; \quad \widehat{p}_\mu = \frac{1}{N} \sum_{n=-N/2}^{N/2} \widehat{P}_{k_n} e^{ik_n \mu a},$$

where $k_n = \frac{2\pi n}{Na}$, $n = \left\{-\frac{N}{2}, -\frac{N}{2} + 1, \ldots, \frac{N}{2}\right\}$.

a) Find similar expressions for the $\widehat{Q}_{k_n}$ and $\widehat{P}_{k_n}$ operators.

b) Evaluate the commutator $\left[\widehat{Q}_{k_n}, \widehat{P}_{k_m}\right]$.

c) Express the Hamiltonian in the new operators $\widehat{Q}_{k_n}$ and $\widehat{P}_{k_n}$.

d) Obtain the infinite chain limit $N \to \infty$ with the interatomic separation kept constant $a = const.$

e) Obtain the continuous limit of this Hamiltonian by taking the limit $a \to 0$ by keeping the length of the chain constant $Na = L$.

---

a) To find an expression for the $\widehat{Q}_{k_n}$ operator, both sides of the definition of $\widehat{q}_\mu$ are multiplied accordingly by $e^{ik_m\mu a}$, and summed for all μ:

$$\widehat{q}_\mu = \frac{1}{N}\sum_{n=-N/2}^{N/2} \widehat{Q}_{k_n} e^{-ik_n\mu a},$$

$$\sum_{\mu=-N/2}^{N/2} \widehat{q}_\mu e^{ik_m\mu a} = \sum_{\mu=-N/2}^{N/2}\left(\frac{1}{N}\sum_{n=-N/2}^{N/2} \widehat{Q}_{k_n} e^{-ik_n\mu a}\right) e^{-ik_m\mu a}$$

$$= \frac{1}{N}\sum_{n=-N/2}^{N/2} \widehat{Q}_{k_n} \sum_{\mu=-N/2}^{N/2} e^{-i(k_n-k_m)\mu a}.$$

Using that $\frac{1}{N}\sum_{\mu=-N/2}^{N/2} e^{-i(k_n-k_m)\mu a} = \delta_{nm}$, we obtain

$$\sum_{\mu=-N/2}^{N/2} \widehat{q}_\mu e^{ik_m\mu a} = \sum_{n=-N/2}^{N/2} \widehat{Q}_{k_n}\delta_{nm} = \widehat{Q}_{k_m}.$$

In a similar way, we find

$$\widehat{P}_{k_m} = \sum_{\mu=-N/2}^{N/2} \widehat{p}_\mu e^{-ik_m\mu a}.$$

b) Next, we evaluate the commutator $\left[\widehat{Q}_{k_n}, \widehat{P}_{k_m}\right]$ using the definitions of the operators directly and the relation $\left[\widehat{q}_\mu, \widehat{p}_\nu\right] = i\delta_{\mu\nu}$:

$$\left[\widehat{Q}_{k_n}, \widehat{P}_{k_m}\right] = \left[\sum_{\mu=-N/2}^{N/2} \widehat{q}_\mu e^{ik_n\mu a}, \sum_{\nu=-N/2}^{N/2} \widehat{p}_\nu e^{-ik_m\nu a}\right] = \sum_{\mu=-N/2}^{N/2}\sum_{\nu=-N/2}^{N/2} e^{ik_n\mu a} e^{-ik_m\nu a}\left[\widehat{q}_\mu, \widehat{p}_\nu\right]$$

$$= \sum_{\mu=-N/2}^{N/2}\sum_{\nu=-N/2}^{N/2} e^{ik_n\mu a} e^{-ik_m\nu a} i\delta_{\mu\nu} = i\sum_{\mu=-N/2}^{N/2} e^{i(k_n-k_m)\mu a} = iN\delta_{nm}.$$

c) To reexpress the Hamiltonian $\widehat{H} = \frac{1}{2}\sum_{\mu=-N/2}^{N/2}\left[\widehat{p}_\mu^2 + \left(\widehat{q}_\mu - \widehat{q}_{\mu-1}\right)^2\right]$, we substitute each operator by the given definitions $\widehat{q}_\mu$ and $\widehat{p}_\mu$:

$$\sum_{\mu=-N/2}^{N/2} \widehat{p}_\mu^2 = \sum_{\mu=-N/2}^{N/2} \frac{1}{N}\sum_{n=-N/2}^{N/2} \widehat{P}_{k_n} e^{ik_n\mu a}\frac{1}{N}\sum_{m=-N/2}^{N/2} \widehat{P}_{k_n} e^{ik_m\mu a}$$

$$= \frac{1}{N^2}\sum_{n=-N/2}^{N/2}\sum_{m=-N/2}^{N/2} \widehat{P}_{k_n}\widehat{P}_{k_m}\sum_{\mu=-N/2}^{N/2} e^{i(k_n+k_m)\mu a}$$

$$= \frac{1}{N}\sum_{n=-N/2}^{N/2}\sum_{m=-N/2}^{N/2} \widehat{P}_{k_n}\widehat{P}_{k_m}\delta_{n(-m)} = \frac{1}{N}\sum_{n=-N/2}^{N/2} \widehat{P}_{k_n}\widehat{P}_{k_{-n}}.$$

For the second term in the Hamiltonian,

$$\sum_{\mu=-N/2}^{N/2} (\widehat{q}_\mu - \widehat{q}_{\mu-1})^2 = \frac{1}{N} \sum_{n=-N/2}^{N/2} \sum_{m=-N/2}^{N/2} \widehat{Q}_{k_n} \widehat{Q}_{k_m} \left(1 - e^{ik_m a} - e^{ik_n a} + e^{i(k_n+k_m)a}\right)$$

$$\times \frac{1}{N} \sum_{\mu=-N/2}^{N/2} e^{-i(k_n+k_m)\mu a}$$

$$= \frac{1}{N} \sum_{n=-N/2}^{N/2} \sum_{m=-N/2}^{N/2} \widehat{Q}_{k_n} \widehat{Q}_{k_m} \left(1 - e^{ik_m a} - e^{ik_n a} + e^{i(k_n+k_m)a}\right) \delta_{(-n)m}$$

$$= \frac{1}{N} \sum_{n=-N/2}^{N/2} \widehat{Q}_{k_n} \widehat{Q}_{k_{-n}} \left(1 - e^{ik_{-n} a} - e^{ik_n a} + e^{i(k_n+k_{-n})a}\right).$$

Note that all remaining exponentials are μ-independent. Using now that $k_{-n} = \frac{2\pi(-n)}{Na} = -k_n$, we can further simplify this result to

$$\sum_{\mu=-N/2}^{N/2} (\widehat{q}_\mu - \widehat{q}_{\mu-1})^2 = \frac{1}{N} \sum_{n=-N/2}^{N/2} \widehat{Q}_{k_n} \widehat{Q}_{k_{-n}} \left(1 - e^{-ik_n a} - e^{ik_n a} + e^{i(k_n-k_n)a}\right)$$

$$= \frac{1}{N} \sum_{n=-N/2}^{N/2} \widehat{Q}_{k_n} \widehat{Q}_{k_{-n}} (2 - 2\cos(k_n a)).$$

Then, the Hamiltonian becomes

$$\widehat{H} = \frac{1}{2N} \sum_{n=-N/2}^{N/2} \left[\widehat{P}_{k_n} \widehat{P}_{-k_n} + \widehat{Q}_{k_n} \widehat{Q}_{-k_n} (2 - 2\cos(k_n a))\right].$$

d) To obtain the infinite chain limit, we start with the Hamiltonian

$$\lim_{N\to\infty} \widehat{H} = \frac{1}{2} \sum_{\mu=-\infty}^{\infty} \left[\widehat{p}_\mu^2 + (\widehat{q}_\mu - \widehat{q}_{\mu-1})^2\right].$$

Applying the $N \to \infty$ limit in $k_n = \frac{2\pi n}{Na}$ for all previous results, it is further found

$$\lim_{N\to\infty} \widehat{q}_\mu = \lim_{N\to\infty} \frac{1}{N} \sum_{n=-N/2}^{N/2} \widehat{Q}_{k_n} e^{-i\frac{2\pi n}{Na}\mu a} = \lim_{N\to\infty} \frac{1}{N} \int_{-N/2}^{N/2} dn \widehat{Q}_{k_n} e^{-i\frac{2\pi n}{Na}\mu a}$$

$$= \lim_{N\to\infty} \frac{1}{N} \frac{Na}{2\pi} \int_{-\frac{\pi}{a}}^{\frac{\pi}{a}} dk \widehat{Q}_k e^{-ik\mu a} = \frac{a}{2\pi} \int_{-\frac{\pi}{a}}^{\frac{\pi}{a}} dk \widehat{Q}_k e^{-ik\mu a},$$

$$\lim_{N\to\infty} \widehat{Q}_{k_n} = \lim_{N\to\infty} \widehat{Q}_k = \lim_{N\to\infty} \sum_{\mu=-N/2}^{N/2} \widehat{q}_\mu e^{ik\mu a} = \sum_{\mu=-\infty}^{\infty} \widehat{q}_\mu e^{ik\mu a}.$$

Similarly, we obtain

$$\lim_{N\to\infty} \widehat{p}_\mu = \frac{a}{2\pi} \int_{-\frac{\pi}{a}}^{\frac{\pi}{a}} dk \widehat{P}_k e^{ik\mu a},$$

$$\lim_{N\to\infty} \widehat{P}_{k_n} = \lim_{N\to\infty} \widehat{P}_k = \sum_{\mu=-\infty}^{\infty} \widehat{p}_\mu e^{-ik\mu a},$$

$$\lim_{N\to\infty}\widehat{H}=\frac{1}{2}\frac{a}{2\pi}\int_{-\frac{\pi}{a}}^{\frac{\pi}{a}}dk\left[\widehat{P}_k\widehat{P}_{-k}+\widehat{Q}_k\widehat{Q}_{-k}\left(2-2\cos\left(ka\right)\right)\right],$$

$$\lim_{N\to\infty}\left[\widehat{q}_\mu,\widehat{p}_\nu\right]=i\delta_{\mu\nu},$$

$$\lim_{N\to\infty}\left[\widehat{Q}_{k_n},\widehat{P}_{k_m}\right]=\left[\widehat{Q}_k,\widehat{P}_u\right]=i\frac{2\pi}{a}\delta\left(k-u\right).$$

Note that the integrals in reciprocal space are carried out over the first Brillouin Zone, and $V_{BZ}=\frac{2\pi}{a}$.

e) Finally, to find the continuous limit of the preceding $\widehat{H}$, using $Na=L$ and $x=\mu a$, we note that $k_n=\frac{2\pi n}{Na}=\frac{2\pi n}{L}$. Thus, the operators can be written as

$$\begin{aligned}\lim_{\substack{N\to\infty\\Na=L}}\widehat{H}&=\lim_{\substack{N\to\infty\\Na=L}}\frac{1}{2}\sum_{\mu=-N/2}^{N/2}\left[\widehat{p}^2(\mu a)+\left(\widehat{q}(\mu a)-\widehat{q}(\mu a-a)\right)^2\right]\\&=\lim_{\substack{N\to\infty\\Na=L}}\frac{1}{2}\int_{-N/2}^{N/2}d\mu\left[\widehat{p}^2(\mu a)+\left(\widehat{q}(x)-\left[\widehat{q}(\mu a)-a\partial_x\widehat{q}(\mu a)+\frac{a^2}{2}\partial_x^2\widehat{q}(\mu a)+\ldots\right]\right)^2\right]\\&=\lim_{\substack{N\to\infty\\Na=L}}\frac{1}{2a}\int_{-aN/2}^{aN/2}dx\left[\widehat{p}^2(x)+\left(a\partial_x\widehat{q}(x)-\frac{a^2}{2}\partial_x^2\widehat{q}(x)+\ldots\right)^2\right]\\&=\frac{1}{2a}\int_{-L/2}^{L/2}dx\left[\widehat{p}^2(x)+a^2\left(\partial_x\widehat{q}(x)\right)^2\right].\end{aligned}$$

If we apply the $N\to\infty$ limit by keeping $Na=L$ to all the results obtained earlier (with $k_n=\frac{2\pi n}{Na}$), we find that the operators tend to singular operators that depend explicitly on the infinitesimal separation of particles a, as we demonstrate in the following:

$$\begin{aligned}\lim_{\substack{N\to\infty\\Na=L}}\widehat{q}_\mu&=\lim_{\substack{N\to\infty\\Na=L}}\widehat{q}(x=\mu a)=\lim_{\substack{N\to\infty\\Na=L}}\frac{1}{N}\sum_{n=-N/2}^{N/2}\widehat{Q}_{k_n}e^{-i\frac{2\pi n}{Na}x}=\lim_{\substack{N\to\infty\\Na=L}}\frac{1}{N}\int_{-N/2}^{N/2}dn\widehat{Q}\left(k_n\right)e^{-i\frac{2\pi n}{L}x}\\&=\lim_{\substack{N\to\infty\\Na=L}}\frac{1}{N}\frac{L}{2\pi}\int_{-\frac{\pi}{a}}^{\frac{\pi}{a}}dk\widehat{Q}(k)e^{-ikx}=\lim_{\substack{N\to\infty\\Na=L}}\frac{a}{2\pi}\int_{-\frac{\pi}{a}}^{\frac{\pi}{a}}dk\widehat{Q}(k)e^{-ikx},\end{aligned}$$

$$\begin{aligned}\lim_{\substack{N\to\infty\\Na=L}}\widehat{Q}_{k_n}&=\lim_{\substack{N\to\infty\\Na=L}}\widehat{Q}\left(k_n\right)=\lim_{\substack{N\to\infty\\Na=L}}\sum_{\mu=-N/2}^{N/2}\widehat{q}(\mu a)e^{i\frac{2\pi n}{Na}\mu a}=\lim_{\substack{N\to\infty\\Na=L}}\int_{-N/2}^{N/2}d\mu\widehat{q}(\mu a)e^{i\frac{2\pi n}{L}\mu a}\\&=\frac{1}{a}\int_{-L/2}^{L/2}dx\widehat{q}(x)e^{ikx}.\end{aligned}$$

To avoid this problem, we redefine the conjugate position as $\widehat{\mathcal{Q}}(k)=a\widehat{Q}(k)$; then we obtain

$$\widehat{\mathcal{Q}}(k)=\lim_{\substack{N\to\infty\\Na=L}}a\widehat{Q}(k)=\int_{-L/2}^{L/2}dx\widehat{q}(x)e^{ikx},$$

$$\lim_{\substack{N\to\infty\\Na=L}}\widehat{q}(x=\mu a)=\int_{-\infty}^{\infty}\frac{dk}{2\pi}\widehat{\mathcal{Q}}(k)e^{-ikx}.$$

Similarly, the momentum operators are obtained as

$$\widehat{\mathscr{P}}(k) = \lim_{\substack{N\to\infty \\ Na=L}} a\widehat{P}(k) = \int_{-L/2}^{L/2} dx\widehat{p}(x)e^{-ikx},$$

$$\lim_{\substack{N\to\infty \\ Na=L}} \widehat{p}(x=\mu a) = \int_{-\infty}^{\infty} \frac{dk}{2\pi}\widehat{\mathscr{P}}(k)e^{ikx},$$

$$\lim_{\substack{N\to\infty \\ Na=L}} \widehat{H} = \frac{1}{2}\int_{-\infty}^{\infty} \frac{dk}{2\pi}\left[\widehat{\mathscr{P}}(k)\widehat{\mathscr{P}}(-k) + \widehat{Q}(k)\widehat{Q}(-k)\,(ka)^2\right],$$

$$\lim_{\substack{N\to\infty \\ Na=L}} \left[\widehat{q}_\alpha, \widehat{p}_\beta\right] = \lim_{\substack{N\to\infty \\ Na=L}} [\widehat{q}(x), \widehat{p}(y)] = i\delta(x-y),$$

$$\lim_{\substack{N\to\infty \\ Na=L}} \left[a\widehat{Q}_{k_n}, a\widehat{P}_{k_m}\right] = \left[\widehat{\mathscr{Q}}(k), \widehat{\mathscr{P}}(u)\right] = iL\operatorname{sinc}\left(\frac{L}{2}(k-u)\right) \xrightarrow[L\to\infty]{} i\frac{\delta(k-u)}{2\pi}.$$

Note that this rescaling transformation is valid because all observables are kept invariant. We could also have used this transformation to symmetrize all the operators obtained in this exercise. Try it!

> *Food for thought:* The atoms of a crystal can be thought of as *d*-dimensional chains of particles separated by a small periodic given distance. Thus, the dynamics of the perturbation from the equilibrium of those chains (called phonons in this context) is given by the *d*-dimensional generalization of the Hamiltonian we have derived here.

Problem 5.18

Consider the many-body density operator in real space given as $\widehat{n}(\boldsymbol{x}) = \sum_{i=1}^{N} \delta(\widehat{\boldsymbol{r}}_i - \boldsymbol{x})$, where $\widehat{\boldsymbol{r}}_i$ is the position operator for an ith particle and $\boldsymbol{x}$ is a spatial vector.

Give its second quantized picture in (a) position $\boldsymbol{x}$ and (b) momentum $\boldsymbol{k}$ spaces.

We first consider the second quantized form in coordinate space, $\boldsymbol{x}$. For this purpose, it is useful to remember that if an operator $\widehat{A}$ consists of many single operators, then the *Fock space* can be built from the eigenstates of these single operators. Specifically, for the operator $\widehat{A} = \sum_{i=1}^{N} \widehat{A}_i$, in the first quantization picture, we have a basis where $\widehat{A}_i|a_1a_2\ldots a_i\ldots a_N\rangle = \lambda_i|a_1a_2\ldots a_i\ldots a_N\rangle$, thus $\widehat{A}_i|a_1a_2\ldots a_i\ldots.a_N\rangle = \sum_{i=1}^{N} \lambda_i|a_1a_2\ldots a_i\ldots a_N\rangle$.

This, on the other hand, can be written in terms of the occupation number operator as

$$\widehat{A}|n_1\ldots n_i\ldots n_R\rangle = \sum_{i=1}^{R} n_i\widehat{A}_i|n_1\ldots n_i\ldots n_R\rangle = \sum_{i=1}^{R} n_i\lambda_i|n_1\ldots n_i\ldots n_R\rangle = \sum_{i=1}^{R} \lambda_i\widehat{c}_i^{\,+}\widehat{c}_i|n_1\ldots n_i\ldots n_R\rangle,$$

$$\widehat{A} = \sum_{i=1}^{R} \lambda_i\widehat{c}_i^{\,+}\widehat{c}_i,$$

where we have used that $\widehat{c}_i^+\widehat{c}_i|n_1\ldots n_i\ldots n_R\rangle = \widehat{n}_i|n_1\ldots n_i\ldots n_R\rangle = n_i|n_1\ldots n_i\ldots n_R\rangle$. Further, using a unitary change of eigenbasis $|a_i\rangle$ with corresponding creation and annihilation operators $\widehat{d}_i^+$, $\widehat{d}_i$, we can write

$$|\alpha_m\rangle = \sum_{i=1}^{N} |a_i\rangle\langle a_i|\alpha_m\rangle.$$

Since $|\alpha_m\rangle = \widehat{c}_m^+|0\rangle$ and $|a_i\rangle = \widehat{d}_i^+|0\rangle$, the creation and annihilation operators of $|\alpha_m\rangle$ states can thus be written as

$$\widehat{c}_m^+ = \sum_{i=1}^{N} \widehat{d}_i^+\langle a_i|\alpha_m\rangle^*,$$

$$\widehat{c}_i = \sum_{m=1}^{N} \langle a_i|\alpha_m\rangle\widehat{d}_m.$$

One can further represent the operator $\widehat{A}$ in Fock space built from the single-particle states $|\alpha_1\alpha_2\ldots\alpha_i\ldots\alpha_N\rangle$ according to

$$\widehat{A} = \sum_{i=1}^{N} \lambda_i\widehat{c}_i^+\widehat{c}_i = \sum_{i=1}^{N} \lambda_i \sum_{m=1}^{N} \widehat{d}_m^+\langle\alpha_m|a_i\rangle \sum_{n=1}^{N} \langle a_i|\alpha_n\rangle\widehat{d}_n = \sum_{m=1}^{N}\sum_{n=1}^{N} \widehat{d}_m^+ \sum_{i=1}^{N} \langle\alpha_m|a_i\rangle\lambda_i\langle a_i|\alpha_n\rangle\widehat{d}_n.$$

We recognize that $\widehat{A}^{(1)} = \sum_{i=1}^{N} |a_i\rangle\lambda_i\langle a_i|$ is an operator composed from the sum of one-particle operators $\widehat{A}_i$ in spectral representation. Thus, in general,

$$\widehat{A} = \sum_{m=1}^{N}\sum_{n=1}^{N} \widehat{d}_m^+\langle\alpha_m|\widehat{A}^{(1)}|\alpha_n\rangle\widehat{d}_n = \sum_{m=1}^{N}\sum_{n=1}^{N} \widehat{d}_m^+ A_{mn}^{(1)}\widehat{d}_n.$$

a) Based on the preceding, the general expression for the many-body density operator $\widehat{n}(\boldsymbol{x}) = \sum_{i=1}^{N} \widehat{n}_i^{(1)}(\boldsymbol{x}) = \sum_{i=1}^{N} \delta(\widehat{\boldsymbol{r}}_i - \boldsymbol{x})$ is

$$\widehat{n}(\boldsymbol{x}) = \sum_{\alpha,\beta} \widehat{a}_\alpha^+\langle\alpha|\widehat{n}^{(1)}(\boldsymbol{x})|\beta\rangle\widehat{a}_\beta,$$

where $\widehat{n}_1^{(1)}(\boldsymbol{x}) = \delta(\widehat{\boldsymbol{r}}_i - \boldsymbol{x})$ is the density operator of the ith particle.

Using the change of basis in real space, we have

$$\widehat{c}^+(\boldsymbol{x}) = \sum_\alpha \widehat{a}_\alpha^+\langle\alpha|\boldsymbol{x}\rangle, \qquad \widehat{c}(\boldsymbol{x}) = \sum_\alpha \langle\boldsymbol{x}|\alpha\rangle\widehat{a}_\alpha,$$

$$\widehat{a}_\alpha^+ = \int d^3\boldsymbol{x}\widehat{c}^+(\boldsymbol{x})\langle\boldsymbol{x}|\alpha\rangle, \qquad \widehat{a}_\alpha = \int d^3\boldsymbol{x}\langle\alpha|\boldsymbol{x}\rangle\widehat{c}(\boldsymbol{x}),$$

$$\begin{aligned}
\widehat{n}(\boldsymbol{x}) &= \sum_{\alpha,\beta} \widehat{a}_\alpha^+\langle\alpha|\widehat{n}^{(1)}(\boldsymbol{x})|\beta\rangle\widehat{a}_\beta = \sum_{\alpha,\beta}\int d^3\boldsymbol{x}_1\widehat{c}^+(\boldsymbol{x}_1)\langle\boldsymbol{x}_1|\alpha\rangle\langle\alpha|\widehat{n}^{(1)}(\boldsymbol{x})|\beta\rangle\int d^3\boldsymbol{x}_2\langle\beta|\boldsymbol{x}_2\rangle\widehat{c}(\boldsymbol{x}_2) \\
&= \int d^3\boldsymbol{x}_1\int d^3\boldsymbol{x}_2\sum_{\alpha,\beta}\widehat{c}^+(\boldsymbol{x}_1)\langle\boldsymbol{x}_1|\alpha\rangle\langle\alpha|\widehat{n}^{(1)}(\boldsymbol{x})|\beta\rangle\langle\beta|\boldsymbol{x}_2\rangle\widehat{c}(\boldsymbol{x}_2) \\
&= \int d^3\boldsymbol{x}_1\int d^3\boldsymbol{x}_2\widehat{c}^+(\boldsymbol{x}_1)\langle\boldsymbol{x}_1|\left[\sum_\alpha|\alpha\rangle\langle\alpha|\right]\widehat{n}^{(1)}(\boldsymbol{x})\left[\sum_\beta|\beta\rangle\langle\beta|\right]|\boldsymbol{x}_2\rangle\widehat{c}(\boldsymbol{x}_2) \\
&= \int d^3\boldsymbol{x}_1\int d^3\boldsymbol{x}_2\widehat{c}^+(\boldsymbol{x}_1)\langle\boldsymbol{x}_1|\widehat{n}^{(1)}(\boldsymbol{x})|\boldsymbol{x}_2\rangle\widehat{c}(\boldsymbol{x}_2).
\end{aligned}$$

In the preceding, we have used the identity operator $\widehat{\mathbb{1}} = \sum_{\alpha} |\alpha\rangle\langle\alpha|$. Also, from the definition $\widehat{n}^{(1)}(\boldsymbol{x}) = \delta(\widehat{\boldsymbol{r}}_i - \boldsymbol{x})$, the action $f(\widehat{\boldsymbol{r}})|\boldsymbol{x}\rangle = f(\boldsymbol{x})|\boldsymbol{x}\rangle$, and the identity $\langle \boldsymbol{x}_1|\boldsymbol{x}_2\rangle = \delta(\boldsymbol{x}_1 - \boldsymbol{x}_2)$, we find

$$\langle \boldsymbol{x}_1|\widehat{n}^{(1)}(\boldsymbol{x})|\boldsymbol{x}_2\rangle = \langle \boldsymbol{x}_1|\delta(\widehat{\boldsymbol{r}} - \boldsymbol{x})|\boldsymbol{x}_2\rangle = \langle \boldsymbol{x}_1|\boldsymbol{x}_2\rangle\delta(\boldsymbol{x}_2 - \boldsymbol{x}) = \delta(\boldsymbol{x}_1 - \boldsymbol{x}_2)\delta(\boldsymbol{x}_2 - \boldsymbol{x}).$$

Then,

$$\widehat{n}(\boldsymbol{x}) = \int d^3\boldsymbol{x}_1 \int d^3\boldsymbol{x}_2 \widehat{c}^+(\boldsymbol{x}_1)\delta(\boldsymbol{x}_1 - \boldsymbol{x}_2)\delta(\boldsymbol{x}_2 - \boldsymbol{x})\widehat{c}(\boldsymbol{x}_2) = \int d^3\boldsymbol{x}_1 \widehat{c}^+(\boldsymbol{x}_1)\delta(\boldsymbol{x}_1 - \boldsymbol{x})\widehat{c}(\boldsymbol{x})$$
$$= \widehat{c}^+(\boldsymbol{x})\widehat{c}(\boldsymbol{x}).$$

Summarizing,

$$\widehat{n}(\boldsymbol{x}) = \sum_{i=1}^{N} \widehat{n}_i^{(1)}(\boldsymbol{x}) = \sum_{i=1}^{N} \delta(\widehat{\boldsymbol{r}}_i - \boldsymbol{x}) - \textit{First quantized coordinate picture.}$$
$$\widehat{n}(\boldsymbol{x}) = \widehat{c}^+(\boldsymbol{x})\widehat{c}(\boldsymbol{x}) - \textit{Second quantized coordinate picture.}$$

b) In momentum space, the one-particle density operator of the ith particle is its Fourier transform

$$\widehat{\widehat{n}}^{(1)}(\boldsymbol{k}) = \int d^3\boldsymbol{x}\langle \boldsymbol{k}|\boldsymbol{x}\rangle\widehat{n}^{(1)}(\boldsymbol{x}) = \int d^3\boldsymbol{x}\frac{e^{i\boldsymbol{k}\cdot\boldsymbol{x}}}{(2\pi)^{3/2}}\delta(\widehat{\boldsymbol{r}} - \boldsymbol{x}) = \frac{e^{i\boldsymbol{k}\cdot\widehat{\boldsymbol{r}}}}{(2\pi)^{3/2}}.$$

Thus, we have

$$\widehat{n}(\boldsymbol{k}) = \sum_{\boldsymbol{p}_1,\boldsymbol{p}_2} \widehat{a}^+_{\boldsymbol{p}_1}\langle \boldsymbol{p}_1|\widehat{\widehat{n}}^{(1)}(\boldsymbol{k})|\boldsymbol{p}_2\rangle\widehat{a}_{\boldsymbol{p}_2},$$

$$\langle \boldsymbol{p}_1|\widehat{\widehat{n}}^{(1)}(\boldsymbol{k})|\boldsymbol{p}_2\rangle = \langle \boldsymbol{p}_1|\left[\int d^3\boldsymbol{x}_1|\boldsymbol{x}_1\rangle\langle \boldsymbol{x}_1|\right]\widehat{\widehat{n}}^{(1)}(\boldsymbol{k})\left[\int d^3\boldsymbol{x}_2|\boldsymbol{x}_2\rangle\langle \boldsymbol{x}_2|\right]|\boldsymbol{p}_2\rangle$$
$$= \int d^3\boldsymbol{x}_1 \int d^3\boldsymbol{x}_2\langle \boldsymbol{p}_1|\boldsymbol{x}_1\rangle\langle \boldsymbol{x}_1|\widehat{\widehat{n}}^{(1)}(\boldsymbol{k})|\boldsymbol{x}_2\rangle\langle \boldsymbol{x}_2|\boldsymbol{p}_2\rangle$$
$$= \int d^3\boldsymbol{x}_1 \int d^3\boldsymbol{x}_2\frac{e^{i\boldsymbol{p}_1\cdot\boldsymbol{x}_1}}{(2\pi)^{3/2}}\langle \boldsymbol{x}_1|\frac{e^{i\boldsymbol{k}\cdot\widehat{\boldsymbol{r}}}}{(2\pi)^{3/2}}|\boldsymbol{x}_2\rangle\frac{e^{-i\boldsymbol{p}_2\cdot\boldsymbol{x}_2}}{(2\pi)^{3/2}}$$
$$= \int d^3\boldsymbol{x}_1 \int d^3\boldsymbol{x}_2\frac{e^{i\boldsymbol{p}_1\cdot\boldsymbol{x}_1}}{(2\pi)^{3/2}}\langle \boldsymbol{x}_1|\boldsymbol{x}_2\rangle\frac{e^{i\boldsymbol{k}\cdot\boldsymbol{x}_2}}{(2\pi)^{3/2}}\frac{e^{-i\boldsymbol{p}_2\cdot\boldsymbol{x}_2}}{(2\pi)^{3/2}}$$
$$= \int d^3\boldsymbol{x}_1 \int d^3\boldsymbol{x}_2\frac{e^{i\boldsymbol{p}_1\cdot\boldsymbol{x}_1}}{(2\pi)^{3/2}}\delta(\boldsymbol{x}_1 - \boldsymbol{x}_2)\frac{e^{i\boldsymbol{x}_2\cdot(\boldsymbol{k}-\boldsymbol{p}_2)}}{(2\pi)^3}$$
$$= \int d^3\boldsymbol{x}_2\frac{e^{i\boldsymbol{p}_1\cdot\boldsymbol{x}_2}}{(2\pi)^{3/2}}\frac{e^{i\boldsymbol{x}_2\cdot(\boldsymbol{k}-\boldsymbol{p}_2)}}{(2\pi)^3} = \int d^3\boldsymbol{x}_2\frac{e^{i\boldsymbol{x}_2\cdot(\boldsymbol{p}_1+\boldsymbol{k}-\boldsymbol{p}_2)}}{(2\pi)^{9/2}} = \frac{\delta(\boldsymbol{p}_1 + \boldsymbol{k} - \boldsymbol{p}_2)}{(2\pi)^{3/2}}.$$

Then,

$$\widehat{n}(\boldsymbol{k}) = \sum_{\boldsymbol{p}_1,\boldsymbol{p}_2} \widehat{a}^+_{\boldsymbol{p}_1}\langle \boldsymbol{p}_1|\widehat{\widehat{n}}^{(1)}(\boldsymbol{k})|\boldsymbol{p}_2\rangle\widehat{a}_{\boldsymbol{p}_2} = \int d^3\boldsymbol{p}_1 \int d^3\boldsymbol{p}_2\widehat{a}^+_{\boldsymbol{p}_1}\langle \boldsymbol{p}_1|\widehat{\widehat{n}}^{(1)}(\boldsymbol{k})|\boldsymbol{p}_2\rangle\widehat{a}_{\boldsymbol{p}_2}$$
$$= \int d^3\boldsymbol{p}_1 \int d^3\boldsymbol{p}_2\widehat{a}^+_{\boldsymbol{p}_1}\frac{\delta(\boldsymbol{p}_1 + \boldsymbol{k} - \boldsymbol{p}_2)}{(2\pi)^{3/2}}\widehat{a}_{\boldsymbol{p}_2} = \int \frac{d^3\boldsymbol{p}_1}{(2\pi)^{3/2}}\widehat{a}^+_{\boldsymbol{p}_1}\widehat{a}_{\boldsymbol{p}_1+\boldsymbol{k}}.$$

Summarizing,

$$\widehat{n}(\boldsymbol{k}) = \sum_{i=1}^{N} \widehat{n}_i^{(1)}(\boldsymbol{k}) = \sum_{i=1}^{N} \frac{e^{i\boldsymbol{k}\cdot\widehat{\boldsymbol{r}}_i}}{(2\pi)^{3/2}} - \textit{First quantized momentum picture.}$$

$$\widehat{n}(\boldsymbol{k}) = \int \frac{d^3\boldsymbol{p}}{(2\pi)^{3/2}} \widehat{a}_{\boldsymbol{p}}^{+} \widehat{a}_{\boldsymbol{p}+\boldsymbol{k}} - \textit{Second quantized momentum picture.}$$

Problem 5.19

a) Consider the momentum operator $\widehat{\boldsymbol{p}} = -i\hbar\boldsymbol{\nabla}_{\boldsymbol{r}}$ and give its second quantized momentum $\boldsymbol{k}$ space picture using corresponding single-particle Fock space operators. Give the real space $\boldsymbol{x}$ picture using corresponding single-particle Fock space operators.

b) Consider the Hamiltonian $\widehat{H} = \sum_{j=11}^{N} \left[\frac{\widehat{\boldsymbol{p}}_j^2}{2m} + V(\widehat{\boldsymbol{r}}_j)\right]$ and give its second quantization picture in momentum space, using corresponding single-particle Fock space operators. Alternatively, give the same Hamiltonian in real space using the single-particle Fock space operators for this Hamiltonian.

---

a) We consider the Fock space built from eigenstates of the position operator $\widehat{\boldsymbol{r}}$, such that $\widehat{\boldsymbol{r}}_j|\boldsymbol{x}_1, \ldots, \boldsymbol{x}_N\rangle = \boldsymbol{x}_j|\boldsymbol{x}_1, \ldots, \boldsymbol{x}_N\rangle$ and $\langle \boldsymbol{x}_1|\boldsymbol{x}_2\rangle = \delta(\boldsymbol{x}_1 - \boldsymbol{x}_2)$.

The Fock space built from the eigenstates of the operator $\widehat{\boldsymbol{p}}$ are such that $\widehat{\boldsymbol{p}}_j|\boldsymbol{p}_1, \ldots, \boldsymbol{p}_N\rangle = \boldsymbol{p}_j|\boldsymbol{p}_1, \ldots, \boldsymbol{p}_N\rangle$. Using that $\langle \boldsymbol{k}|\boldsymbol{x}\rangle = \frac{e^{i\boldsymbol{k}\cdot\boldsymbol{x}}}{(2\pi)^{d/2}}$, we find that $\langle \boldsymbol{p}_1|\boldsymbol{p}_2\rangle = \langle \boldsymbol{p}_1 \left|\left[\int d\boldsymbol{x}|\boldsymbol{x}\rangle\langle\boldsymbol{x}|\right]\right| \boldsymbol{p}_2\rangle = \int d\boldsymbol{x}\langle \boldsymbol{p}_1|\boldsymbol{x}\rangle\langle\boldsymbol{x}|\boldsymbol{p}_2\rangle = \int d\boldsymbol{x} \frac{e^{i\boldsymbol{p}_1\cdot\boldsymbol{x}}}{(2\pi)^{d/2}} \frac{e^{i\boldsymbol{p}_2\cdot\boldsymbol{x}}}{(2\pi)^{d/2}} = \frac{1}{(2\pi)^d} \int d\boldsymbol{x} e^{i(\boldsymbol{p}_1-\boldsymbol{p}_2)\cdot\boldsymbol{x}} = \delta(\boldsymbol{p}_1 - \boldsymbol{p}_2)$.

Next, we want to evaluate the matrix elements

$$\langle \boldsymbol{x}_1|\widehat{\boldsymbol{p}}^{(1)}|\boldsymbol{x}_2\rangle = \langle \boldsymbol{x}_1| - i\hbar\boldsymbol{\nabla}_{\widehat{\boldsymbol{r}}}|\boldsymbol{x}_2\rangle = -i\hbar\boldsymbol{\nabla}_{\boldsymbol{x}_2}\langle \boldsymbol{x}_1|\boldsymbol{x}_2\rangle = -i\hbar\boldsymbol{\nabla}_{\boldsymbol{x}_2}\delta(\boldsymbol{x}_1 - \boldsymbol{x}_2),$$

$$\begin{aligned}
\langle \boldsymbol{k}_1|\widehat{\boldsymbol{p}}^{(1)}|\boldsymbol{k}_2\rangle &= \langle \boldsymbol{k}_1| \left[\int d\boldsymbol{x}_1|\boldsymbol{x}_1\rangle\langle\boldsymbol{x}_1|\right] \widehat{\boldsymbol{p}}^{(1)} \left[\int d\boldsymbol{x}_2|\boldsymbol{x}_2\rangle\langle\boldsymbol{x}_2|\right] |\boldsymbol{k}_2\rangle \\
&= \int d\boldsymbol{x}_1 \int d\boldsymbol{x}_2 \langle \boldsymbol{k}_1|\boldsymbol{x}_1\rangle\langle\boldsymbol{x}_1|\widehat{\boldsymbol{p}}^{(1)}|\boldsymbol{x}_2\rangle\langle\boldsymbol{x}_2|\boldsymbol{k}_2\rangle \\
&= \int d\boldsymbol{x}_1 \int d\boldsymbol{x}_2 \frac{e^{i\boldsymbol{k}_1\cdot\boldsymbol{x}_1}}{(2\pi)^{d/2}} \left[-i\hbar\boldsymbol{\nabla}_{\boldsymbol{x}_2}\delta(\boldsymbol{x}_1 - \boldsymbol{x}_2)\right] \frac{e^{-i\boldsymbol{k}_2\cdot\boldsymbol{x}_2}}{(2\pi)^{d/2}} \\
&= -i\hbar \int d\boldsymbol{x}_1 \int d\boldsymbol{x}_2 \left[\boldsymbol{\nabla}_{\boldsymbol{x}_2}\delta(\boldsymbol{x}_1 - \boldsymbol{x}_2)\right] \frac{e^{i(\boldsymbol{k}_1\cdot\boldsymbol{x}_1 - \boldsymbol{k}_2\cdot\boldsymbol{x}_2)}}{(2\pi)^d} \\
&= i\hbar \int d\boldsymbol{x}_1 \int d\boldsymbol{x}_2 \delta(\boldsymbol{x}_1 - \boldsymbol{x}_2) \left(\boldsymbol{\nabla}_{\boldsymbol{x}_2} \frac{e^{i(\boldsymbol{k}_1\cdot\boldsymbol{x}_1 - \boldsymbol{k}_2\cdot\boldsymbol{x}_2)}}{(2\pi)^d}\right) \\
&= i\hbar \int d\boldsymbol{x}_1 \int d\boldsymbol{x}_2 \delta(\boldsymbol{x}_1 - \boldsymbol{x}_2) \left((-i\boldsymbol{k}_2) \frac{e^{i(\boldsymbol{k}_1\cdot\boldsymbol{x}_1 - \boldsymbol{k}_2\cdot\boldsymbol{x}_2)}}{(2\pi)^d}\right) \\
&= \hbar\boldsymbol{k}_2 \int d\boldsymbol{x}_1 \int d\boldsymbol{x}_2 \delta(\boldsymbol{x}_1 - \boldsymbol{x}_2) \frac{e^{i(\boldsymbol{k}_1\cdot\boldsymbol{x}_1 - \boldsymbol{k}_2\cdot\boldsymbol{x}_2)}}{(2\pi)^d} \\
&= \hbar\boldsymbol{k}_2 \int d\boldsymbol{x}_1 \frac{e^{i(\boldsymbol{k}_1 - \boldsymbol{k}_2)\cdot\boldsymbol{x}_1}}{(2\pi)^d} = \hbar\boldsymbol{k}_2\delta(\boldsymbol{k}_1 - \boldsymbol{k}_2).
\end{aligned}$$

We further write, the momentum operator in the second quantization picture as

$$\widehat{p}=\int dx_1\int dx_2\widehat{c}^+(x_1)\langle x_1|-i\hbar\nabla_{x_2}|x_2\rangle\widehat{c}(x_2),$$

where the quantized (field) operators are defined as

$$\widehat{c}(x)=\sum_k\langle x|k\rangle\widehat{C}(k)=\int dk\frac{e^{-ik\cdot x}}{(2\pi)^{d/2}}\widehat{C}(k),$$

$$\widehat{c}^+(x)=\sum_k C^+(k)\langle k|x\rangle=\int dk\widehat{C}^+(k)\frac{e^{-ik\cdot x}}{(2\pi)^{d/2}},$$

$$\widehat{C}(k)=\int dx\langle k|x\rangle\widehat{c}(x)=\int dx\frac{e^{ik\cdot x}}{(2\pi)^{d/2}}\widehat{c}(x),$$

$$\widehat{C}^+(k)=\int dx\widehat{c}^+(x)\langle x|k\rangle=\int dx\widehat{c}^+(x)\frac{e^{-ik\cdot x}}{(2\pi)^{d/2}},$$

with the commutation relations

$$\left[\widehat{C}(k_1),\widehat{C}^+(k_2)\right]=\delta(k_1-k_2)\text{ for bosons,}$$
$$\left\{\widehat{C}(k_1),\widehat{C}^+(k_2)\right\}=\delta(k_1-k_2)\text{ for fermions.}$$

Then, the second quantization picture of the momentum operator in position $\widehat{r}$ space is:

$$\begin{aligned}\widehat{p}&=\int dx_1\int dx_2\widehat{c}^+(x_1)\langle x_1|\widehat{p}^{(1)}|x_2\rangle\widehat{c}(x_2)\\&=\int dx_1\int dx_2\widehat{c}^+(x_1)\left[-i\hbar\delta(x_1-x_2)\right]\left[\nabla_{x_2}\widehat{c}(x_2)\right]\\&=\int dx_1\widehat{c}^+(x_1)\left[-i\hbar\nabla_{x_1}\widehat{c}(x_1)\right].\end{aligned}$$

The second quantization picture of the momentum operator in momentum k space is:

$$\begin{aligned}\widehat{p}&=\int dx_1\int dx_2\widehat{C}^+(k_1)\langle k_1|\widehat{p}^{(1)}|k_2\rangle\widehat{C}(k_2)\\&=\int dk_1\int dk_2\widehat{C}^+(k_1)\hbar k_2\delta(k_1-k_2)\widehat{C}(k_2)\\&=\int dk_2\widehat{C}^+(k_2)\hbar k_2\widehat{C}(k_2).\end{aligned}$$

b) For the Hamiltonian in momentum space, we realize that

$$\langle k_1|\widehat{p}^2|k_2\rangle=\langle k_1|(-i\hbar\nabla_x)^2|k_2\rangle=(\hbar k_2)^2\delta(k_1-k_2),$$

$$\begin{aligned}\langle k_1|V(\widehat{r})|k_2\rangle&=\int dx_1\int dx_2\langle k_1|x_1\rangle\langle x_1|V(\widehat{r})|x_2\rangle\langle x_2|k_2\rangle\\&=\int dx_1\int dx_2\frac{e^{ik_1\cdot x_1}}{(2\pi)^{d/2}}\left[V(x_2)\delta(x_1-x_2)\right]\frac{e^{-ik_2\cdot x_2}}{(2\pi)^{d/2}}\end{aligned}$$

$$= \int d\boldsymbol{x}_1 \int d\boldsymbol{x}_2 V(\boldsymbol{x}_2)\,\delta(\boldsymbol{x}_1 - \boldsymbol{x}_2)\,\frac{e^{i(\boldsymbol{k}_1\cdot\boldsymbol{x}_1 - \boldsymbol{k}_2\cdot\boldsymbol{x}_2)}}{(2\pi)^d}$$
$$= \int d\boldsymbol{x}_1 V(\boldsymbol{x}_1)\,\frac{e^{i(\boldsymbol{k}_1 - \boldsymbol{k}_2)\cdot\boldsymbol{x}_1}}{(2\pi)^d} = \frac{\tilde{V}(\boldsymbol{k}_1 - \boldsymbol{k}_2)}{(2\pi)^{d/2}}.$$

From here, the second quantization picture of the Hamiltonian in momentum space is

$$\widehat{H} = \int d\boldsymbol{k}_1 \int d\boldsymbol{k}_2 \widehat{C}^+(\boldsymbol{k}_1)\left\langle \boldsymbol{k}_1 \middle| \widehat{H} \middle| \boldsymbol{k}_2 \right\rangle \widehat{C}(\boldsymbol{k}_2)$$
$$= \int d\boldsymbol{k}_1 \int d\boldsymbol{k}_2 \widehat{C}^+(\boldsymbol{k}_1)\langle \boldsymbol{k}_1 | \left[\frac{\widehat{\boldsymbol{p}}^2}{2m} + V(\widehat{\boldsymbol{r}})\right] | \boldsymbol{k}_2 \rangle \widehat{C}(\boldsymbol{k}_2)$$
$$= \frac{1}{2m}\int d\boldsymbol{k}_1 \int d\boldsymbol{k}_2 \widehat{C}^+(\boldsymbol{k}_1)\left\langle \boldsymbol{k}_1 \middle| \widehat{\boldsymbol{p}}^2 \middle| \boldsymbol{k}_2 \right\rangle \widehat{C}(\boldsymbol{k}_2)$$
$$+ \int d\boldsymbol{k}_1 \int d\boldsymbol{k}_2 \widehat{C}^+(\boldsymbol{k}_1)\langle \boldsymbol{k}_1 | V(\widehat{\boldsymbol{r}}) | \boldsymbol{k}_2 \rangle \widehat{C}(\boldsymbol{k}_2)$$
$$= \frac{1}{2m}\int d\boldsymbol{k}_1 \int d\boldsymbol{k}_2 \widehat{C}^+(\boldsymbol{k}_1)(\hbar \boldsymbol{k}_2)^2\,\delta(\boldsymbol{k}_1 - \boldsymbol{k}_2)\,\widehat{C}(\boldsymbol{k}_2)$$
$$+ \int d\boldsymbol{k}_1 \int d\boldsymbol{k}_2 \widehat{C}^+(\boldsymbol{k}_1)\frac{\tilde{V}(\boldsymbol{k}_1 - \boldsymbol{k}_2)}{(2\pi)^{d/2}}\widehat{C}(\boldsymbol{k}_2)$$
$$= \int d\boldsymbol{k}_1 \int d\boldsymbol{k}_2 \widehat{C}^+(\boldsymbol{k}_1)\left[\frac{\hbar^2 \boldsymbol{k}_2^2}{2m}\delta(\boldsymbol{k}_1 - \boldsymbol{k}_2) + \frac{\tilde{V}(\boldsymbol{k}_1 - \boldsymbol{k}_2)}{(2\pi)^{d/2}}\right]\widehat{C}(\boldsymbol{k}_2).$$

Therefore, the second quantized form of the Hamiltonian in momentum $\boldsymbol{k}$ space becomes

$$\widehat{H} = \int d\boldsymbol{k}_1 \widehat{C}^+(\boldsymbol{k}_1)\frac{\hbar^2 \boldsymbol{k}_1^2}{2m}\widehat{C}(\boldsymbol{k}_1) + \int d\boldsymbol{k}_1 \int d\boldsymbol{k}_2 \widehat{C}^+(\boldsymbol{k}_1)\frac{\tilde{V}(\boldsymbol{k}_1 - \boldsymbol{k}_2)}{(2\pi)^{d/2}}\widehat{C}(\boldsymbol{k}_2).$$

Problem 5.20

Consider a two-particle interaction potential given in real space $\tilde{V} = \frac{1}{2}\sum_{\ell \neq m}^{N} V^{(2)}(\widehat{\boldsymbol{r}}_\ell - \widehat{\boldsymbol{r}}_m)$, where N denotes the number of particles in the system.

a) Write this potential in the *second quantization picture* in *real space* and in *momentum space* when the spatial coordinates are $x_i \in \left[-\frac{L}{2}, \frac{L}{2}\right]$.
b) Write this potential in the *second quantization picture* in *real space* and in *momentum space* when the spatial coordinates are $x_i \in (-\infty, \infty)$.

For a two-body operator $\widehat{A} = \sum_{i=1}^{N}\sum_{j\neq i}^{N} \widehat{A}^{(2)}(\boldsymbol{r}_i, \boldsymbol{r}_j)$, we remember that the second quantized picture in an arbitrary basis can be given as

$$\widehat{A} = \frac{1}{2}\sum_{\alpha\beta\gamma\delta} \widehat{a}_\alpha^+ \widehat{a}_\beta^+ A_{\alpha\beta,\gamma\delta}^{(2)} \widehat{a}_\gamma \widehat{a}_\delta,$$

where $A^{(2)}_{\alpha\beta,\gamma\delta} = \left\langle \alpha_i\beta_j \left| \widehat{A}^{(2)}\left(\boldsymbol{r}_i,\boldsymbol{r}_j\right) \right| \gamma_i\delta_j \right\rangle$. The states $|\alpha_i\rangle$ and $|\gamma_i\rangle$ correspond to the ith particle, while the states $|\beta_j\rangle$ and $|\delta_j\rangle$ correspond to the jth particle.

Using the change of basis applied in Problem 5.18 and 5.19, for example, the $\widehat{a}$ operators can be replaced by the field operators $\widehat{c}(\boldsymbol{r})$ in real space,

$$\widehat{V} = \frac{1}{2}\int d\boldsymbol{x}_1 \int d\boldsymbol{x}_2 \int d\boldsymbol{x}_3 \int d\boldsymbol{x}_4 \widehat{c}^+(\boldsymbol{x}_1)\widehat{c}^+(\boldsymbol{x}_2)\left\langle \boldsymbol{x}_1\boldsymbol{x}_2 \left| V^{(2)}\left(\widehat{\boldsymbol{r}}_i - \widehat{\boldsymbol{r}}_j\right) \right| \boldsymbol{x}_3\boldsymbol{x}_4 \right\rangle \widehat{c}(\boldsymbol{x}_3)\widehat{c}(\boldsymbol{x}_4),$$

$$\widehat{V} = \frac{1}{2}\int d\boldsymbol{x}_1 \int d\boldsymbol{x}_2 \int d\boldsymbol{x}_3 \int d\boldsymbol{x}_4 \widehat{c}^+(\boldsymbol{x}_1)\widehat{c}^+(\boldsymbol{x}_2) V^{(2)}(\boldsymbol{x}_3 - \boldsymbol{x}_4)\langle \boldsymbol{x}_1\boldsymbol{x}_2 \,|\, \boldsymbol{x}_3\boldsymbol{x}_4\rangle \widehat{c}(\boldsymbol{x}_3)\widehat{c}(\boldsymbol{x}_4).$$

We further use that $\boldsymbol{x}_1$ and $\boldsymbol{x}_3$ correspond to the ith particle, while $\boldsymbol{x}_2$ and $\boldsymbol{x}_4$ correspond to the jth particle

$$\langle \boldsymbol{x}_1\boldsymbol{x}_2 \,|\, \boldsymbol{x}_3\boldsymbol{x}_4\rangle = \delta(\boldsymbol{x}_1 - \boldsymbol{x}_3)\,\delta(\boldsymbol{x}_2 - \boldsymbol{x}_4),$$

$$\begin{aligned}\widehat{V} &= \frac{1}{2}\int d\boldsymbol{x}_1 \int d\boldsymbol{x}_2 \int d\boldsymbol{x}_3 \int d\boldsymbol{x}_4 \widehat{c}^+(\boldsymbol{x}_1)\widehat{c}^+(\boldsymbol{x}_2) V^{(2)}(\boldsymbol{x}_3 - \boldsymbol{x}_4)\,\delta(\boldsymbol{x}_1 - \boldsymbol{x}_3) \\ &\quad\times \delta(\boldsymbol{x}_2 - \boldsymbol{x}_4)\,\widehat{c}(\boldsymbol{x}_3)\widehat{c}(\boldsymbol{x}_4) \\ &= \frac{1}{2}\int d\boldsymbol{x}_1 \int d\boldsymbol{x}_2 \widehat{c}^+(\boldsymbol{x}_1)\widehat{c}^+(\boldsymbol{x}_2) V^{(2)}(\boldsymbol{x}_1 - \boldsymbol{x}_2)\,\widehat{c}(\boldsymbol{x}_1)\widehat{c}(\boldsymbol{x}_2).\end{aligned}$$

a) In momentum space, for a system in a finite volume V_0, we use the relation $\widehat{c}(\boldsymbol{x}) = \sum_{\boldsymbol{k}} \frac{e^{-i\boldsymbol{k}\cdot\boldsymbol{x}}}{\sqrt{V_0}}\widehat{a}_{\boldsymbol{k}}$. Thus, $\widehat{c}^+(\boldsymbol{x}) = \sum_{\boldsymbol{k}} \widehat{a}^+_{\boldsymbol{k}} \frac{e^{+i\boldsymbol{k}\cdot\boldsymbol{x}}}{\sqrt{V_0}}$ and

$$\begin{aligned}\widehat{V} &= \frac{1}{2}\int_{V_0} d\boldsymbol{x}_1 \int_{V_0} d\boldsymbol{x}_2 \widehat{c}^+(\boldsymbol{x}_1)\widehat{c}^+(\boldsymbol{x}_2) V^{(2)}(\boldsymbol{x}_1 - \boldsymbol{x}_2)\,\widehat{c}(\boldsymbol{x}_1)\widehat{c}(\boldsymbol{x}_2) \\ &= \frac{1}{2}\int_{V_0} d\boldsymbol{x}_1 \int_{V_0} d\boldsymbol{x}_2 \sum_{\boldsymbol{k}_1} \widehat{a}^+_{\boldsymbol{k}_1}\frac{e^{i\boldsymbol{k}_1\cdot\boldsymbol{x}_1}}{\sqrt{V_0}} \sum_{\boldsymbol{k}_2} \widehat{a}^+_{\boldsymbol{k}_2}\frac{e^{i\boldsymbol{k}_2\cdot\boldsymbol{x}_2}}{\sqrt{V_0}} V^{(2)}(\boldsymbol{x}_1 - \boldsymbol{x}_2) \sum_{\boldsymbol{k}_3} \frac{e^{-i\boldsymbol{k}_3\cdot\boldsymbol{x}_1}}{\sqrt{V_0}}\widehat{a}_{\boldsymbol{k}_3} \sum_{\boldsymbol{k}_4} \frac{e^{-i\boldsymbol{k}_4\cdot\boldsymbol{x}_2}}{\sqrt{V_0}}\widehat{a}_{\boldsymbol{k}_4} \\ &= \frac{1}{2V_0^2} \sum_{\boldsymbol{k}_1,\boldsymbol{k}_2,\boldsymbol{k}_3,\boldsymbol{k}_4} \widehat{a}^+_{\boldsymbol{k}_1}\widehat{a}^+_{\boldsymbol{k}_2}\widehat{a}_{\boldsymbol{k}_3}\widehat{a}_{\boldsymbol{k}_4} \int_{V_0} d\boldsymbol{x}_1 \int_{V_0} d\boldsymbol{x}_2 e^{i(\boldsymbol{k}_1-\boldsymbol{k}_3)\cdot\boldsymbol{x}_1} V^{(2)}(\boldsymbol{x}_1 - \boldsymbol{x}_2)\, e^{i(\boldsymbol{k}_2-\boldsymbol{k}_4)\cdot\boldsymbol{x}_2}.\end{aligned}$$

On the other hand,

$$V^{(2)}(\boldsymbol{x}_1 - \boldsymbol{x}_2) = \frac{1}{\sqrt{V_0}}\sum_{\boldsymbol{k}} \tilde{V}^{(2)}(\boldsymbol{k}) e^{-i\boldsymbol{k}\cdot(\boldsymbol{x}_1-\boldsymbol{x}_2)}.$$

Therefore,

$$\begin{aligned}\widehat{V} &= \frac{1}{2V_0^2} \sum_{\boldsymbol{k}_1,\boldsymbol{k}_2,\boldsymbol{k}_3,\boldsymbol{k}_4} \widehat{a}^+_{\boldsymbol{k}_1}\widehat{a}^+_{\boldsymbol{k}_2}\widehat{a}_{\boldsymbol{k}_3}\widehat{a}_{\boldsymbol{k}_4} \int_{V_0} d\boldsymbol{x}_1 \int_{V_0} d\boldsymbol{x}_2 e^{i(\boldsymbol{k}_1-\boldsymbol{k}_3)\cdot\boldsymbol{x}_1} \frac{1}{\sqrt{V_0}} \\ &\quad\times \sum_{\boldsymbol{k}_0} \tilde{V}^{(2)}(\boldsymbol{k}_0)\, e^{-i\boldsymbol{k}_0\cdot(\boldsymbol{x}_1-\boldsymbol{x}_2)} e^{i(\boldsymbol{k}_2-\boldsymbol{k}_4)\cdot\boldsymbol{x}_2} \\ &= \frac{1}{2V_0^{5/2}} \sum_{\boldsymbol{k}_0,\boldsymbol{k}_1,\boldsymbol{k}_2,\boldsymbol{k}_3,\boldsymbol{k}_4} \widehat{a}^+_{\boldsymbol{k}_1}\widehat{a}^+_{\boldsymbol{k}_2}\widehat{a}_{\boldsymbol{k}_3}\widehat{a}_{\boldsymbol{k}_4} \tilde{V}^{(2)}(\boldsymbol{k}_0) \int_{V_0} d\boldsymbol{x}_1 e^{i(\boldsymbol{k}_1-\boldsymbol{k}_3-\boldsymbol{k}_0)\cdot\boldsymbol{x}_1} \int_{V_0} d\boldsymbol{x}_2 e^{i(\boldsymbol{k}_2-\boldsymbol{k}_4+\boldsymbol{k}_0)\cdot\boldsymbol{x}_2} \\ &= \frac{1}{2V_0^{5/2}} \sum_{\boldsymbol{k}_0,\boldsymbol{k}_1,\boldsymbol{k}_2,\boldsymbol{k}_3,\boldsymbol{k}_4} \widehat{a}^+_{\boldsymbol{k}_1}\widehat{a}^+_{\boldsymbol{k}_2}\widehat{a}_{\boldsymbol{k}_3}\widehat{a}_{\boldsymbol{k}_4} \tilde{V}^{(2)}(\boldsymbol{k}_0)\, V_0\delta_{\boldsymbol{k}_3,\boldsymbol{k}_1-\boldsymbol{k}_0} V_0\delta_{\boldsymbol{k}_4,\boldsymbol{k}_2+\boldsymbol{k}_0} \\ &= \frac{1}{2\sqrt{V_0}} \sum_{\boldsymbol{k}_0,\boldsymbol{k}_1,\boldsymbol{k}_2} \widehat{a}^+_{\boldsymbol{k}_1}\widehat{a}^+_{\boldsymbol{k}_2}\tilde{V}^{(2)}(\boldsymbol{k}_0)\,\widehat{a}_{\boldsymbol{k}_1-\boldsymbol{k}_0}\widehat{a}_{\boldsymbol{k}_2+\boldsymbol{k}_0}.\end{aligned}$$

b) In position space, for the momentum we use the relation $\widehat{c}(\boldsymbol{x}) = \sum_{\boldsymbol{k}} \langle \boldsymbol{x}|\boldsymbol{k}\rangle \widehat{c}(\boldsymbol{k}) = \int d\boldsymbol{k} \frac{e^{-i\boldsymbol{k}\cdot\boldsymbol{x}}}{(2\pi)^{d/2}} \widehat{c}(\boldsymbol{k})$. Thus, $\widehat{c}^{+}(\boldsymbol{x}) = \sum_{\boldsymbol{k}} \widehat{c}(\boldsymbol{k})\langle \boldsymbol{k}|\boldsymbol{x}\rangle = \int d\boldsymbol{k} \widehat{c}^{+}(\boldsymbol{k}) \frac{e^{i\boldsymbol{k}\cdot\boldsymbol{x}}}{(2\pi)^{d/2}}$. From here,

$$
\begin{aligned}
\widehat{V} &= \frac{1}{2}\int d\boldsymbol{x}_1 \int d\boldsymbol{x}_2 \widehat{c}^{+}(\boldsymbol{x}_1)\widehat{c}^{+}(\boldsymbol{x}_2) V^{(2)}(\boldsymbol{x}_1-\boldsymbol{x}_2)\widehat{c}(\boldsymbol{x}_1)\widehat{c}(\boldsymbol{x}_2) \\
&= \frac{1}{2}\int d\boldsymbol{x}_1 \int d\boldsymbol{x}_2 \int d\boldsymbol{k}_1 \widehat{c}^{+}(\boldsymbol{k}_1)\frac{e^{i\boldsymbol{k}_1\cdot\boldsymbol{x}_1}}{(2\pi)^{d/2}} \int d\boldsymbol{k}_2 \widehat{c}^{+}(\boldsymbol{k}_2)\frac{e^{i\boldsymbol{k}_2\cdot\boldsymbol{x}_2}}{(2\pi)^{d/2}} V^{(2)}(\boldsymbol{x}_1-\boldsymbol{x}_2) \\
&\quad \times \int d\boldsymbol{k}_3 \frac{e^{-i\boldsymbol{k}_3\cdot\boldsymbol{x}_1}}{(2\pi)^{d/2}}\widehat{c}(\boldsymbol{k}_3) \int d\boldsymbol{k}_4 \frac{e^{-i\boldsymbol{k}_4\cdot\boldsymbol{x}_2}}{(2\pi)^{d/2}}\widehat{c}(\boldsymbol{k}_4) \\
&= \frac{1}{2(2\pi)^{2d}} \int d\boldsymbol{k}_1 \int d\boldsymbol{k}_2 \int d\boldsymbol{k}_3 \int d\boldsymbol{k}_4 \widehat{c}^{+}(\boldsymbol{k}_1)\widehat{c}^{+}(\boldsymbol{k}_2)\widehat{c}(\boldsymbol{k}_3)\widehat{c}(\boldsymbol{k}_4) \\
&\quad \times \int d\boldsymbol{x}_1 \int d\boldsymbol{x}_2 e^{i(\boldsymbol{k}_1-\boldsymbol{k}_3)\cdot\boldsymbol{x}_1} V^{(2)}(\boldsymbol{x}_1-\boldsymbol{x}_2) e^{i(\boldsymbol{k}_2-\boldsymbol{k}_4)\cdot\boldsymbol{x}_2}.
\end{aligned}
$$

On the other hand,

$$
V^{(2)}(\boldsymbol{x}_1-\boldsymbol{x}_2) = \int d\boldsymbol{k}\tilde{V}^{(2)}(\boldsymbol{k})\frac{e^{-i\boldsymbol{k}\cdot(\boldsymbol{x}_1-\boldsymbol{x}_2)}}{(2\pi)^{d/2}}.
$$

Therefore,

$$
\begin{aligned}
\widehat{V} &= \frac{1}{2(2\pi)^{2d}} \int d\boldsymbol{k}_1 \int d\boldsymbol{k}_2 \int d\boldsymbol{k}_3 \int d\boldsymbol{k}_4 \widehat{c}^{+}(\boldsymbol{k}_1)\widehat{c}^{+}(\boldsymbol{k}_2)\widehat{c}(\boldsymbol{k}_3)\widehat{c}(\boldsymbol{k}_4) \int d\boldsymbol{x}_1 \\
&\quad \times \int d\boldsymbol{x}_2 e^{i(\boldsymbol{k}_1-\boldsymbol{k}_3)\cdot\boldsymbol{x}_1} \int d\boldsymbol{k}_0 \tilde{V}^{(2)}(\boldsymbol{k}_0)\frac{e^{-i\boldsymbol{k}_0\cdot(\boldsymbol{x}_1-\boldsymbol{x}_2)}}{(2\pi)^{d/2}} e^{i(\boldsymbol{k}_2-\boldsymbol{k}_4)\cdot\boldsymbol{x}_2}, \\
\widehat{V} &= \frac{1}{2(2\pi)^{5d/2}} \int d\boldsymbol{k}_0 \int d\boldsymbol{k}_1 \int d\boldsymbol{k}_2 \int d\boldsymbol{k}_3 \int d\boldsymbol{k}_4 \widehat{c}^{+}(\boldsymbol{k}_1)\widehat{c}^{+}(\boldsymbol{k}_2)\widehat{c}(\boldsymbol{k}_3)\widehat{c}(\boldsymbol{k}_4)\tilde{V}^{(2)}(\boldsymbol{k}_0) \\
&\quad \times \int d\boldsymbol{x}_1 e^{i(\boldsymbol{k}_1-\boldsymbol{k}_3-\boldsymbol{k}_0)\cdot\boldsymbol{x}_1} \int d\boldsymbol{x}_2 e^{i(\boldsymbol{k}_2-\boldsymbol{k}_4+\boldsymbol{k}_0)\cdot\boldsymbol{x}_2}, \\
\widehat{V} &= \frac{1}{2(2\pi)^{5d/2}} \int d\boldsymbol{k}_0 \int d\boldsymbol{k}_1 \int d\boldsymbol{k}_2 \int d\boldsymbol{k}_3 \int d\boldsymbol{k}_4 \widehat{c}^{+}(\boldsymbol{k}_1)\widehat{c}^{+}(\boldsymbol{k}_2)\widehat{c}(\boldsymbol{k}_3)\widehat{c}(\boldsymbol{k}_4)\tilde{V}^{(2)}(\boldsymbol{k}_0) \\
&\quad \times (2\pi)^d \delta(\boldsymbol{k}_1-\boldsymbol{k}_3-\boldsymbol{k}_0)(2\pi)^d \delta(\boldsymbol{k}_2-\boldsymbol{k}_4+\boldsymbol{k}_0), \\
\widehat{V} &= \frac{1}{2(2\pi)^{d/2}} \int d\boldsymbol{k}_0 \int d\boldsymbol{k}_1 \int d\boldsymbol{k}_2 \widehat{c}^{+}(\boldsymbol{k}_1)\widehat{c}^{+}(\boldsymbol{k}_2)\tilde{V}^{(2)}(\boldsymbol{k}_0)\widehat{c}(\boldsymbol{k}_1-\boldsymbol{k}_0)\widehat{c}(\boldsymbol{k}_2+\boldsymbol{k}_0).
\end{aligned}
$$

Consider the two-particle interaction written as $\widehat{V} = \frac{1}{2}\int d\boldsymbol{x}_1 \int d\boldsymbol{x}_2 V^{(2)}(\boldsymbol{x}_1-\boldsymbol{x}_2)\widehat{\rho}(\boldsymbol{x}_1)\widehat{\rho}(\boldsymbol{x}_2)$. After substituting the density operators and using the appropriate commutation relations, you should get

$$
\begin{aligned}
\widehat{V} &= \frac{1}{2}\int d\boldsymbol{x}_1 \int d\boldsymbol{x}_2 \widehat{c}^{+}(\boldsymbol{x}_1)\widehat{c}^{+}(\boldsymbol{x}_2) V^{(2)}(\boldsymbol{x}_1-\boldsymbol{x}_2)\widehat{c}(\boldsymbol{x}_1)\widehat{c}(\boldsymbol{x}_2) \\
&= \frac{1}{2}\int d\boldsymbol{x}_1 \int d\boldsymbol{x}_2 V^{(2)}(\boldsymbol{x}_1-\boldsymbol{x}_2)\widehat{\rho}(\boldsymbol{x}_1)\widehat{\rho}(\boldsymbol{x}_2) - \frac{1}{2}\int d\boldsymbol{x}_1 V^{(2)}(0)\widehat{c}^{+}(\boldsymbol{x}_1)\widehat{c}(\boldsymbol{x}_1),
\end{aligned}
$$

where $\widehat{\rho}(\boldsymbol{x}_1) = \widehat{c}^{+}(\boldsymbol{x}_1)\widehat{c}(\boldsymbol{x}_1)$. The last term is the (infinite) self-energy of the interacting system.

Problem 5.21
Consider the ground state of a gas composed of N identical noninteracting particles contained in some finite volume V. In the case of fermions, this is a *Fermi gas*; in the case of bosons, this is a *Bose gas*. Calculate the mean particle number density for each type of gas (Taylor & Heinonen, 2002).

The Hamiltonian of a noninteracting quantum many-body system is $\widehat{H} = \sum_{n=1}^{N} \widehat{h}_n$, where $\widehat{h}_n$ is the one-particle Hamiltonian. In the case of noninteracting particles, $\widehat{h}_n = K(\widehat{\boldsymbol{p}}_n) + V(\widehat{\boldsymbol{r}}_n)$, with kinetic and potential energies, respectively,

$$K(\widehat{\boldsymbol{p}}_n) = \frac{\widehat{\boldsymbol{p}}_n^2}{2m},$$

$$V(\widehat{x}_i) = \begin{cases} 0, & 0 < x_i < L_i \\ \infty, & otherwise \end{cases}.$$

The eigenstates of such a Hamiltonian are denoted as $|a_1 \ldots a_N\rangle$, where $\widehat{h}_n |a_n\rangle = E_n |a_n\rangle$. To find the mean particle number density, we have to evaluate

$$\langle \widehat{n} \rangle = \frac{\langle \widehat{N} \rangle}{V},$$

where $\langle \widehat{N} \rangle$ is the expectation value of the density operator. In Problem 5.16, we showed that the density operator for one particle in the *first quantization picture* is $\widehat{n}(\boldsymbol{x}) = \sum_{n=1}^{N} \widehat{n}_n^{(1)}(\boldsymbol{x}) = \sum_{n=1}^{N} \delta(\widehat{\boldsymbol{r}}_n - \boldsymbol{x})$. Therefore, taking the $|a_1 \ldots a_N\rangle$ eigenbasis,

$$\begin{aligned}
\langle \widehat{n} \rangle &= \frac{\langle \widehat{N} \rangle}{V} = \frac{1}{V} \int_V d\boldsymbol{x} \langle a_1 \ldots a_N | \widehat{n}(\boldsymbol{x}) | a_1 \ldots a_N \rangle \\
&= \frac{1}{V} \int_V d\boldsymbol{x} \int_V \prod_{a=1}^{N} d\boldsymbol{x}_a \int_V \prod_{\beta=1}^{N} d\boldsymbol{y}_\beta \langle a_1 \ldots a_N | \boldsymbol{x}_1 \ldots \boldsymbol{x}_N \rangle \\
&\quad \times \left\langle \boldsymbol{x}_1 \ldots \boldsymbol{x}_N \left| \sum_{n=1}^{N} \delta(\widehat{\boldsymbol{r}}_n - \boldsymbol{x}) \right| \boldsymbol{y}_1 \ldots \boldsymbol{y}_N \right\rangle \langle \boldsymbol{y}_1 \ldots \boldsymbol{y}_N | a_1 \ldots a_N \rangle \\
&= \frac{1}{V} \int_V d\boldsymbol{x} \int_V \prod_{a=1}^{N} d\boldsymbol{x}_a \int_V \prod_{\beta=1}^{N} d\boldsymbol{y}_\beta \langle a_1 \ldots a_N | \boldsymbol{x}_1 \ldots \boldsymbol{x}_N \rangle \\
&\quad \times \sum_{n=1}^{N} \delta(\boldsymbol{x}_n - \boldsymbol{x}) \langle \boldsymbol{x}_1 \ldots \boldsymbol{x}_N | \boldsymbol{y}_1 \ldots \boldsymbol{y}_N \rangle \langle \boldsymbol{y}_1 \ldots \boldsymbol{y}_N | a_1 \ldots a_N \rangle \\
&= \frac{1}{V} \int_V d\boldsymbol{x} \int_V \prod_{a=1}^{N} d\boldsymbol{x}_a \langle a_1 \ldots a_N | \boldsymbol{x}_1 \ldots \boldsymbol{x}_N \rangle \sum_{n=1}^{N} \delta(\boldsymbol{x}_n - \boldsymbol{x}) \langle \boldsymbol{x}_1 \ldots \boldsymbol{x}_N | a_1 \ldots a_N \rangle \\
&= \frac{1}{V} \int_V d\boldsymbol{x} \sum_{n=1}^{N} \langle a_n | \boldsymbol{x} \rangle \langle \boldsymbol{x} | a_n \rangle \prod_{\alpha \neq n}^{N} \langle a_\alpha | a_\alpha \rangle = \frac{1}{V} \sum_{n=1}^{N} \int d\boldsymbol{x} \left| \psi_{a_n}(\boldsymbol{x}) \right|^2 = \frac{1}{V} \sum_{n=1}^{N} 1 = \frac{N}{V}.
\end{aligned}$$

For deriving $\langle \widehat{n} \rangle$ in the *second quantization picture*, we use that $\widehat{n}(\boldsymbol{x}) = \widehat{c}^+(\boldsymbol{x}) \widehat{c}(\boldsymbol{x})$ and the eigenstates of the Hamiltonian are labeled as $|n_1(\boldsymbol{x}_1) \ldots n_R(\boldsymbol{x}_R)\rangle$, where R is the number of different one-particle states. For N fermions, $R = N$ due to the Pauli exclusion principle; for N bosons, $R \leqslant N$ two or more particles can occupy the same state. Then we write

$$\begin{aligned}\langle\widehat{n}\rangle &= \frac{\langle\widehat{N}\rangle}{V} = \frac{1}{V}\int_V dx\,\langle n_1\ldots n_R\,|\widehat{c}^{+}(x)\widehat{c}(x)|\,n_1\ldots n_R\rangle \\ &= \frac{1}{V}\int_V dx\,\langle n_1(x_1)\ldots n_R(x_R)\,|\widehat{c}^{+}(x)\widehat{c}(x)|\,n_1(x_1)\ldots n_R(x_R)\rangle \\ &= \frac{1}{V}\int_V dx\,\langle n_1(x_1)\ldots n_R(x_R)\,|\widehat{n}(x)|n_1(x_1)\ldots n_R(x_R)\rangle \\ &= \frac{1}{V}\int_V dxn(x)\,\langle n_1(x_1)\ldots n_R(x_R)\,|\,n_1(x_1)\ldots n_R(x_R)\rangle = \frac{1}{V}\int_V dxn(x) = \frac{N}{V},\end{aligned}$$

where we have used $N = \int_V drn(r)$. These results are valid for both bosonic and fermionic noninteracting particles confined in a finite volume V.

Problem 5.22

In physics problems we often have to deal with multiparticle systems at nonzero temperature T. In such cases, one may need to deal with correlation properties taken into account via *Green's functions* (Mahan, 2000).

Consider the case of free electrons whose Hamiltonian can be written as $\widehat{H} = \sum_\alpha E_\alpha \widehat{a}_\alpha^{+}\widehat{a}_\alpha$, where $\widehat{a}_\alpha^{+}$ and $\widehat{a}_\alpha$ are the creation and annihilation operators for the particle with properties labeled by α. The Green's function is defined as

$$\mathcal{G}_{\alpha\gamma}(t,t_0) = \frac{1}{i\hbar}Tr\left(\widehat{\rho}\mathcal{T}_t\left\{\widehat{a}_\alpha(t)\,\widehat{a}_\gamma^{+}(t_0)\right\}\right),$$

where $\mathcal{T}_t$ is the time ordering meta-operator,

$$\mathcal{T}_t\left\{\widehat{a}_\alpha(t)\widehat{a}_\gamma^{+}(t_0)\right\} = \begin{cases}\widehat{a}_\alpha(t)\widehat{a}_\gamma^{+}(t_0), & t > t_0 \\ e^{i\theta}\widehat{a}_\gamma^{+}(t_0)\,\widehat{a}_\alpha(t), & t < t_0\end{cases}.$$

In the preceding, a phase factor $e^{i\theta}$ acquired upon changing the order of the creation and annihilation operators is also included in the definition. Note that, for bosons $\theta = 0$ while, for fermions, $\theta = \pi$.

Based on the preceding definitions:

a) Obtain $\langle\widehat{a}_\alpha(t)\widehat{a}_\gamma^{+}(t_0)\rangle$ for the canonical ensemble and $t < t_0$; then $\widehat{\rho} = \widehat{\rho}_C = \frac{e^{-\beta\widehat{H}}}{Z}$, $Z = Tr(\widehat{\rho}_C) = e^{-\beta\mathcal{F}}$, and $\beta = 1/k_BT$ with $\mathcal{F}$ being the Helmholtz free energy.
b) Obtain an expression for the Green's function from the definition working in real time.

---

a) Starting from the definition, and assuming first that $t < t_0$, in the canonical ensemble we have

$$\begin{aligned}\mathcal{G}_{\alpha\gamma}(t,t_0) &= \frac{1}{i\hbar}\langle\mathcal{T}_t\left\{\widehat{a}_\alpha(t)\widehat{a}_\gamma^{+}(t_0)\right\}\rangle = \frac{1}{i\hbar}Tr\left(\widehat{\rho}_C\mathcal{T}_t\left\{\widehat{a}_\alpha(t)\widehat{a}_\gamma^{+}(t_0)\right\}\right) \\ &\underset{t<t_0}{=} \frac{1}{i\hbar}Tr\left(\frac{e^{-\beta\widehat{H}}}{Z}e^{i\theta}\widehat{a}_\gamma^{+}(t_0)\,\widehat{a}_\alpha(t)\right) = \frac{1}{i\hbar}e^{i\theta}\langle\widehat{a}_\gamma^{+}(t_0)\,\widehat{a}_\alpha(t)\rangle.\end{aligned}$$

Here we remember that the t-dependent operators in the Heisenberg picture are related to the t-independent operators in the Schrödinger picture as

$$\widehat{a}_\alpha(t) = e^{i\frac{t}{\hbar}\widehat{H}}\widehat{a}_\alpha e^{-i\frac{t}{\hbar}\widehat{H}} = e^{-i\frac{t}{\hbar}E_\alpha}\widehat{a}_\alpha,$$
$$\widehat{a}_\alpha^+(t) = e^{i\frac{t}{\hbar}\widehat{H}}\widehat{a}_\alpha^+ e^{-i\frac{t}{\hbar}\widehat{H}} = e^{i\frac{t}{\hbar}E_\alpha}\widehat{a}_\alpha^+.$$

Thus, by using the property $Tr(ABC) = Tr(CAB)$, we obtain

$$\begin{aligned}
\langle \widehat{a}_\gamma^+(t_0)\widehat{a}_\alpha(t)\rangle &\underset{t<t_0}{=} \frac{1}{Z}Tr\left(e^{-\beta\widehat{H}}e^{i\frac{t_0}{\hbar}\widehat{H}}\widehat{a}_\gamma^+ e^{-i\frac{t_0}{\hbar}\widehat{H}}e^{i\frac{t}{\hbar}\widehat{H}}\widehat{a}_\alpha e^{-i\frac{t}{\hbar}\widehat{H}}\right)\\
&= \frac{1}{Z}Tr\left(e^{i\frac{t_0}{\hbar}\widehat{H}}e^{-\beta\widehat{H}}\widehat{a}_\gamma^+ e^{-i\frac{t_0}{\hbar}\widehat{H}}e^{i\frac{t}{\hbar}\widehat{H}}\widehat{a}_\alpha e^{-i\frac{t}{\hbar}\widehat{H}}\right)\\
&= \frac{1}{Z}Tr\left(e^{-\beta\widehat{H}}\widehat{a}_\gamma^+ e^{i\frac{(t-t_0)}{\hbar}\widehat{H}}\widehat{a}_\alpha e^{-i\frac{(t-t_0)}{\hbar}\widehat{H}}\right)\\
&\underset{t<t_0}{=} \frac{1}{Z}Tr\left(e^{-\beta\widehat{H}}\widehat{a}_\gamma^+ e^{\beta\widehat{H}}e^{-\beta\widehat{H}}e^{i\frac{(t-t_0)}{\hbar}\widehat{H}}\widehat{a}_\alpha e^{-i\frac{(t-t_0)}{\hbar}\widehat{H}}\right)\\
&= \langle \widehat{a}_\gamma^+(0)\widehat{a}_\alpha(t-t_0)\rangle.
\end{aligned}$$

We evaluate $e^{-\beta\widehat{H}}\widehat{a}_\gamma^+ e^{\beta\widehat{H}}$ and $e^{-\beta\widehat{H}}\widehat{a}_\alpha e^{\beta\widehat{H}}$ by using the relation $e^{\widehat{A}}\widehat{B}e^{-\widehat{A}} = \sum_{n=0}^{\infty}\frac{\widehat{C}_n}{n!}$, with $\widehat{C}_0 = \widehat{B}$, and $\widehat{C}_{n+1} = \left[\widehat{A},\widehat{C}_n\right]$ and the commutation relations for fermionic creation and annihilation operators, obtaining

$$e^{-\beta\widehat{H}}\widehat{a}_\gamma^+ e^{\beta\widehat{H}} = \sum_{n=0}^{\infty}\frac{(-\beta E_\gamma)}{n!}\widehat{a}_\gamma^+ = e^{-\beta E_\gamma}\widehat{a}_\gamma^+,$$
$$e^{-\beta\widehat{H}}\widehat{a}_\alpha e^{\beta\widehat{H}} = \sum_{n=0}^{\infty}\frac{(\beta E_\alpha)^n}{n!}\widehat{a}_\alpha = e^{\beta E_\alpha}\widehat{a}_\alpha.$$

Substituting these results and using the property $Tr(ABC) = Tr(CAB)$, we obtain

$$\langle \widehat{a}_\gamma^+(t_0)\widehat{a}_\alpha(t)\rangle \underset{t<t_0}{=} e^{-\beta E_\gamma}\frac{1}{Z}Tr\left(e^{-\beta\widehat{H}}e^{i\frac{(t-t_0)}{\hbar}\widehat{H}}\widehat{a}_\alpha e^{-i\frac{(t-t_0)}{\hbar}\widehat{H}}\widehat{a}_\gamma^+\right) = e^{-\beta E_\gamma}\langle \widehat{a}_\alpha(t-t_0)\widehat{a}_\gamma^+(0)\rangle.$$

Then, we have the following relations for $t < t_0$:

$$\langle \widehat{a}_\gamma^+(t_0)\widehat{a}_\alpha(t)\rangle \underset{t<t_0}{=} \langle \widehat{a}_\gamma^+(0)\widehat{a}_\alpha(t-t_0)\rangle = e^{-\beta E_\gamma}\langle \widehat{a}_\alpha(t-t_0)\widehat{a}_\gamma^+(0)\rangle.$$

Further using $\widehat{a}_\alpha(t-t_0) = e^{-i\frac{(t-t_0)}{\hbar}E_\alpha}\widehat{a}_\alpha$, we find $\langle \widehat{a}_\gamma^+\widehat{a}_\alpha\rangle = e^{-\beta E_\gamma}\langle \widehat{a}_\alpha\widehat{a}_\gamma^+\rangle$. In addition, because of the (anti)commutation relation $\left[\widehat{a}_\alpha,\widehat{a}_\gamma^+\right]_\theta = \widehat{a}_\alpha\widehat{a}_\gamma^+ - e^{i\theta}\widehat{a}_\gamma^+\widehat{a}_\alpha = \delta_{\alpha\gamma}$, we also have that $\widehat{a}_\alpha\widehat{a}_\gamma^+ = e^{i\theta}\widehat{a}_\gamma^+\widehat{a}_\alpha + \delta_{\alpha\gamma}$.

Therefore

$$e^{\beta E_\gamma}\langle \widehat{a}_\gamma^+\widehat{a}_\alpha\rangle = \langle \widehat{a}_\alpha\widehat{a}_\gamma^+\rangle = \langle e^{i\theta}\widehat{a}_\gamma^+\widehat{a}_\alpha + \delta_{\alpha\gamma}\rangle = e^{i\theta}\langle \widehat{a}_\gamma^+\widehat{a}_\alpha\rangle + \delta_{\alpha\gamma}\langle 1\rangle,$$

where

$$\langle \widehat{a}_\gamma^+\widehat{a}_\alpha\rangle = \frac{\delta_{\alpha\gamma}}{e^{\beta E_\alpha} - e^{i\theta}} = \frac{\delta_{\alpha\gamma}}{e^{\beta E_\alpha} - 1} \text{ for bosons } (\theta = 0),$$
$$\langle \widehat{a}_\gamma^+\widehat{a}_\alpha\rangle = \frac{\delta_{\alpha\gamma}}{e^{\beta E_\alpha} - e^{i\theta}} = \frac{\delta_{\alpha\gamma}}{e^{\beta E_\alpha} + 1} \text{ for fermions } (\theta = \pi).$$

Finally,

$$\langle \hat{a}_\gamma^+(t_0)\hat{a}_\alpha(t)\rangle \underset{t<t_0}{=} \langle \hat{a}_\gamma^+(0)\hat{a}_\alpha(t-t_0)\rangle = e^{-i\frac{(t-t_0)}{\hbar}E_\alpha}\langle \hat{a}_\gamma^+\hat{a}_\alpha\rangle = e^{-i\frac{(t-t_0)}{\hbar}E_\alpha}\frac{\delta_{\alpha\gamma}}{e^{\beta E_\alpha}-e^{i\theta}}.$$

b) From the definition of the Green's function, we have

$$\begin{aligned}\mathcal{G}_{\alpha\gamma}(t,t_0) &= \frac{1}{i\hbar}\left\langle \mathcal{T}_t\left\{\hat{a}_\alpha(t)\,\hat{a}_\gamma^+(t_0)\right\}\right\rangle\\ &= \frac{1}{i\hbar}\Theta(t-t_0)\langle \hat{a}_\alpha(t)\hat{a}_\gamma^+(t_0)\rangle + \frac{1}{i\hbar}\Theta(t_0-t)e^{i\theta}\langle \hat{a}_\gamma^+(t_0)\hat{a}_\alpha(t)\rangle\\ &= \frac{1}{i\hbar}\Theta(t-t_0)e^{i\frac{t}{\hbar}E_\alpha}e^{-i\frac{t_0}{\hbar}E_\gamma}\langle \hat{a}_\alpha\hat{a}_\gamma^+\rangle + \frac{1}{i\hbar}\Theta(t_0-t)e^{i\theta}e^{i\frac{t}{\hbar}E_\alpha}e^{-i\frac{t_0}{\hbar}E_\gamma}\langle \hat{a}_\gamma^+\hat{a}_\alpha\rangle.\end{aligned}$$

By using the (anti)commutation relation $\left[\hat{a}_\alpha,\hat{a}_\gamma^+\right]_\theta = \delta_{\alpha\gamma}$ and $\langle \hat{a}_\gamma^+\hat{a}_\alpha\rangle = n_\alpha\delta_{\alpha\gamma} = \frac{\delta_{\alpha\gamma}}{e^{\beta E_\alpha}-e^{i\theta}}$,

$$\mathcal{G}_{\alpha\gamma}(t,t_0) = \frac{1}{i\hbar}e^{i\frac{(t-t_0)}{\hbar}E_\alpha}\left(\Theta(t-t_0)\left(1+e^{i\theta}n_\alpha\right)+e^{i\theta}\Theta\left(t_0-t\right)n_\alpha\right)\delta_{\alpha\gamma}.$$

Problem 5.23

In quantum thermal statistical physics, it is useful to work in imaginary time ($t=-i\tau$) instead in real time. This is the so-called Matsubara formalism (Mahan, 2000). Green's functions are defined similarly as for real time, but new properties become evident. For example, the Green's function can be expanded in a Fourier series at imaginary time ($t=i\tau$),

$$\tilde{\mathcal{G}}_{\alpha\gamma}(i\omega_n) = \int_0^{\hbar\beta} d\tau e^{i\omega_n\tau}\mathcal{G}_{\alpha\gamma}(i\tau,0),$$

$$\mathcal{G}_{\alpha\gamma}(i\tau,0) = \frac{1}{\hbar\beta}\sum_{n\in\mathbb{Z}} e^{-i\omega_n\tau}\tilde{\mathcal{G}}_{\alpha\gamma}(i\omega_n),$$

where $i\omega_n$ are imaginary (Matsubara) frequencies. The preceding formalism requires that we know how to treat the Green's functions from the previous problem in imaginary time. Let us then consider the following problems:

a) How are the Green's function and the time-ordering meta-operator modified in the imaginary time domain? To answer this question, apply the change $t=i\tau$ to the Schrödinger equation for the two-point Green's function, check how the Heisenberg picture of the creation and annihilation operators change, and write the Green function for imaginary time.
b) Obtain a similar expression for the Green's function, but in the imaginary (Matsubara) formalism by finding $\langle \hat{a}_\alpha(i\tau)\,\hat{a}_\gamma^-(i\tau_0)\rangle$ for $\tau<\tau_0$ where $\hat{a}_\gamma^-(i\tau_0) = e^{\frac{\tau_0}{\hbar}\hat{H}}\hat{a}_\gamma^+e^{-\frac{\tau_0}{\hbar}\hat{H}}$.
c) Show that the Green's function is periodic in imaginary time.
d) By imposing the periodicity of the Green's function $\mathcal{G}_{\alpha\gamma}(i\tau+i\hbar\beta,0) = e^{i\theta}\mathcal{G}_{\alpha\gamma}(i\tau,0)$, find explicit expressions for $\mathcal{G}_{\alpha\gamma}(i\tau)$ and $\tilde{\mathcal{G}}_{\alpha\gamma}(i\omega_n)$ in the case of bosonic and fermionic particles.
e) How does this Green's function change in the limit $T\to 0$?

a) To work in the imaginary time domain, we make the substitution $t = i\tau$ (to avoid subsequent nonconvergent solutions) in the Schrödinger equation for the two-point Green's function, and we obtain

$$\left(-\hbar\partial_\tau - \widehat{h}_\alpha(\boldsymbol{x})\right)\mathcal{G}_{\alpha\gamma}(\boldsymbol{x}, i\tau; \boldsymbol{x}_0, i\tau_0) = \delta_{\alpha\gamma}\delta(t-\tau_0)\delta(\boldsymbol{x}-\boldsymbol{x}_0),$$

where we have used $\delta(t-s) = \delta(i\tau - i\tau_0) = \frac{\delta(t-t_0)}{|i|} = \delta(t-\tau_0)$ and $i\hbar\partial_t f(t) = -\hbar\partial_\tau f(i\tau)$. The retarded Green's function is now

$$\mathcal{G}_{\alpha\gamma}(i\tau, i\tau_0) = \frac{-1}{\hbar}\langle\mathcal{T}_{i\tau}\left\{\widehat{a}_\alpha(i\tau)\,\widehat{a}_\gamma^-(i\tau_0)\right\}\rangle = \frac{-1}{\hbar}Tr\left(\widehat{\rho}_C\mathcal{T}_{i\tau}\left\{\widehat{a}_\alpha(i\tau)\,\widehat{a}_\gamma^-(i\tau_0)\right\}\right),$$

$$\widehat{a}_\alpha(i\tau) = e^{\frac{\tau}{\hbar}\widehat{H}}\widehat{a}_\alpha e^{-\frac{\tau}{\hbar}\widehat{H}} = e^{-\frac{\tau}{\hbar}E_\alpha}\widehat{a}_\alpha,$$

$$\widehat{a}_\gamma^-(i\tau_0) = e^{\frac{\tau_0}{\hbar}\widehat{H}}\widehat{a}_\gamma^+ e^{-\frac{\tau_0}{\hbar}\widehat{H}} = e^{\frac{\tau_0}{\hbar}E_\gamma}\widehat{a}_\gamma^+.$$

Here, $\widehat{a}_\gamma^-(i\tau_0)$ is no longer the Hermitian conjugate of $\widehat{a}_\gamma(i\tau_0)$. The time-ordering meta-operator now orders time in the imaginary axis

$$\mathcal{T}_{i\tau}\left\{\widehat{a}_\alpha(i\tau)\,\widehat{a}_\gamma^-(i\tau_0)\right\} = \begin{cases}\widehat{a}_\alpha(i\tau)\,\widehat{a}_\gamma^-(i\tau_0), & \tau > \tau_0 \\ e^{i\theta}\widehat{a}_\gamma^-(i\tau_0)\,\widehat{a}_\alpha(i\tau), & \tau < \tau_0\end{cases}.$$

Therefore, the Green's function becomes

$$\begin{aligned}\mathcal{G}_{\alpha\gamma}(i\tau, i\tau_0) &= \frac{-1}{\hbar}\langle\mathcal{T}_{i\tau}\{\widehat{a}_\alpha(i\tau)\,\widehat{a}_\gamma^-(i\tau_0)\}\rangle \\ &= \frac{-1}{\hbar}\Theta(\tau-\tau_0)\,\langle\widehat{a}_\alpha(i\tau)\,\widehat{a}_\gamma^-(i\tau_0)\rangle - \frac{1}{\hbar}e^{i\theta}\Theta(\tau_0-\tau)\langle\widehat{a}_\gamma^-(i\tau_0)\widehat{a}_\alpha(i\tau)\rangle.\end{aligned}$$

b) Here we rely on the result from part a). We also use the expressions for $\widehat{a}_\alpha(i\tau)$, $\widehat{a}_\gamma^-(i\tau_0)$, and $\widehat{\rho}_C = \frac{e^{-\beta\widehat{H}}}{Z}$. Thus,

$$\begin{aligned}\mathcal{G}_{\alpha\gamma}(i\tau, i\tau_0) \underset{\tau<\tau_0}{=} &-\frac{1}{\hbar}e^{i\theta}\Theta(\tau_0-\tau)\langle\widehat{a}_\gamma^-(i\tau_0)\widehat{a}_\alpha(i\tau)\rangle \\ &= \frac{-1}{\hbar}e^{i\theta}\frac{1}{Z}Tr\left(e^{-\beta\widehat{H}}e^{\frac{\tau_0}{\hbar}\widehat{H}}\widehat{a}_\gamma^+ e^{-\frac{\tau_0}{\hbar}\widehat{H}}e^{\frac{\tau}{\hbar}\widehat{H}}\widehat{a}_\alpha e^{-\frac{\tau}{\hbar}\widehat{H}}\right).\end{aligned}$$

Using the cyclic property of the trace $Tr(ABC) = Tr(CAB)$, and the fact that $\widehat{H}$ commutes with itself, we further write

$$\begin{aligned}\mathcal{G}_{\alpha\gamma}(i\tau, i\tau_0) \underset{\tau<\tau_0}{=} &\frac{-1}{\hbar}e^{i\theta}\frac{1}{Z}Tr\left(e^{-\beta\widehat{H}}e^{-\frac{\tau}{\hbar}\widehat{H}}e^{\frac{\tau_0}{\hbar}\widehat{H}}\widehat{a}_\gamma^+ e^{-\frac{\tau_0}{\hbar}\widehat{H}}e^{\frac{\tau}{\hbar}\widehat{H}}\widehat{a}_\alpha\right) \\ &= \frac{-1}{\hbar}e^{i\theta}\frac{1}{Z}Tr\left(e^{-\beta\widehat{H}}e^{\frac{(\tau_0-\tau)}{\hbar}}\widehat{a}_\gamma^+ e^{-\frac{(\tau_0-\tau)}{\hbar}\widehat{H}}\widehat{a}_\alpha\right) \\ &= \langle\widehat{a}_\gamma^-(i\tau_0 - i\tau)\widehat{a}_\alpha(0)\rangle = \mathcal{G}_{\alpha\gamma}(0, i\tau_0 - i\tau).\end{aligned}$$

The result for $\tau > \tau_0$ is

$$\mathcal{G}_{\alpha\gamma}(i\tau, i\tau_0) \underset{\tau>\tau_0}{=} \frac{-1}{\hbar}\Theta(\tau-\tau_0)\,\langle\widehat{a}_\alpha(i\tau)\widehat{a}_\gamma^-(i\tau_0)\rangle = \frac{-1}{\hbar}\langle\widehat{a}_\alpha(i\tau - i\tau_0)\widehat{a}_\gamma^-(0)\rangle = \mathcal{G}_{\alpha\gamma}(i\tau - i\tau_0, 0).$$

c) To check the periodicity of the Green's function, let's consider that for $\Delta\tau=\tau-\tau_0<0$, while one has $\Delta\tau+\hbar\beta>0$. Thus,

$$\begin{aligned}
\mathcal{G}_{\alpha\gamma}(i\Delta\tau+i\hbar\beta,0) &\underset{\Delta\tau+\hbar\beta>0}{=} \frac{-1}{\hbar}\langle\mathcal{T}_{i\tau}\{\widehat{a}_\alpha(i\Delta\tau+i\hbar\beta)\,\widehat{a}_\gamma^{+}(0)\}\rangle \\
&\underset{\Delta\tau+\hbar\beta>0}{=} \frac{-1}{\hbar}\langle\widehat{a}_\alpha(i\Delta\tau+i\hbar\beta)\,\widehat{a}_\gamma^{+}(0)\rangle,
\end{aligned}$$

$$\begin{aligned}
\langle\widehat{a}_\alpha(i\Delta\tau+i\hbar\beta)\,\widehat{a}_\gamma^{+}(0)\rangle &= Tr\left(\widehat{\rho}_C\widehat{a}_\alpha(i\Delta\tau+i\hbar\beta)\widehat{a}_\gamma^{+}(0)\right) \\
&= Tr\left(\frac{e^{-\beta\widehat{H}}}{Z}e^{\frac{(\Delta\tau+\hbar\beta)}{\hbar}\widehat{H}}\widehat{a}_\alpha e^{-\frac{(\Delta\tau+\hbar\beta)}{\hbar}\widehat{H}}e^{\frac{0}{\hbar}\widehat{H}}\widehat{a}_\gamma^{+}e^{-\frac{0}{\hbar}\widehat{H}}\right) \\
&= \frac{1}{Z}Tr\left(e^{-\beta\widehat{H}}e^{\beta\widehat{H}}e^{\frac{\Delta\tau}{\hbar}\widehat{H}}\widehat{a}_\alpha e^{-\frac{\Delta\tau}{\hbar}\widehat{H}}e^{-\beta\widehat{H}}\widehat{a}_\gamma^{+}\right) \\
&= \frac{1}{Z}Tr\left(\widehat{a}_\alpha(i\Delta\tau)e^{-\beta\widehat{H}}\widehat{a}_\gamma^{+}(0)\right) = \frac{1}{Z}Tr\left(e^{-\beta\widehat{H}}\widehat{a}_\gamma^{+}(0)\widehat{a}_\alpha(i\Delta\tau)\right).
\end{aligned}$$

Substituting back into the Green's function,

$$\mathcal{G}_{\alpha\gamma}(i\Delta\tau+i\hbar\beta,0) \underset{\Delta\tau+\hbar\beta>0}{=} \frac{-1}{\hbar Z}Tr\left(e^{-\beta\widehat{H}}\widehat{a}_\gamma^{+}(0)\widehat{a}_\alpha(i\Delta\tau)\right).$$

Rewriting this using the time-ordering operator and realizing that $-\hbar\beta<\Delta\tau<0<\hbar\beta$,

$$\mathcal{T}_{i\tau}\left\{\widehat{a}_\alpha(i\Delta\tau)\,\widehat{a}_\gamma^{+}(0)\right\} \underset{\Delta\tau<0}{=} e^{i\theta}\widehat{a}_\gamma^{+}(0)\widehat{a}_\alpha(i\Delta\tau),$$

$$\begin{aligned}
\mathcal{G}_{\alpha\gamma}(i\Delta\tau+i\hbar\beta,0) &\underset{\Delta\tau+\hbar\beta>0}{=} \frac{-1}{\hbar}\frac{1}{Z}Tr\left(e^{-\beta\widehat{H}}\widehat{a}_\gamma^{+}(0)\widehat{a}_\alpha(i\Delta\tau)\right) \\
&\underset{\Delta\tau<0}{=} e^{-i\theta}\frac{-1}{\hbar}\frac{1}{Z}Tr\left(e^{-\beta\widehat{H}}\mathcal{T}_{i\tau}\left\{\widehat{a}_\alpha(i\Delta\tau)\widehat{a}_\gamma^{+}(0)\right\}\right) \\
&= e^{-i\theta}\mathcal{G}_{\alpha\gamma}(i\Delta\tau,0).
\end{aligned}$$

The preceding relation demonstrates the periodicity of the Green's function for $\Delta\tau+\hbar\beta>0$. The case of $\Delta\tau-\hbar\beta<0$ is left as an exercise.

d) To obtain explicit expressions of the Green's function $\mathcal{G}_{\alpha\gamma}(i\tau)$, we start with

$$\begin{aligned}
\mathcal{G}_{\alpha\gamma}(i\tau,i\tau_0) &= \frac{-1}{\hbar}\left\langle\mathcal{T}_{i\tau}\left\{\widehat{a}_\alpha(i\tau)\widehat{a}_\gamma^{+}(i\tau_0)\right\}\right\rangle \\
&= \frac{-1}{\hbar}\Theta(\tau-\tau_0)\left\langle\widehat{a}_\alpha(i\tau)\widehat{a}_\gamma^{+}(i\tau_0)\right\rangle - \frac{1}{\hbar}e^{i\theta}\Theta(\tau_0-\tau)\left\langle\widehat{a}_\gamma^{+}(i\tau_0)\widehat{a}_\alpha(i\tau)\right\rangle,
\end{aligned}$$

and use that $\left\langle\widehat{a}_\alpha(i\tau)\widehat{a}_\gamma^{+}(i\tau_0)\right\rangle = e^{-\frac{\tau}{\hbar}E_\alpha}e^{\frac{\tau_0}{\hbar}E_\gamma}\left\langle\widehat{a}_\alpha\widehat{a}_\gamma^{+}\right\rangle$ and $\left\langle\widehat{a}_\alpha\widehat{a}_\gamma^{+}\right\rangle = \frac{1}{Z}Tr\left(e^{-\beta\widehat{H}}\widehat{a}_\alpha\widehat{a}_\gamma^{+}\right)$ $= e^{\beta E_\alpha}\left\langle\widehat{a}_\gamma^{+}\widehat{a}_\alpha\right\rangle$.

On the other hand, from the commutation relations $\left[\widehat{a}_\alpha,\widehat{a}_\gamma^{+}\right]_\theta = \delta_{\alpha\gamma}$, then $\widehat{a}_\alpha\widehat{a}_\gamma^{+} = e^{i\theta}\widehat{a}_\gamma^{+}\widehat{a}_\alpha+\delta_{\alpha\gamma}$ for fermions ($\theta=\pi$) and bosons ($\theta=0$), we find that

$$e^{\beta E_\alpha}\langle\widehat{a}_\gamma^{+}\widehat{a}_\alpha\rangle = \langle\widehat{a}_\alpha\widehat{a}_\gamma^{+}\rangle = \langle e^{i\theta}\widehat{a}_\gamma^{+}\widehat{a}_\alpha+\delta_{\alpha\gamma}\rangle = e^{i\theta}\langle\widehat{a}_\gamma^{+}\widehat{a}_\alpha\rangle+\delta_{\alpha\gamma},$$

$$\langle\widehat{a}_\gamma^{+}\widehat{a}_\alpha\rangle = \frac{\delta_{\alpha\gamma}}{e^{\beta E_\alpha}-e^{i\theta}} = n_\alpha\delta_{\alpha\gamma},$$

where n_α is the occupation number derived earlier. Then, the Matsubara Green's function becomes

$$\begin{aligned}\mathscr{G}_{\alpha\gamma}(i\tau,i\tau_0) &= \frac{-1}{\hbar}\Theta(\tau-\tau_0)e^{-\frac{\tau}{\hbar}E_\alpha}e^{\frac{\tau_0}{\hbar}E_\gamma}\langle\hat{a}_\alpha\hat{a}_\gamma^+\rangle - \frac{1}{\hbar}e^{i\theta}\Theta(\tau_0-\tau)e^{-\frac{\tau}{\hbar}E_\alpha}e^{\frac{\tau_0}{\hbar}E_\gamma}\langle\hat{a}_\gamma^+\hat{a}_\alpha\rangle \\ &= \frac{-1}{\hbar}\Theta(\tau-\tau_0)e^{-\frac{\tau}{\hbar}E_\alpha}e^{\frac{\tau_0}{\hbar}E_\gamma}(e^{i\theta}n_\alpha\delta_{\alpha\gamma}+\delta_{\alpha\gamma}) - \frac{1}{\hbar}e^{i\theta}\Theta(\tau_0-\tau)e^{-\frac{\tau}{\hbar}E_\alpha}e^{\frac{\tau_0}{\hbar}E_\gamma}n_\alpha\delta_{\alpha\gamma} \\ &= \frac{-1}{\hbar}e^{-\frac{(\tau-\tau_0)}{\hbar}E_\alpha}\left[\Theta(\tau-\tau_0)(1+e^{i\theta}n_\alpha)+e^{i\theta}\Theta(\tau_0-\tau)n_\alpha\right]\delta_{\alpha\gamma}.\end{aligned}$$

Now, to obtain explicit expressions of the Green's function $\mathscr{G}_{\alpha\gamma}(i\omega_n)$, we start with the Fourier series relations

$$\tilde{\mathscr{G}}_{\alpha\gamma}(i\omega_n) = \frac{1}{2}\int_{-\hbar\beta}^{\hbar\beta} d\tau e^{i\omega_n\tau}\mathscr{G}_{\alpha\gamma}(i\tau,0),$$

$$\mathscr{G}_{\alpha\gamma}(i\tau,0) = \frac{1}{\hbar\beta}\sum_{n\in\mathbb{Z}} e^{-i\omega_n\tau}\tilde{\mathscr{G}}_{\alpha\gamma}(i\omega_n).$$

Imposing the periodicity in imaginary times $\mathscr{G}_{\alpha\gamma}(i\tau+i\hbar\beta,0) = e^{i\theta}\mathscr{G}_{\alpha\gamma}(i\tau,0)$ onto the Fourier expansion,

$$\frac{1}{\hbar\beta}\sum_{n\in\mathbb{Z}} e^{-i\omega_n(\tau+\hbar\beta)}\tilde{\mathscr{G}}_{\alpha\gamma}(i\omega_n) = e^{i\theta}\frac{1}{\hbar\beta}\sum_{n\in\mathbb{Z}} e^{-i\omega_n\tau}\tilde{\mathscr{G}}_{\alpha\gamma}(i\omega_n),$$

$$0 = \frac{1}{\hbar\beta}\sum_{n\in\mathbb{Z}}(e^{-i\omega_n(\tau+\hbar\beta)}-e^{i\theta}e^{-i\omega_n\tau})\tilde{\mathscr{G}}_{\alpha\gamma}(i\omega_n) = \frac{1}{\hbar\beta}\sum_{n\in\mathbb{Z}}(e^{i\omega_n\hbar\beta}-e^{i\theta})e^{-i\omega_n\tau}\tilde{\mathscr{G}}_{\alpha\gamma}(i\omega_n).$$

Thus, for every n we derive the Matsubara frequencies

$$e^{i\omega_n\hbar\beta} = e^{i\theta} = e^{i(\theta+2\pi n)} \rightarrow \omega_n = \frac{2\pi n+\theta}{\hbar\beta},$$

meaning that for bosons ($\theta=0$), $\omega_n = \frac{\pi}{\hbar\beta}(2n)$, while for fermions ($\theta=\pi$), $\omega_n = \frac{\pi}{\hbar\beta}(2n+1)$ for $n\in\mathbb{Z}$. Continuing for the Green's function:

$$\begin{aligned}\tilde{\mathscr{G}}_{\alpha\gamma}(i\omega_n) &= \int_0^{\hbar\beta} d\tau e^{i\omega_n\tau}\mathscr{G}_{\alpha\gamma}(i\tau,0) = \frac{-1}{\hbar}\int_0^{\hbar\beta} d\tau e^{i\omega_n\tau}e^{\frac{-\tau}{\hbar}E_\alpha}(1+e^{i\theta}n_\alpha)\delta_{\alpha\gamma} \\ &= \frac{-1}{\hbar}(1+e^{i\theta}n_\alpha)\delta_{\alpha\gamma}\int_0^{\hbar\beta} d\tau e^{\left(i\omega_n-\frac{E_\alpha}{\hbar}\right)\tau} \\ &= \frac{-1}{\hbar}(1+e^{i\theta}n_\alpha)\delta_{\alpha\gamma}\frac{e^{\left(i\omega_n-\frac{E_\alpha}{\hbar}\right)\hbar\beta}-e^{\left(i\omega_n-\frac{E_\alpha}{\hbar}\right)0}}{i\omega_n-\frac{E_\alpha}{\hbar}} \\ &= \left(1+e^{i\theta}n_\alpha\right)\delta_{\alpha\gamma}\frac{1-e^{i\theta}e^{-\beta E_\alpha}}{i\hbar\omega_n-E_\alpha}.\end{aligned}$$

Finally, using $n_\alpha = \frac{1}{e^{\beta E_\alpha}-e^{i\theta}}$, the last expression can be simplified further into

$$\tilde{\mathscr{G}}_{\alpha\gamma}(i\omega_n) = \frac{\delta_{\alpha\gamma}}{i\hbar\omega_n-E_\alpha}.$$

e) In the limit $T \to 0$, we have to take the limit $\beta = \frac{1}{k_B T} \to \infty$ in the appropriate expressions. The Matsubara frequencies $\omega_n = \frac{2\pi n+\theta}{\hbar\beta}$ now fill the entire imaginary axis, becoming continuous for all $\omega \in (0,\infty)$.

$$\lim_{\beta\to\infty} \mathcal{G}_{\alpha\gamma}(i\tau,0) = \lim_{\beta\to\infty} \frac{1}{\hbar\beta} \sum_{n\in\mathbb{Z}} e^{-i\omega_n\tau} \tilde{\mathcal{G}}_{\alpha\gamma}(i\omega_n) = \int_{-\infty}^{\infty} \frac{d\omega}{2\pi} e^{-i\omega\tau} \tilde{\mathcal{G}}^{T=0}_{\alpha\gamma}(i\omega),$$

$$\lim_{\beta\to\infty} \tilde{\mathcal{G}}_{\alpha\gamma}(i\omega_n) = \lim_{\beta\to\infty} \int_0^{\hbar\beta} d\tau e^{i\omega_n} \mathcal{G}_{\alpha\gamma}(i\tau,0) = \int_0^{\infty} d\tau e^{i\omega\tau} \mathcal{G}_{\alpha\gamma}(i\tau,0) = \tilde{\mathcal{G}}^{T=0}_{\alpha\gamma}(i\omega),$$

thus recovering the Green's function at $T = 0$, which is $\tilde{\mathcal{G}}^{T=0}_{\alpha\gamma}(i\omega) = \frac{\delta_{\alpha\gamma}}{i\hbar\omega - E_\alpha}$.

Note that we can safely rotate back all expressions to real time (and real frequencies), obtaining

$$\mathcal{G}^{T=0}_{\alpha\gamma}(t,0) = \int_{-\infty}^{\infty} \frac{d\omega}{2\pi} e^{-i\omega t} \tilde{\mathcal{G}}^{T=0}_{\alpha\gamma}(\omega),$$

$$\tilde{\mathcal{G}}^{T=0}_{\alpha\gamma}(\omega) = \int_0^{\infty} dt e^{i\omega t} \mathcal{G}_{\alpha\gamma}(t,0) = \frac{\delta_{\alpha\gamma}}{\hbar\omega - E_\alpha}.$$

Food for thought: The limit of $\lim_{\beta\to\infty} \mathcal{G}_{\alpha\gamma}(i\tau,0)$ is a direct application of the Abel–Plana formula, a result derived from the Argument Principle,

$$\sum_{n\in\mathbb{Z}} f(\Lambda n) = \int_{-\infty}^{\infty} dn f(\Lambda n) + 2i \int_0^{\infty} \frac{dy}{e^{2\pi y} - 1} [f(i\Lambda y) - f(-i\Lambda y)].$$

The possibility of applying the Argument Principle to Matsubara sums is one of the best advantages of using this formalism in equilibrium quantum thermodynamics (Coleman, 2015; Bordag et al., 2009).

Problem 5.24

The retarded Green's function is defined as

$$\mathcal{G}_{\alpha\gamma}(\boldsymbol{x},t;\boldsymbol{x}_0,t_0) = \frac{1}{i\hbar}\Theta(t-t_0)\left\langle \left[\widehat{a}_\alpha(\boldsymbol{x},t), \widehat{a}^\dagger_\gamma(\boldsymbol{x}_0,t_0)\right]_\theta \right\rangle = \frac{1}{i\hbar}\Theta(t-t_0) Tr\left(\widehat{\rho}\left[\widehat{a}_\alpha(\boldsymbol{x},t), \widehat{a}^\dagger_\gamma(\boldsymbol{x}_0,t_0)\right]_\theta\right).$$

Show that, for the linear Hamiltonian $\widehat{H} = \int d\boldsymbol{x} \sum_\alpha \widehat{a}^\dagger_\alpha(\boldsymbol{x}) \widehat{h}_\alpha(\boldsymbol{x}) \widehat{a}_\alpha(\boldsymbol{x})$, the retarded Green's function satisfies the following (Schrödinger) equation:

$$(i\hbar\partial_t - \widehat{h}_\alpha(\boldsymbol{x}))\mathcal{G}_{\alpha\gamma}(\boldsymbol{x},t;\boldsymbol{x}_0,t_0) = \delta_{\alpha\gamma}\delta(t-t_0)\delta(\boldsymbol{x}-\boldsymbol{x}_0).$$

We begin with the temporal derivative of the retarded Green's function:

$$\begin{aligned}
i\hbar\partial_t \mathcal{G}_{\alpha\gamma}(\boldsymbol{x},t;\boldsymbol{x}_0,t_0) &= i\hbar\partial_t \frac{1}{i\hbar}\Theta(t-t_0)\left\langle \left[\widehat{a}_\alpha(\boldsymbol{x},t), \widehat{a}^\dagger_\gamma(\boldsymbol{x}_0,t_0)\right]_\theta \right\rangle \\
&= (\partial_t\Theta(t-t_0))\left\langle \left[\widehat{a}_\alpha(\boldsymbol{x},t), \widehat{a}^\dagger_\gamma(\boldsymbol{x}_0,t_0)\right]_\theta \right\rangle + \Theta(t-t_0)\left\langle \left[\partial_t\widehat{a}_\alpha(\boldsymbol{x},t), \widehat{a}^\dagger_\gamma(\boldsymbol{x}_0,t_0)\right]_\theta \right\rangle \\
&\quad + \Theta(t-t_0)\left\langle \left[\widehat{a}_\alpha(\boldsymbol{x},t), \partial_t\widehat{a}^\dagger_\gamma(\boldsymbol{x}_0,t_0)\right]_\theta \right\rangle \\
&= \delta(t-t_0)\left\langle \left[\widehat{a}_\alpha(\boldsymbol{x},t), \widehat{a}^\dagger_\gamma(\boldsymbol{x}_0,t_0)\right]_\theta \right\rangle + \Theta(t-t_0)\left\langle \left[\partial_t\widehat{a}_\alpha(\boldsymbol{x},t), \widehat{a}^\dagger_\gamma(\boldsymbol{x}_0,t_0)\right]_\theta \right\rangle,
\end{aligned}$$

since $\partial_t \Theta(t-t_0) = \delta(t-t_0)$, and $\partial_t \hat{a}_\gamma^+(\boldsymbol{x}_0, t_0) = 0$. We focus on the second term first, for which we write

$$\partial_t \hat{a}_\alpha(\boldsymbol{x},t) = e^{\frac{i}{\hbar}\hat{H}t} \frac{1}{i\hbar} \left[\hat{a}_\alpha(\boldsymbol{x}), \hat{H}\right] e^{-\frac{i}{\hbar}\hat{H}t},$$

$$\left[\hat{a}_\alpha(\boldsymbol{x}), \hat{H}\right] = \int d\boldsymbol{x}_1 \sum_n \hat{h}_n(\boldsymbol{x}_1) \left(\{e^{i\theta}\hat{a}_n^+(\boldsymbol{x}_1)\hat{a}_\alpha(\boldsymbol{x}) + \delta_{\alpha n}\delta(\boldsymbol{x}-\boldsymbol{x}_1)\}\hat{a}_n(\boldsymbol{x}_1)\right.$$
$$\left. - \hat{a}_n^+(\boldsymbol{x}_1)\{e^{i\theta}\hat{a}_\alpha(\boldsymbol{x})\hat{a}_n(\boldsymbol{x}_1)\}\right)$$
$$= \int d\boldsymbol{x}_1 \sum_n \hat{h}_n(\boldsymbol{x}_1)\delta_{\alpha n}\delta(\boldsymbol{x}-\boldsymbol{x}_1)\hat{a}_n(\boldsymbol{x}_1) = \hat{h}_\alpha(\boldsymbol{x})\hat{a}_\alpha(\boldsymbol{x}),$$

$$\partial_t \hat{a}_\alpha(\boldsymbol{x},t) = e^{\frac{i}{\hbar}\hat{H}t} \frac{1}{i\hbar} \left[\hat{a}_\alpha(\boldsymbol{x}), \hat{H}\right] e^{-\frac{i}{\hbar}\hat{H}t} = e^{\frac{1}{i\hbar}\hat{H}t} \frac{1}{i\hbar}\hat{h}_\alpha(\boldsymbol{x})\hat{a}_\alpha(\boldsymbol{x})e^{-\frac{i}{\hbar}\hat{H}t} = \frac{1}{i\hbar}\hat{h}_\alpha(\boldsymbol{x})\hat{a}_\alpha(\boldsymbol{x},t).$$

Let us now consider the first term in the temporal derivative of the Green's function,

$$\left\langle \left[\hat{a}_\alpha(\boldsymbol{x},t), \hat{a}_\gamma^+(\boldsymbol{x}_0,t)\right]_\theta \right\rangle = Tr\left(\hat{\rho}\left[e^{\frac{i}{\hbar}\hat{H}t}\hat{a}_\alpha(\boldsymbol{x})e^{-\frac{i}{\hbar}\hat{H}t}, e^{\frac{i}{\hbar}\hat{H}t}\hat{a}_\gamma^+(\boldsymbol{x}_0)e^{-\frac{i}{\hbar}\hat{H}t0}\right]_\theta\right)$$
$$= Tr\left(\hat{\rho}\left(\hat{a}_\alpha(\boldsymbol{x})e^{\frac{i}{\hbar}\hat{H}(t-t)}\hat{a}_\gamma^+(\boldsymbol{x}_0)e^{\frac{-i}{\hbar}\hat{H}(t-t)} - e^{i\theta}\hat{a}_\gamma^+(\boldsymbol{x}_0)e^{\frac{i}{\hbar}\hat{H}(t-t)}\hat{a}_\alpha(\boldsymbol{x})e^{\frac{-i}{\hbar}\hat{H}(t-t)}\right)\right)$$
$$= Tr\left(\hat{\rho}\left(\hat{a}_\alpha(\boldsymbol{x})\hat{a}_\gamma^+(\boldsymbol{x}_0) - e^{i\theta}\hat{a}_\gamma^+(\boldsymbol{x}_0)\hat{a}_\alpha(\boldsymbol{x})\right)\right)$$
$$= Tr\left(\hat{\rho}\left[\hat{a}_\alpha(\boldsymbol{x}), \hat{a}_\gamma^+(\boldsymbol{x}_0)\right]_\theta\right) = \delta_{\alpha\gamma}\delta(\boldsymbol{x}-\boldsymbol{x}_0),$$

where $\hat{\rho}$ is the density operator for which $Tr(\hat{\rho}) = 1$. After putting everything together, we obtain the desired relation,

$$\left(i\hbar\partial_t - \hat{h}_\alpha(\boldsymbol{x})\right)\mathcal{G}_{\alpha\gamma}(\boldsymbol{x},t;\boldsymbol{x}_0,t_0) = \delta_{\alpha\gamma}\delta(t-t_0)\delta(\boldsymbol{x}-\boldsymbol{x}_0).$$

> We have used here implicitly the so-called *Zubarev* formalism, a method to obtain the equation of motion of Green functions for general problems (Zubarev, 1960).

Problem 5.25

Superconductivity in layered YBCO perovskite is often modeled to be carried by its CuO_2 monolayer whose properties are characterized by the electronic states of the Cu positioned in the vertices of a square lattice, while the role of the O atoms laying in the middle of the edges is neglected. Assume that the Cu atoms are specified by $\boldsymbol{R}_{n_1,n_2} = (n_1\boldsymbol{a}_1 + n_2\boldsymbol{a}_2)$, where $\boldsymbol{a}_1 = a\hat{x}$ and $\boldsymbol{a}_2 = a\hat{y}$ are unit vectors and a is the distance between Cu–Cu atoms (see Figure 5.5). For such a covalently bonded material, the real-space Hamiltonian involving only nearest-neighbor coupling can be written as

$$\hat{H} = \sum_n \left(t\hat{a}_{\boldsymbol{R}_n}^+\hat{a}_{\boldsymbol{R}_n+\boldsymbol{a}_1} + t\hat{a}_{\boldsymbol{R}_n}^+\hat{a}_{\boldsymbol{R}_n+\boldsymbol{a}_2} - \mu\hat{a}_{\boldsymbol{R}_n}^+\hat{a}_{\boldsymbol{R}_n} + H.c.\right),$$

where t is the effective oxygen-mediated hopping parameter between Cu atoms, and $\hat{a}_{\boldsymbol{R}_n}^+(\hat{a}_{\boldsymbol{R}_n})$ are the creation (annihilation) operators for electrons on Cu atoms placed at the position $\boldsymbol{R}_n$.

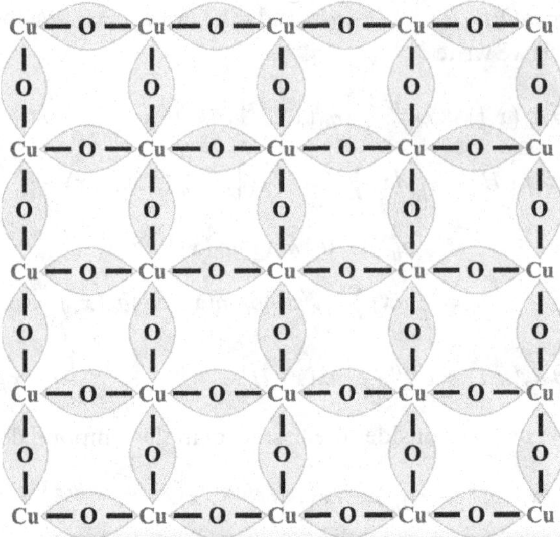

Figure 5.5 Schematic representation of the CuO lattice.

a) Obtain the basis vectors of the reciprocal lattice and determine the first Brillouin zone.
b) Write the preceding Hamiltonian in its second quantized form in reciprocal space.
c) Examine the Hamiltonian in the continuous limit $\boldsymbol{k} \to \boldsymbol{0}$.

---

a) The unit cell of this *square* lattice consists of one atom, effectively bonded to its immediate neighbors. For the Hamiltonian given in the problem, the distance between the nearest neighbor atoms is needed. For the atom in the origin, this can be found immediately as

$$\boldsymbol{a}_1 = \boldsymbol{R}_{n_1+1,n_2} - \boldsymbol{R}_{n_1,n_2} = a\begin{pmatrix} 1 \\ 0 \end{pmatrix} = a\widehat{\boldsymbol{x}},$$

$$\boldsymbol{a}_2 = \boldsymbol{R}_{n_1,n_2+1} - \boldsymbol{R}_{n_1,n_2} = a\begin{pmatrix} 0 \\ 1 \end{pmatrix} = a\widehat{\boldsymbol{y}}.$$

The basis vectors of the reciprocal lattice are found by using the relation $\boldsymbol{a}_i \cdot \boldsymbol{b}_j = 2\pi\delta_{ij}$; therefore, we obtain

$$\boldsymbol{b}_1 = \frac{2\pi}{a}\begin{pmatrix} 1 \\ 0 \end{pmatrix}, \quad \boldsymbol{b}_2 = \frac{2\pi}{a}\begin{pmatrix} 0 \\ 1 \end{pmatrix}.$$

Therefore, the first Brillouin zone consists of all points of the reciprocal space that fulfil

$$\boldsymbol{k} \in BZ \Rightarrow \boldsymbol{k} = k_1\boldsymbol{b}_1 + k_2\boldsymbol{b}_2,$$

with $k_1, k_2 \in \left(\frac{-1}{2}, \frac{+1}{2}\right)$. The given Hamiltonian can then be rewritten as

$$\widehat{H} = t\sum_n \left(\widehat{a}^+_{\boldsymbol{R}_n}\widehat{a}_{\boldsymbol{R}_n+\boldsymbol{a}_1} + \widehat{a}^+_{\boldsymbol{R}_n}\widehat{a}_{\boldsymbol{R}_n+\boldsymbol{a}_2} + \widehat{a}^+_{\boldsymbol{R}_n}\widehat{a}_{\boldsymbol{R}_n-\boldsymbol{a}_1} + \widehat{a}^+_{\boldsymbol{R}_n}\widehat{a}_{\boldsymbol{R}_n-\boldsymbol{a}_2}\right) - \mu\sum_n \widehat{a}^+_{\boldsymbol{R}_n}\widehat{a}_{\boldsymbol{R}_n}.$$

b) To transform this Hamiltonian to the reciprocal space, we use the relations

$$\widehat{a}_{\boldsymbol{R}_n} = \int_{BZ}\frac{d\boldsymbol{k}}{\sqrt{V_{BZ}}}\widehat{A}_{\boldsymbol{k}}e^{-i\boldsymbol{k}\cdot\boldsymbol{R}_n} \Leftrightarrow \widehat{A}_{\boldsymbol{k}} = \frac{1}{\sqrt{V_{BZ}}}\sum_n e^{i\boldsymbol{k}\cdot\boldsymbol{R}_n}\widehat{a}_{\boldsymbol{R}_n},$$
$$\widehat{a}^+_{\boldsymbol{R}_n} = \int_{BZ}\frac{d\boldsymbol{k}}{\sqrt{V_{BZ}}}e^{i\boldsymbol{k}\cdot\boldsymbol{R}_n}\widehat{A}^+_{\boldsymbol{k}} \Leftrightarrow \widehat{A}^+_{\boldsymbol{k}} = \frac{1}{\sqrt{V_{BZ}}}\sum_n \widehat{a}^+_{\boldsymbol{R}_n}e^{-i\boldsymbol{k}\cdot\boldsymbol{R}_n},$$

where V_{BZ} is the 2-volume of the Brillouin zone, and $\boldsymbol{k}$ is the wave vector in reciprocal space. After substitution, the Hamiltonian transforms to

$$\begin{aligned}\widehat{H} = t\sum_n \Bigg(&\int_{BZ}\frac{d\boldsymbol{k}_1}{\sqrt{V_{BZ}}}e^{i\boldsymbol{k}_1\cdot\boldsymbol{R}_n}\widehat{A}^+_{\boldsymbol{k}_1}\int_{BZ}\frac{d\boldsymbol{k}_2}{\sqrt{V_{BZ}}}\widehat{A}_{\boldsymbol{k}_2}e^{-i\boldsymbol{k}_2\cdot(\boldsymbol{R}_n+\boldsymbol{a}_1)} + \\ &+\int_{BZ}\frac{d\boldsymbol{k}_2}{\sqrt{V_{BZ}}}e^{i\boldsymbol{k}_1\cdot\boldsymbol{R}_n}\widehat{A}^+_{\boldsymbol{k}_1}\int_{BZ}\frac{d\boldsymbol{k}_1}{\sqrt{V_{BZ}}}A_{\boldsymbol{k}_2}e^{-i\boldsymbol{k}_2\cdot(\boldsymbol{R}_n+\boldsymbol{a}_2)} + \int_{BZ}\frac{d\boldsymbol{k}_1}{\sqrt{V_{BZ}}}e^{i\boldsymbol{k}_1\cdot\boldsymbol{R}_n}\widehat{A}^+_{\boldsymbol{k}_1} \\ &\times\int_{BZ}\frac{d\boldsymbol{k}_2}{\sqrt{V_{BZ}}}\widehat{A}_{\boldsymbol{k}_2}e^{-i\boldsymbol{k}_2\cdot(\boldsymbol{R}_n-\boldsymbol{a}_1)} + \int_{BZ}\frac{d\boldsymbol{k}_1}{\sqrt{V_{BZ}}}e^{-i\boldsymbol{k}_1\cdot\boldsymbol{R}_n}\widehat{A}^+_{\boldsymbol{k}_1}\int_{BZ}\frac{d\boldsymbol{k}_2}{\sqrt{V_{BZ}}}\widehat{A}_{\boldsymbol{k}_2}e^{-i\boldsymbol{k}_2\cdot(\boldsymbol{R}_n-\boldsymbol{a}_2)}\Bigg) \\ &-\mu\sum_n\int_{BZ}\frac{d\boldsymbol{k}_1}{\sqrt{V_{BZ}}}e^{i\boldsymbol{k}_1\cdot\boldsymbol{R}_n}\widehat{A}^+_{\boldsymbol{k}_1}\int_{BZ}\frac{d\boldsymbol{k}_2}{\sqrt{V_{BZ}}}\widehat{A}^+_{\boldsymbol{k}_2}e^{i\boldsymbol{k}_2\cdot\boldsymbol{R}_n}.\end{aligned}$$

$$\begin{aligned}\widehat{H} = &\frac{1}{V_{BZ}}\int_{BZ}d\boldsymbol{k}_1\int_{BZ}d\boldsymbol{k}_2\widehat{A}^+_{\boldsymbol{k}_1}\widehat{A}_{\boldsymbol{k}_2}\left[t\left(e^{-i\boldsymbol{k}_2\cdot\boldsymbol{a}_1} + e^{-i\boldsymbol{k}_2\cdot\boldsymbol{a}_2} + e^{i\boldsymbol{k}_2\cdot\boldsymbol{a}_1} + e^{i\boldsymbol{k}_2\cdot\boldsymbol{a}_2}\right) - \mu\right] \\ &\times\sum_n e^{i(\boldsymbol{k}_1-\boldsymbol{k}_2)\cdot\boldsymbol{R}_n}.\end{aligned}$$

Using that $\sum_n e^{i(\boldsymbol{k}_1-\boldsymbol{k}_2)\cdot\boldsymbol{R}_n} = V_{BZ}\delta(\boldsymbol{k}_1-\boldsymbol{k}_2)$ leads to further simplifications:

$$\begin{aligned}\widehat{H} &= \int_{BZ}d\boldsymbol{k}_1\int_{BZ}d\boldsymbol{k}_2\widehat{A}^+_{\boldsymbol{k}_1}\widehat{A}_{\boldsymbol{k}_2}\left[t\left(e^{-i\boldsymbol{k}_2\cdot\boldsymbol{a}_1} + e^{-i\boldsymbol{k}_2\cdot\boldsymbol{a}_2} + e^{i\boldsymbol{k}_2\cdot\boldsymbol{a}_1} + e^{i\boldsymbol{k}_2\cdot\boldsymbol{a}_2}\right) - \mu\right]\delta(\boldsymbol{k}_1-\boldsymbol{k}_2) \\ &= \int_{BZ}d\boldsymbol{k}_1\widehat{A}^+_{\boldsymbol{k}_1}\widehat{A}_{\boldsymbol{k}_1}\left[t(e^{-i\boldsymbol{k}_1\cdot\boldsymbol{a}_1} + e^{-i\boldsymbol{k}_1\cdot\boldsymbol{a}_2} + e^{i\boldsymbol{k}_1\cdot\boldsymbol{a}_1} + e^{i\boldsymbol{k}_1\cdot\boldsymbol{a}_2}) - \mu\right] \\ &= \int_{BZ}d\boldsymbol{k}\widehat{A}^+_{\boldsymbol{k}}\widehat{A}_{\boldsymbol{k}}\left[2t(\cos(\boldsymbol{k}\cdot\boldsymbol{a}_1) + \cos(\boldsymbol{k}\cdot\boldsymbol{a}_2)) - \mu\right].\end{aligned}$$

c) In the $\boldsymbol{k}\to\boldsymbol{0}$ limit (using $\cos(x) = 1 - \frac{x^2}{2}$), we obtain in the continuous limit:

$$\begin{aligned}\widehat{H} &= \int_{\mathbb{R}^2}d\boldsymbol{k}\widehat{A}^+(\boldsymbol{k})\left[t(4 - (\boldsymbol{k}\cdot\boldsymbol{a}_1)^2 - (\boldsymbol{k}\cdot\boldsymbol{a}_2)^2) - \mu\right]\widehat{A}(\boldsymbol{k}) \\ &= \int_{\mathbb{R}^2}d\boldsymbol{k}\widehat{A}^+(\boldsymbol{k})\left[(4t-\mu) - a^2t(k_x^2 + k_y^2)\right]\widehat{A}(\boldsymbol{k}).\end{aligned}$$

Food for thought: Take the preceding Hamiltonian and transform it back into position space. We obtain

$$\widehat{H} = \int dx \widehat{a}^{+}(\boldsymbol{x}) \left[a^2 t(\partial_x^2 + \partial_y^2) + (4t - \mu)\right] \widehat{a}(\boldsymbol{x}) = \int dx \widehat{a}^{+}(\boldsymbol{x}) \widehat{h} \widehat{a}(\boldsymbol{x}).$$

This equation immediately identifies the Hamiltonian of the Schrödinger equation,

$$\widehat{h} = a^2 t(\partial_x^2 + \partial_y^2) + (4t - \mu) = \frac{\widehat{p}^2}{2m^*} + V(\widehat{r}),$$

where the effective mass of the particles is found as $m^* = \frac{-\hbar^2}{2a^2 t}$, while the effective potential is $V(\widehat{r}) = (4t - \mu)$. This result is quite interesting, as it shows that the collective behavior of the Cu electrons is equivalent to quasi-particles with an effective mass m^* under an effective potential $V(\widehat{r})$.

Problem 5.26

Graphene is a 2D carbon layer with a honeycomb lattice, in which the location of each atom can be specified by $\boldsymbol{R} = (n_1 \boldsymbol{a}_1 + n_2 \boldsymbol{a}_2)$, where $\boldsymbol{a}_1 = a\left(\frac{3}{2}\widehat{x} + \frac{\sqrt{3}}{3}\widehat{y}\right)$ and $\boldsymbol{a}_2 = a\left(\frac{3}{2}\widehat{x} - \frac{\sqrt{3}}{3}\widehat{y}\right)$ are unit vectors and $a = 1.42$ Å is the distance between first neighbors (see Figure 5.6). For such a covalently bonded material, the real-space Hamiltonian involving only nearest-neighbor coupling can be written as

$$\widehat{H} = -t \sum_{n} (\widehat{a}_{\boldsymbol{R}_n}^{+} \widehat{b}_{\boldsymbol{R}_n}^{+} + \widehat{a}_{\boldsymbol{R}_n}^{+} \widehat{b}_{\boldsymbol{R}_n + \boldsymbol{a}_1}^{+} + \widehat{a}_{\boldsymbol{R}_n}^{+} \widehat{b}_{\boldsymbol{R}_n + \boldsymbol{a}_2}^{+} + H.c.),$$

where t is a parameter and $\widehat{a}_{\boldsymbol{R}}^{+} (\widehat{b}_{\boldsymbol{R}}^{+})$, $\widehat{a}_{\boldsymbol{R}} (\widehat{b}_{\boldsymbol{R}})$ are the creation and annihilation operators for electrons on atoms $A(B)$ in the unit cell labelled by $\boldsymbol{R}$.

a) Obtain the basis vectors of the reciprocal lattice and determine the first Brillouin zone.
b) Write the preceding Hamiltonian in its second quantized form in reciprocal space.
c) Find its eigenvalues and eigenstates.
d) Examine the Hamiltonian and its eigenenergies and eigenstates in the vicinity of the special points, $\boldsymbol{K}_{\pm} = \frac{2\pi}{3a}\left(1, \frac{\pm 1}{\sqrt{3}}\right)$, such that $\boldsymbol{k} = \boldsymbol{K}_{\pm} + \boldsymbol{q}$ in the limit of $\boldsymbol{q} \to \boldsymbol{0}$.

---

a) The unit cell of this *honeycomb* lattice consists of two inequivalent atoms, shown in dark and clear gray colors in Figure 5.6. For the Hamiltonian given in the problem, the distance between the nearest neighbor atoms is needed. For the atom in the origin, these distances can be found from geometry considerations,

$$\boldsymbol{\delta}_1 = \frac{a}{2}\begin{pmatrix} 1 \\ \sqrt{3} \end{pmatrix}, \ \boldsymbol{\delta}_2 = \frac{a}{2}\begin{pmatrix} 1 \\ -\sqrt{3} \end{pmatrix}, \ \boldsymbol{\delta}_3 = a\begin{pmatrix} -1 \\ 0 \end{pmatrix}.$$

The basis vectors of the reciprocal lattice are found by using the relation $\boldsymbol{a}_i \cdot \boldsymbol{b}_j = 2\pi\delta_{ij}$; therefore, we obtain

$$\boldsymbol{b}_1 = \frac{2\pi}{3a}\begin{pmatrix} 1 \\ \sqrt{3} \end{pmatrix}, \quad \boldsymbol{b}_2 = \frac{2\pi}{3a}\begin{pmatrix} 1 \\ -\sqrt{3} \end{pmatrix}.$$

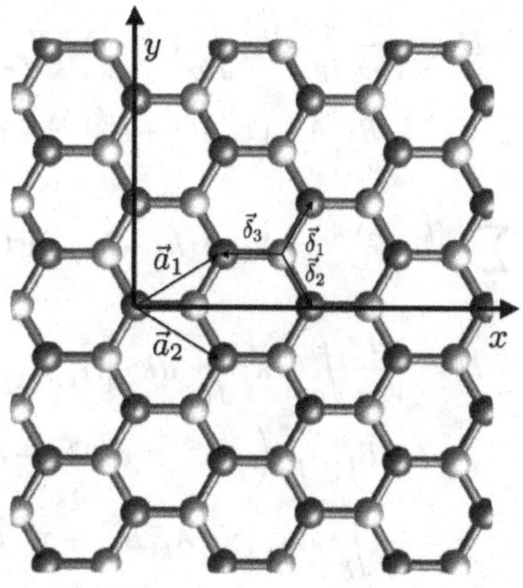

Figure 5.6 Lattice structure of 2D graphene with its unit lattice vector $\boldsymbol{a}_{1,2}$ and nearest-neighbor vector $\boldsymbol{\delta}_{1,2,3}$.

The given Hamiltonian can then be rewritten as

$$\widehat{H} = -t\sum_n \left(\widehat{a}^+_{\boldsymbol{R}_n}\widehat{b}_{\boldsymbol{R}_n+\delta_1} + \widehat{a}^+_{\boldsymbol{R}_n}\widehat{b}_{\boldsymbol{R}_n+\delta_2} + \widehat{a}^+_{\boldsymbol{R}_n}\widehat{b}_{\boldsymbol{R}_n+\delta_3} + \widehat{b}^+_{\boldsymbol{R}_n+\delta_1}\widehat{a}_{\boldsymbol{R}_n} + \widehat{b}^+_{\boldsymbol{R}_n+\delta_2}\widehat{a}_{\boldsymbol{R}_n} + \widehat{b}^+_{\boldsymbol{R}_n+\delta_3}\widehat{a}_{\boldsymbol{R}_n}\right).$$

b) To transform this Hamiltonian to the reciprocal space, we use the relations

$$\widehat{a}_{\boldsymbol{R}_n} = \int_{BZ} \frac{d\boldsymbol{k}}{\sqrt{V_{BZ}}} \widehat{A}_{\boldsymbol{k}} e^{-i\boldsymbol{k}\cdot\boldsymbol{R}_n} \Leftrightarrow \widehat{A}_{\boldsymbol{k}} = \frac{1}{\sqrt{V_{BZ}}} \sum_n e^{i\boldsymbol{k}\cdot\boldsymbol{R}_n} \widehat{a}_{\boldsymbol{R}_n},$$

$$\widehat{b}_{\boldsymbol{R}_n} = \int_{BZ} \frac{d\boldsymbol{k}}{\sqrt{V_{BZ}}} \widehat{B}_{\boldsymbol{k}} e^{-i\boldsymbol{k}\cdot\boldsymbol{R}_n} \Leftrightarrow \widehat{B}_{\boldsymbol{k}} = \frac{1}{\sqrt{V_{BZ}}} \sum_n e^{i\boldsymbol{k}\cdot\boldsymbol{R}_n} \widehat{b}_{\boldsymbol{R}_n},$$

$$\widehat{a}^+_{\boldsymbol{R}_n} = \int_{BZ} \frac{d\boldsymbol{k}}{\sqrt{V_{BZ}}} e^{i\boldsymbol{k}\cdot\boldsymbol{R}_n} \widehat{B}^+_{\boldsymbol{k}} \Leftrightarrow \widehat{B}^+_{\boldsymbol{k}} = \frac{1}{\sqrt{V_{BZ}}} \sum_n \widehat{a}^+_{\boldsymbol{R}_n} e^{-i\boldsymbol{k}\cdot\boldsymbol{R}_n},$$

$$\widehat{b}^+_{\boldsymbol{R}_n} = \int_{BZ} \frac{d\boldsymbol{k}}{\sqrt{V_{BZ}}} e^{i\boldsymbol{k}\cdot\boldsymbol{R}_n} \widehat{B}^+_{\boldsymbol{k}} \Leftrightarrow \widehat{B}^+_{\boldsymbol{k}} = \frac{1}{\sqrt{V_{BZ}}} \sum_n \widehat{b}^+_{\boldsymbol{R}_n} e^{-i\boldsymbol{k}\cdot\boldsymbol{R}_n},$$

where V_{BZ} is the 2-volume of the Brillouin zone, and $\boldsymbol{k}$ is the wave vector in reciprocal space. After substitution, the Hamiltonian transforms to

$$\begin{aligned}
\widehat{H} = -t\sum_n \Bigg(& \int_{BZ} \frac{d\boldsymbol{k}_1}{\sqrt{V_{BZ}}} e^{i\boldsymbol{k}_1\cdot\boldsymbol{R}_n} \widehat{A}^+_{\boldsymbol{k}_1} \int_{BZ} \frac{d\boldsymbol{k}_2}{\sqrt{BZ}} \widehat{B}_{\boldsymbol{k}_2} e^{-i\boldsymbol{k}_2\cdot(\boldsymbol{R}_n+\delta_1)} \\
& + \int_{BZ} \frac{d\boldsymbol{k}_1}{\sqrt{V_{BZ}}} e^{i\boldsymbol{k}_1\cdot\boldsymbol{R}_n} \widehat{A}^+_{\boldsymbol{k}_1} \int_{BZ} \frac{d\boldsymbol{k}_2}{\sqrt{V_{BZ}}} \widehat{B}_{\boldsymbol{k}_2} e^{-i\boldsymbol{k}_2\cdot(\boldsymbol{R}_n+\delta_2)} \\
& + \int_{BZ} \frac{d\boldsymbol{k}_1}{\sqrt{V_{BZ}}} e^{i\boldsymbol{k}_1\cdot\boldsymbol{R}_n} \widehat{A}^+_{\boldsymbol{k}_1} \int_{BZ} \frac{d\boldsymbol{k}_2}{\sqrt{V_{BZ}}} \widehat{B}_{\boldsymbol{k}_2} e^{-i\boldsymbol{k}_2\cdot(\boldsymbol{R}_n+\delta_3)} \\
& + \int_{BZ} \frac{d\boldsymbol{k}_1}{\sqrt{V_{BZ}}} e^{i\boldsymbol{k}_1\cdot(\boldsymbol{R}_n+\delta_1)} \widehat{B}^+_{\boldsymbol{k}_1} \int_{BZ} \frac{d\boldsymbol{k}_2}{\sqrt{V_{BZ}}} \widehat{A}_{\boldsymbol{k}_2} e^{-i\boldsymbol{k}_2\cdot\boldsymbol{R}_n} \\
& + \int_{BZ} \frac{d\boldsymbol{k}_1}{\sqrt{V_{BZ}}} e^{i\boldsymbol{k}_1\cdot(\boldsymbol{R}_n+\delta_2)} \widehat{B}^+_{\boldsymbol{k}_1} \int_{BZ} \frac{d\boldsymbol{k}_2}{\sqrt{V_{BZ}}} \widehat{A}_{\boldsymbol{k}_2} e^{-i\boldsymbol{k}_2\cdot\boldsymbol{R}_n} \\
& + \int_{BZ} \frac{d\boldsymbol{k}_1}{\sqrt{V_{BZ}}} e^{i\boldsymbol{k}_1\cdot(\boldsymbol{R}_n+\delta_3)} \widehat{B}^+_{\boldsymbol{k}_1} \int_{BZ} \frac{d\boldsymbol{k}_2}{\sqrt{V_{BZ}}} \widehat{A}_{\boldsymbol{k}_2} e^{-i\boldsymbol{k}_2\cdot\boldsymbol{R}_n} \Bigg)
\end{aligned}$$

$$\widehat{H} = \frac{-1}{V_{BZ}} \int_{BZ} d\boldsymbol{k}_1 \int_{BZ} d\boldsymbol{k}_2 \left[\widehat{A}^+_{\boldsymbol{k}_1} \widehat{B}_{\boldsymbol{k}_2} \left(e^{-i\boldsymbol{k}_2\cdot\boldsymbol{\delta}_1} + e^{-i\boldsymbol{k}_2\cdot\boldsymbol{\delta}_2} + e^{-i\boldsymbol{k}_2\cdot\boldsymbol{\delta}_3} \right) + \widehat{B}^+_{\boldsymbol{k}_1} \widehat{A}_{\boldsymbol{k}_2} \left(e^{i\boldsymbol{k}_1\cdot\boldsymbol{\delta}_1} + e^{i\boldsymbol{k}_1\cdot\boldsymbol{\delta}_2} + e^{i\boldsymbol{k}_1\cdot\boldsymbol{\delta}_3} \right) \right] \sum_n e^{i(\boldsymbol{k}_1-\boldsymbol{k}_2)\cdot\boldsymbol{R}_n}.$$

Using $\sum_n e^{i(\boldsymbol{k}_1-\boldsymbol{k}_2)\cdot\boldsymbol{R}_n} = V_{BZ}\delta(\boldsymbol{k}_1-\boldsymbol{k}_2)$ further simplifies the preceding expression:

$$\begin{aligned}\widehat{H} &= \frac{-t}{V_{BZ}} \int_{BZ} d\boldsymbol{k}_1 \int_{BZ} d\boldsymbol{k}_2 \left[\widehat{A}^+_{\boldsymbol{k}_1} \widehat{B}_{\boldsymbol{k}_2} \left(e^{-i\boldsymbol{k}_2\cdot\boldsymbol{\delta}_1} + e^{-i\boldsymbol{k}_2\cdot\boldsymbol{\delta}_2} + e^{-i\boldsymbol{k}_2\cdot\boldsymbol{\delta}_3} \right) + \widehat{B}^+_{\boldsymbol{k}_1} \widehat{A}_{\boldsymbol{k}_2} \left(e^{i\boldsymbol{k}_1\cdot\boldsymbol{\delta}_1} + e^{i\boldsymbol{k}_1\cdot\boldsymbol{\delta}_2} + e^{i\boldsymbol{k}_1\cdot\boldsymbol{\delta}_3} \right) \right] V_{BZ}\delta(\boldsymbol{k}_1-\boldsymbol{k}_2) \\ &= -t \int_{BZ} d\boldsymbol{k}_1 \left[\gamma_{\boldsymbol{k}_1} \widehat{A}^+_{\boldsymbol{k}_1} \widehat{B}_{\boldsymbol{k}_1} + \gamma^*_{\boldsymbol{k}_1} \widehat{B}^+_{\boldsymbol{k}_1} \widehat{A}_{\boldsymbol{k}_1} \right],\end{aligned}$$

where $\gamma_{\boldsymbol{k}} = \sum_{n=1}^{3} e^{-i\boldsymbol{k}\cdot\boldsymbol{\delta}_n}$.

c) To find the eigenvalues and eigenstates, we give $\widehat{H}$ in a matrix form,

$$\widehat{H} = -t \int_{BZ} d\boldsymbol{k} \left(\widehat{A}^+_{\boldsymbol{k}}, \widehat{B}^+_{\boldsymbol{k}} \right) \begin{pmatrix} 0 & \gamma_{\boldsymbol{k}} \\ \gamma^*_{\boldsymbol{k}} & 0 \end{pmatrix} \begin{pmatrix} \widehat{A}_{\boldsymbol{k}} \\ \widehat{B}_{\boldsymbol{k}} \end{pmatrix} = -t \int_{BZ} d\boldsymbol{k} \left(\widehat{A}^+_{\boldsymbol{k}}, \widehat{B}^+_{\boldsymbol{k}} \right) \widehat{h}(\boldsymbol{k}) \begin{pmatrix} \widehat{A}_{\boldsymbol{k}} \\ \widehat{B}_{\boldsymbol{k}} \end{pmatrix},$$

from which it is easy to obtain

$$\begin{aligned} E_\pm &= \pm t\,|\gamma_{\boldsymbol{k}}|;\ \chi_1 = \frac{1}{\sqrt{2}} \begin{pmatrix} \frac{\gamma_{\boldsymbol{k}}}{|\gamma_{\boldsymbol{k}}|} \\ \mp 1 \end{pmatrix}, \\ |\gamma_{\boldsymbol{k}}| &= \sqrt{3 + 2\cos(\boldsymbol{k}\cdot(\boldsymbol{\delta}_1-\boldsymbol{\delta}_2)) + 2\cos(\boldsymbol{k}\cdot(\boldsymbol{\delta}_1-\boldsymbol{\delta}_3)) + 2\cos(\boldsymbol{k}\cdot(\boldsymbol{\delta}_2-\boldsymbol{\delta}_3))} \\ &= \sqrt{3 + 4\cos\left(\frac{3ak_x}{2}\right)\cos\left(\frac{\sqrt{3}ak_y}{2} + 2\cos^2\left(\sqrt{3}ak_y\right)\right)}. \end{aligned}$$

The energy dispersion is schematically given here (Figure 5.7), and we see that the energy bands touch at the special $\boldsymbol{K}_\pm = \frac{2\pi}{3a}\left(1, \frac{\pm 1}{\sqrt{3}}\right)$ points.

d) In the vicinity of those special points, we expand the wave vector $\boldsymbol{k} = \boldsymbol{K}_\pm + \boldsymbol{q}$ and consider the limit $\boldsymbol{q} \to \boldsymbol{0}$, for which we find

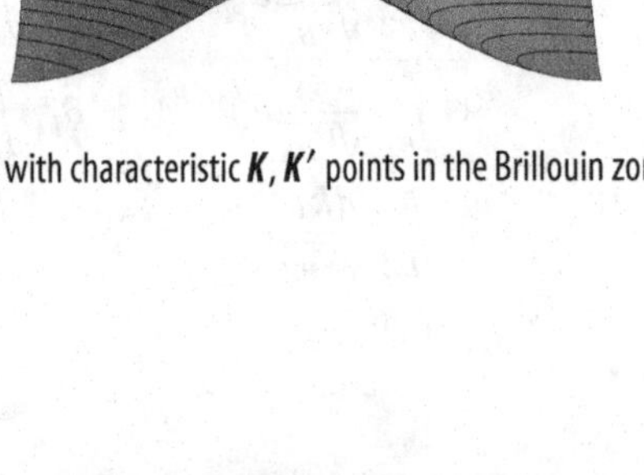

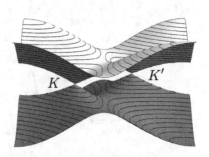

Figure 5.7 The graphene lowest energy bands with characteristic $\boldsymbol{K}, \boldsymbol{K'}$ points in the Brillouin zone.

$$\gamma_{K_\pm+q} = \frac{3ta}{2}e^{-i\frac{\pi}{6}}\left(q_x \pm iq_y\right) = \frac{3ta}{2}e^{-i\frac{\pi}{6}}q(\cos(\theta) \pm i\sin(\theta))$$
$$= \frac{3taq}{2}e^{-i\frac{\pi}{6}}e^{\pm i\theta} = \frac{3taq}{2}e^{i(-\frac{\pi}{6}\pm\theta)},$$

$$\widehat{h}_+(q) = -\frac{3ta}{2}\begin{pmatrix} 0 & e^{-i\frac{\pi}{6}}\left(q_x - iq_y\right) \\ e^{i\frac{\pi}{6}}\left(q_x + iq_y\right) & 0 \end{pmatrix},$$
$$\widehat{h}_-(q) = -\frac{3ta}{2}\begin{pmatrix} 0 & e^{-i\frac{\pi}{6}}\left(q_x + iq_y\right) \\ e^{i\frac{\pi}{6}}\left(q_x - iq_y\right) & 0 \end{pmatrix}.$$

The phase $\left(e^{\pm i\frac{\pi}{6}}\right)$ can be ignored since it only reflects a global $\frac{\pi}{6}$ rotation of q_x and q_y. Thus, the low-energy Hamiltonian and its eigenvalues and eigenvectors at K_+ are

$$\widehat{h}_+(q) = \frac{3ta}{2}\begin{pmatrix} 0 & q_x - iq_y \\ q_x + iq_y & 0 \end{pmatrix} = \frac{3ta}{2}\boldsymbol{\sigma}\cdot\boldsymbol{q},$$
$$E_\pm = \pm\frac{3ta}{2}q; \quad \chi_\pm = \frac{1}{\sqrt{2}}\begin{pmatrix} \frac{q_x - iq_y}{q} \\ \pm 1 \end{pmatrix} = \frac{1}{\sqrt{2}}\begin{pmatrix} e^{-i\theta} \\ \pm 1 \end{pmatrix},$$

where $\boldsymbol{\sigma} = (\sigma_1, \sigma_2)$ are two Pauli matrices. Similarly, the low-energy Hamiltonian and its eigenvalues and eigenvectors at K_- are

$$\widehat{h}_-(q) = \frac{3ta}{2}\begin{pmatrix} 0 & q_x + iq_y \\ q_x - iq_y & 0 \end{pmatrix} = \frac{3ta}{2}\boldsymbol{\sigma}^*\cdot\boldsymbol{q},$$
$$E_\pm = \pm\frac{3ta}{2}q; \; \chi_\pm = \frac{1}{\sqrt{2}}\begin{pmatrix} \frac{q_x + iq_y}{q} \\ \pm 1 \end{pmatrix} = \frac{1}{\sqrt{2}}\begin{pmatrix} e^{i\theta} \\ \pm 1 \end{pmatrix},$$

where $\boldsymbol{\sigma}^* = (\sigma_1, -\sigma_2)$.

Food for thought: The Hamiltonian for small wave vector around the special K-point is often written as $h = \hbar v_F \boldsymbol{\sigma}\cdot\boldsymbol{q}$, which has exactly the same form as the Dirac Hamiltonian (considered in the next chapter). The fact that the energy dispersion in the special K-point is linear is one of the main reasons for the unique properties of graphene, which is often referred to as a Dirac 2D system (Castro Neto et al., 2009).

6 Relativistic Effects in Quantum Mechanics: The Dirac Equation

In quantum chemistry, relativistic effects are usually considered as small corrections to dominating nonrelativistic quantum mechanics as described by the Schrödinger equation. Typically, one applies perturbation theory to explain properties of heavy elements in the periodic table. The relativistic correction to the electron kinetic energy in atoms, spin–orbit coupling, and Zeeman splitting are some examples, which can be found in many textbooks. Taking into account concepts from special theory of relativity, however, has led to the development of relativistic quantum mechanics, which has been successful in areas beyond atoms and chemical elements. For example, the prediction of antiparticles, dynamics of charged particles in electromagnetic fields, and spin textures of fermions among others are unattainable without fully relativistic description. Much of the current research in topologically nontrivial materials has also shown that relativistic quantum mechanics is needed to understand such physics.

In available textbooks, relativistic quantum mechanics is left as the last chapter with a few examples worked out in detail and a handful of problems left for practice. One has to make a big jump by filling out a big gap of needed technical skills. The Klein–Gordon and Dirac equations are Lorentz-invariant relations, which set the foundation of relativistic quantum mechanics. In this chapter we primarily focus on the Dirac equation, which is linear in spatial and time derivatives whose wave function can still be interpreted as a probability density. The Dirac equation gives an accurate description of the hydrogen spectrum and predicts unusual effects, such as the Klein paradox, for example. Some of the problems we offer are similar to those in previous chapters, while others are adopted from the research literature.

6.1 The Klein–Gordon Equation

The Klein–Gordon equation is defined as

$$-\hbar^2\partial_t^2\psi(\boldsymbol{x},t) = \widehat{H}^2\psi(\boldsymbol{x},t),$$
$$\widehat{H}^2 = c^2\widehat{p}^i\widehat{p}_i + m^2c^4 + V(\widehat{\boldsymbol{r}}) = [-\hbar^2c^2\boldsymbol{\nabla}^2 + m^2c^4 + V(\boldsymbol{x})].$$

where $i = 1,2,3$ and the Einstein summation is implied.

6.2 The Dirac Equation

The Dirac equation is defined as

$$i\hbar\alpha^0\partial_t\psi(\boldsymbol{x},t) = \widehat{H}\psi(\boldsymbol{x},t),$$
$$\widehat{H} = c\boldsymbol{\alpha}\cdot\widehat{\boldsymbol{p}} + mc^2\alpha^4 + \alpha^0 V(\widehat{\boldsymbol{r}}) = c\alpha^i p_i + mc^2\alpha^4 + \alpha^0 V(\boldsymbol{x}).$$

where

$$\alpha^0 = \left(\begin{array}{c|c} \sigma^0 & 0 \\ \hline 0 & \sigma^0 \end{array}\right), \alpha^i = \left(\begin{array}{c|c} 0 & \sigma^i \\ \hline \sigma^i & 0 \end{array}\right), \alpha^4 = \left(\begin{array}{c|c} \sigma^0 & 0 \\ \hline 0 & -\sigma^0 \end{array}\right).$$

Note that $\{\alpha^\mu,\alpha^\nu\} = 2\delta^{\mu\nu}\alpha^0$, $\left[\alpha^0,\alpha^\nu\right] = 0$ for $\mu,\nu = 1,\ldots,4$ and σ^μ are the Pauli matrices:

$$\sigma^0 = \begin{pmatrix} 1 & 0 \\ 0 & 1 \end{pmatrix}; \sigma^1 = \begin{pmatrix} 0 & 1 \\ 1 & 0 \end{pmatrix}; \sigma^2 = \begin{pmatrix} 0 & -i \\ i & 0 \end{pmatrix}; \sigma^3 = \begin{pmatrix} 1 & 0 \\ 0 & -1 \end{pmatrix}.$$

The Hamiltonian is given in the Dirac representation, but using α^4 instead of β.

6.3 Symmetry Operations for Dirac Particles

Time-Reversal Operator:	$\widehat{\Theta} = \eta\alpha^1\alpha^3\widehat{K}$
Parity Operator:	$\widehat{\Pi} = \eta\alpha^4\widehat{\pi}$
Particle-Hole Symmetry Operator:	$\widehat{C} = \eta\alpha^2\alpha^4\widehat{K}$

In the preceding definitions: η – arbitrary phase, $\widehat{K}$ – complex conjugation, $\pi\widehat{r}\pi^{-1} = -\widehat{r}$ – parity operation for the position operator $\widehat{\boldsymbol{r}}$.

Problem 6.1

The Dirac Hamiltonian for a massless particle is $\widehat{H} = c\boldsymbol{\sigma}\cdot\widehat{\boldsymbol{p}} + V(\widehat{\boldsymbol{r}})\sigma_z$, where c is the speed of light, $\boldsymbol{\sigma} = (\sigma_x,\sigma_y,\sigma_z)$ are the Pauli matrices, $\widehat{\boldsymbol{p}}$ is the momentum operator, and $V(\widehat{\boldsymbol{r}})$ is an arbitrary potential (Berry & Mondragon, 1987).

a) Obtain the eigenspinor $|\psi(\boldsymbol{x})\rangle$ and eigenenergy for this massless particle constrained to move in 2D for the case of $V(\widehat{\boldsymbol{r}}) = 0$.
b) Consider also the local electric current $\boldsymbol{j}(\boldsymbol{x}) = \langle\psi(\boldsymbol{x})|\boldsymbol{\nabla}_{\boldsymbol{p}}\widehat{H}|\psi(\boldsymbol{x})\rangle$ and give its explicit representation.

---

a) We begin with the eigenvalue equation for the given Hamiltonian,

$$\widehat{H}|\psi(\boldsymbol{x})\rangle = E|\psi(\boldsymbol{x})\rangle, \quad c\boldsymbol{\sigma}\cdot\widehat{\boldsymbol{p}}|\psi(\boldsymbol{x})\rangle = -i\hbar c\boldsymbol{\sigma}\cdot\boldsymbol{\nabla}|\psi(\boldsymbol{x})\rangle$$

$$-i\hbar c\begin{pmatrix} 0 & \partial_x - i\partial_y \\ \partial_x + i\partial_y & 0 \end{pmatrix}\begin{pmatrix} \psi_1(\boldsymbol{x}) \\ \psi_2(\boldsymbol{x}) \end{pmatrix} = E\begin{pmatrix} \psi_1(\boldsymbol{x}) \\ \psi_2(\boldsymbol{x}) \end{pmatrix},$$

The energy for this massless particle is taken as $E = \hbar ck$ with $\boldsymbol{k} = (k_x, k_y)$ being the 2D wave vector. Writing the preceding equation in a matrix as a linear system, we obtain

$$\begin{cases} -i\left(\partial_x - i\partial_y\right)\psi_2(\boldsymbol{x}) = k\psi_1(\boldsymbol{x}) \\ -i\left(\partial_x + i\partial_y\right)\psi_1(\boldsymbol{x}) = k\psi_2(\boldsymbol{x}) \end{cases}.$$

After applying the conjugate operator to each equation, $+i\left(\partial_x + i\partial_y\right)$ to the first equation, and $-i\left(\partial_x - i\partial_y\right)$ to the second, we obtain

$$\left[i\left(\partial_x + i\partial_y\right)\right]\left[-i\left(\partial_x - i\partial_y\right)\right]\psi_n(\boldsymbol{x}) = \left(\partial_x^2 + \partial_y^2\right)\psi_n(\boldsymbol{x}) \to \Delta\psi_n(x) = -k^2\psi_n(\boldsymbol{x})$$

for $n = \{1,2\}$. Therefore, each spinor is a harmonic function in the $\boldsymbol{x} = (x,y)$ plane $\psi_n(\boldsymbol{x}) = \tilde{\psi}_n(\boldsymbol{k})e^{i\boldsymbol{k}\cdot\boldsymbol{x}}$.

Let us take $\psi_2(\boldsymbol{x}) = \frac{-i\left(\partial_x + i\partial_y\right)}{k}\psi_1(\boldsymbol{x})$, from the second equation, for example. It is easy to see that

$$\tilde{\psi}_2(\boldsymbol{k}) = \frac{k_x + ik_y}{k}\tilde{\psi}_1(\boldsymbol{k}),$$

$$|\tilde{\psi}(\boldsymbol{k})\rangle = A\begin{pmatrix} \tilde{\psi}_1(\boldsymbol{k}) \\ \tilde{\psi}_2(\boldsymbol{k}) \end{pmatrix} = A\begin{pmatrix} \tilde{\psi}_1(\boldsymbol{k}) \\ \frac{k_x+ik_y}{k}\tilde{\psi}_1(\boldsymbol{k}) \end{pmatrix}$$
$$= A\begin{pmatrix} 1 \\ \frac{k_x+ik_y}{k} \end{pmatrix}\tilde{\psi}_1(\boldsymbol{k}) = A\begin{pmatrix} 1 \\ e^{i\varphi_{\boldsymbol{k}}} \end{pmatrix}\tilde{\psi}_1(\boldsymbol{k}),$$

where $k^2 = k_x^2 + k_y^2$, and $\frac{k_x+ik_y}{k} = e^{i\varphi_{\boldsymbol{k}}}$ in polar coordinates. The spinor must also be normalized:

$$1 = \langle\tilde{\psi}(\boldsymbol{k})|\tilde{\psi}(\boldsymbol{k})\rangle = \int\frac{d\boldsymbol{k}}{(2\pi)^{2/2}}\tilde{\psi}^+(\boldsymbol{k})\tilde{\psi}(\boldsymbol{k}) = 2|A|^2\int\frac{d\boldsymbol{k}}{2\pi}\tilde{\psi}_1^+(\boldsymbol{k})\tilde{\psi}_1(\boldsymbol{k})$$
$$= 2|A|^2\left\langle\tilde{\psi}_1(\boldsymbol{k})|\tilde{\psi}_1(\boldsymbol{k})\right\rangle.$$

Since $\left\langle\tilde{\psi}_1(\boldsymbol{k})|\tilde{\psi}_1(\boldsymbol{k})\right\rangle = 1$, we obtain $|A|^2 = \frac{1}{2}$, and $A = \frac{e^{i\theta}}{\sqrt{2}}$, with θ being an arbitrary phase. Thus, the eigenspinor for the massless 2D Dirac Hamiltonian in free space is

$$|\psi(\boldsymbol{x})\rangle = \begin{pmatrix} \psi_1(\boldsymbol{x}) \\ \psi_2(\boldsymbol{x}) \end{pmatrix} = \frac{e^{i\theta}}{\sqrt{2}}\begin{pmatrix} 1 \\ e^{i\varphi_{\boldsymbol{k}}} \end{pmatrix}e^{i\boldsymbol{k}\cdot\boldsymbol{x}}.$$

b) Next, let us consider the current $\boldsymbol{j}(\boldsymbol{x}) = \langle\psi(\boldsymbol{x})|\boldsymbol{\nabla}_{\boldsymbol{p}}\widehat{H}|\psi(\boldsymbol{x})\rangle$. Since $\boldsymbol{\nabla}_{\boldsymbol{p}}\widehat{H} = c\boldsymbol{\sigma}$, we find

$$j(\boldsymbol{x}) = \langle\psi(\boldsymbol{x})|c\boldsymbol{\sigma}|\psi(\boldsymbol{x})\rangle = c\langle\psi(\boldsymbol{x})|\left(\sigma_x\widehat{\boldsymbol{x}} + \sigma_y\widehat{\boldsymbol{y}}\right)|\psi(\boldsymbol{x})\rangle$$
$$= \frac{c}{2}e^{-i\boldsymbol{k}\cdot\boldsymbol{x}}\left(1, e^{-i\varphi_{\boldsymbol{k}}}\right)\begin{pmatrix} 0 & 1 \\ 1 & 0 \end{pmatrix}\begin{pmatrix} 1 \\ e^{i\varphi_{\boldsymbol{k}}} \end{pmatrix}e^{i\boldsymbol{k}\cdot\boldsymbol{x}}\widehat{\boldsymbol{x}}$$
$$+ \frac{c}{2}e^{-i\boldsymbol{k}\cdot\boldsymbol{x}}\left(1, e^{-i\varphi_{\boldsymbol{k}}}\right)\begin{pmatrix} 0 & -i \\ i & 0 \end{pmatrix}\begin{pmatrix} 1 \\ e^{i\varphi_{\boldsymbol{k}}} \end{pmatrix}e^{i\boldsymbol{k}\cdot\boldsymbol{x}}\widehat{\boldsymbol{y}}$$
$$= \frac{c}{2}\left(1, e^{-i\varphi_{\boldsymbol{k}}}\right)\begin{pmatrix} e^{i\varphi_{\boldsymbol{k}}} \\ 1 \end{pmatrix}\widehat{\boldsymbol{x}} - \frac{c}{2}i\left(1, e^{-i\varphi_{\boldsymbol{k}}}\right)\begin{pmatrix} e^{i\varphi_{\boldsymbol{k}}} \\ -1 \end{pmatrix}\widehat{\boldsymbol{y}} = c\left(\cos\left(\varphi_{\boldsymbol{k}}\right), \sin\left(\varphi_{\boldsymbol{k}}\right)\right).$$

Problem 6.2
The Hamiltonian for a relativistic particle constrained to move in 2D is $\hat{H} = c\boldsymbol{\sigma}\cdot\hat{\boldsymbol{p}} + V(\hat{\boldsymbol{r}})\sigma_z$. Here, $\boldsymbol{\sigma} = (\sigma_x, \sigma_y, \sigma_z)$ are the Pauli matrices, $\hat{\boldsymbol{p}}$ is the momentum operator, and $V(\hat{\boldsymbol{r}})$ is an arbitrary real potential (Berry & Mondragon, 1987).

a) When is $\hat{H}$ invariant under time reversal?
b) Consider the antiunitary operator $\hat{A} = \sigma_z \hat{R}_y \hat{K}$, where $\hat{R}_y$ represents the reflection $\hat{R}_y(x,y) = (x,-y)$ and $\hat{K}$ is the complex conjugation operator. When is $\hat{H}$ invariant under $\hat{A}$?
c) Consider the antiunitary operator $\hat{C} = i\sigma_y \hat{R}_{2\pi}\hat{K}$, where the rotation by 2π is $\hat{R}_{2\pi}(r,\phi) = (r, \phi + 2\pi)$. When is $\hat{H}$ invariant under $\hat{C}$?

---

a) The time-reversal operator for a spin ½ particle is $\hat{\Theta} = \eta e^{i\sigma_2\pi/2}\hat{K} = \eta i\sigma_2\hat{K}$, where η is an arbitrary phase (see Problems 1.24 and 1.25). Transforming the Hamiltonian gives

$$
\begin{aligned}
\hat{H}_\Theta = \hat{\Theta}\hat{H}\hat{\Theta}^{-1} &= \eta i\sigma_2\hat{K}\left(c\boldsymbol{\sigma}\cdot\hat{\boldsymbol{p}} + V(\hat{\boldsymbol{r}})\sigma_z\right)\hat{K}^{-1}\sigma_2^{-1}i^{-1}\eta^{-1} \\
&= \eta i\sigma_2\hat{K}\begin{pmatrix} V(\hat{\boldsymbol{r}}) & c(\hat{p}_x - i\hat{p}_y) \\ c(\hat{p}_x + i\hat{p}_y) & -V(\hat{\boldsymbol{r}}) \end{pmatrix}\hat{K}^{-1}\sigma_2^{-1}i^{-1}\eta^{-1} \\
&= \eta(i\sigma_2)\begin{pmatrix} V(\hat{\boldsymbol{r}}) & c(\hat{p}_x^* + i\hat{p}_y^*) \\ c(\hat{p}_x^* - i\hat{p}_y^*) & -V(\hat{\boldsymbol{r}}) \end{pmatrix}(\hat{K}\hat{K}^{-1})(i\sigma_2)^{-1}\eta^{-1} \\
&= \eta\begin{pmatrix} 0 & -1 \\ 1 & 0 \end{pmatrix}\begin{pmatrix} V(\hat{\boldsymbol{r}}) & c(\hat{p}_x^* + i\hat{p}_y^*) \\ c(\hat{p}_x^* - i\hat{p}_y^*) & -V(\hat{\boldsymbol{r}}) \end{pmatrix}\begin{pmatrix} 0 & 1 \\ -1 & 0 \end{pmatrix}\eta^{-1} \\
&= \begin{pmatrix} -V(\hat{\boldsymbol{r}}) & c(\hat{p}_x - i\hat{p}_y) \\ c(\hat{p}_x + i\hat{p}_y) & -V(\hat{\boldsymbol{r}}) \end{pmatrix} \neq \hat{H}.
\end{aligned}
$$

The preceding result, obtained by using $\hat{p}_i^* = -\hat{p}_i$, shows that the given Hamiltonian is not invariant under time reversal. We realize that the Hamiltonian is invariant under time inversion only when $V(\hat{\boldsymbol{r}}) = 0$.

b) Transforming the Hamiltonian under $\hat{A} = \sigma_z\hat{R}_y\hat{K}$ operation gives

$$
\begin{aligned}
\hat{H}_{\mathrm{A}} = \hat{\mathrm{A}}\hat{H}\hat{\mathrm{A}}^{-1} &= \sigma_z\hat{R}_y\hat{K}\begin{pmatrix} V(\hat{\boldsymbol{r}}) & c(\hat{p}_x - i\hat{p}_y) \\ c(\hat{p}_x + i\hat{p}_y) & -V(\hat{\boldsymbol{r}}) \end{pmatrix}\hat{K}^{-1}\hat{R}_y^{-1}\sigma_z^{-1} \\
&= \sigma_z\hat{R}_y\begin{pmatrix} V(\hat{x},\hat{y}) & c(\hat{p}_x^* + i\hat{p}_y^*) \\ c(\hat{p}_x^* - i\hat{p}_y^*) & -V(\hat{x},\hat{y}) \end{pmatrix}(\hat{K}\hat{K}^{-1})\hat{R}_y^{-1}\sigma_z^{-1} \\
&= \sigma_z\begin{pmatrix} V(\hat{x},-\hat{y}) & c(\hat{p}_x^* - i\hat{p}_y^*) \\ c(\hat{p}_x^* + i\hat{p}_y^*) & -V(\hat{x},-\hat{y}) \end{pmatrix}\left(\hat{R}_y\hat{R}_y^{-1}\right)\sigma_z^{-1} \\
&= \begin{pmatrix} V(\hat{x},-\hat{y}) & -c(\hat{p}_x^* - i\hat{p}_y^*) \\ -c(\hat{p}_x^* + i\hat{p}_y^*) & -V(\hat{x},-\hat{y}) \end{pmatrix} = \begin{pmatrix} V(\hat{x},-\hat{y}) & c(\hat{p}_x - i\hat{p}_y) \\ c(\hat{p}_x + i\hat{p}_y) & -V(\hat{x},-\hat{y}) \end{pmatrix}.
\end{aligned}
$$

The Hamiltonian is invariant under the A operation only if $\hat{R}_y V(\hat{\boldsymbol{r}})\hat{R}_y^{-1} = V(\hat{\boldsymbol{r}})$, while in general $\hat{H}_A \neq \hat{H}$.

c) For the operator $\widehat{C} = i\sigma_y \widehat{R}_{2\pi}\widehat{K}$, the rotation matrix operator is $\widehat{R}_{\theta,\widehat{z}} = e^{\frac{i}{2}\theta\widehat{z}\cdot\sigma} = e^{\pi i\sigma_z} = \begin{pmatrix} e^{i\pi} & 0 \\ 0 & e^{-i\pi} \end{pmatrix} = \begin{pmatrix} -1 & 0 \\ 0 & -1 \end{pmatrix}$. The transformed Hamiltonian then becomes

$$\widehat{H}_C = \widehat{C}\widehat{H}\widehat{C}^{-1} = (i\sigma_y)\,\widehat{R}_{2\pi}\widehat{K}\begin{pmatrix} V(\widehat{\boldsymbol{r}}) & c(\widehat{p}_x - i\widehat{p}_y) \\ c(\widehat{p}_x + i\widehat{p}_y) & -V(\widehat{\boldsymbol{r}}) \end{pmatrix}\widehat{K}^{-1}\widehat{R}_{2\pi}^{-1}(i\sigma_y)^{-1}$$

$$= \begin{pmatrix} 0 & 1 \\ -1 & 0 \end{pmatrix}\begin{pmatrix} V(\widehat{r},\widehat{\phi}+2\pi) & c(\widehat{p}_x^* + i\widehat{p}_y^*) \\ c(\widehat{p}_x^* - i\widehat{p}_y^*) & -V(\widehat{r},\widehat{\phi}+2\pi) \end{pmatrix}\begin{pmatrix} 0 & -1 \\ 1 & 0 \end{pmatrix}$$

$$= \begin{pmatrix} -V(\widehat{\boldsymbol{r}}) & -c(\widehat{p}_x^* - i\widehat{p}_y^*) \\ -c(\widehat{p}_x^* + i\widehat{p}_y^*) & V(\widehat{\boldsymbol{r}}) \end{pmatrix} = \begin{pmatrix} -V(\widehat{\boldsymbol{r}}) & c(\widehat{p}_x - i\widehat{p}_y) \\ c(\widehat{p}_x + i\widehat{p}_y) & V(\widehat{\boldsymbol{r}}) \end{pmatrix}.$$

In the preceding Hamiltonian, we have used that $\widehat{R}_{2\pi}V(\widehat{\boldsymbol{r}})\widehat{R}_{2\pi}^{-1} = V(\widehat{r},\phi+2\pi) = V(\widehat{\boldsymbol{r}})$. While in general $\widehat{H}_C \neq \widehat{H}$, in the case of $V(\widehat{\boldsymbol{r}}) = -V(\widehat{\boldsymbol{r}})$ (meaning $V(\widehat{\boldsymbol{r}}) = 0$) the given Hamiltonian is invariant under the $\widehat{C}$ operation.

Problem 6.3

The Klein–Gordon equation $-\hbar^2\frac{\partial^2}{\partial t^2}\psi(\boldsymbol{x},t) = (-\hbar^2c^2\boldsymbol{\nabla}^2 + m^2c^4)\psi(\boldsymbol{x},t)$ incorporates relativistic effects in the quantum mechanical framework. This is a second-order differential equation with respect to time and it is consistent with the special theory of relativity since the equation is invariant under inertial frame transformations. Is the Klein–Gordon equation consistent with the meaning of probability density as given by standard quantum mechanics?

In quantum mechanics, the probability density must be real non-negative, and the norm of the wave function must be time independent.

Note that the solution to the Klein–Gordon equation with no potential, given in the problem, is satisfied by $\psi(\boldsymbol{x},t) = \frac{e^{iEt-i\boldsymbol{k}\cdot\boldsymbol{x}}}{N}$, where the energy $E = \pm\sqrt{(\hbar ck)^2 + (mc^2)^2}$ can be either positive or negative.

Let us remind ourselves that the probability density ρ and the probability current density $\boldsymbol{j}$ must satisfy the continuity equation, such that

$$\frac{d\rho}{dt} + \boldsymbol{\nabla}\cdot\boldsymbol{j} = 0,$$

where

$$\rho(\boldsymbol{x}) = i\left[\psi^*(\boldsymbol{x},t)\partial^0\psi(\boldsymbol{x},t) - \partial^0\psi^*(\boldsymbol{x},t)\psi(\boldsymbol{x},t)\right],$$

$$j^\mu(\boldsymbol{x}) = i\left[\psi^*(\boldsymbol{x},t)\partial^\mu\psi(\boldsymbol{x},t) - \partial^\mu\psi^*(\boldsymbol{x},t)\psi(\boldsymbol{x},t)\right].$$

Substituting $\psi(\boldsymbol{x},t) = \frac{e^{iEt-i\boldsymbol{k}\cdot\boldsymbol{x}}}{\sqrt{V}}$ in $\rho(\boldsymbol{x})$, we find $\rho(\boldsymbol{x}) = \frac{2E}{V}$. Given that E can be either positive or negative, this quantity $\rho(\boldsymbol{x})$ cannot be interpreted as a probability density in the quantum mechanical sense which must always be non-negative.

Problem 6.4
The Dirac equation for a free particle can be written in the familiar form $i\hbar\alpha^0\frac{\partial}{\partial t}\psi(\boldsymbol{x},t) = \widehat{H}\psi(\boldsymbol{x},t) = \left[c\boldsymbol{\alpha}\cdot\widehat{\boldsymbol{p}} + mc^2\alpha^4\right]\psi(\boldsymbol{x},t)$. Is this equation consistent with the meaning of probability density defined as $\rho(\boldsymbol{x},t) = \langle\psi(\boldsymbol{x},t)\,|\psi(\boldsymbol{x},t)\rangle$?

We must check to see if the norm of the wave function satisfying the Dirac equation $i\hbar\alpha^0\frac{\partial}{\partial t}\psi(\boldsymbol{x},t) = \widehat{H}\psi(\boldsymbol{x},t)$ is conserved in time. Thus,

$$i\hbar\frac{d}{dt}\langle\psi(\boldsymbol{x},t)\,|\psi(\boldsymbol{x},t)\rangle = i\hbar\frac{d}{dt}\int d^3\boldsymbol{x}\psi^+(\boldsymbol{x},t)\psi(\boldsymbol{x},t)$$
$$= \int d^3\boldsymbol{x}\left[\left(i\hbar\alpha^0\frac{d}{dt}\psi^+(\boldsymbol{x},t)\right)\psi(\boldsymbol{x},t) + \psi^+(\boldsymbol{x},t)\left(i\hbar\alpha^0\frac{d}{dt}\psi(\boldsymbol{x},t)\right)\right]$$
$$= \int d^3\boldsymbol{x}\left[\psi^+(\boldsymbol{x},t)\left(\widehat{H}\psi(\boldsymbol{x},t)\right) - \left(\widehat{H}\psi(\boldsymbol{x},t)\right)^+\psi(\boldsymbol{x},t)\right] = 0.$$

Given that $\langle\psi(\boldsymbol{x},t)\,|\psi(\boldsymbol{x},t)\rangle > 0\forall x$ and t, $\langle\psi(\boldsymbol{x},t)\,|\psi(\boldsymbol{x},t)\rangle$ can be considered a probability density in quantum mechanical sense.

Problem 6.5
Consider the Dirac equation for a free particle $i\hbar\alpha^0\frac{\partial}{\partial t}\psi(\boldsymbol{x},t) = \widehat{H}\psi(\boldsymbol{x},t) = \left[c\boldsymbol{\alpha}\cdot\widehat{\boldsymbol{p}} + mc^2\alpha^4\right]\psi(\boldsymbol{x},t)$, where the $\boldsymbol{\alpha}$ matrices were defined at the beginning of the chapter. Show that the eigenenergies are $E = \pm\sqrt{p^2c^2 + m^2c^4}$ with corresponding eigenspinors expressed in the eigenbasis of the $\boldsymbol{\sigma}\cdot\widehat{\boldsymbol{p}}$ operator with $\chi_+ = \begin{pmatrix}\cos\left(\frac{\theta}{2}\right)\\ \sin\left(\frac{\theta}{2}\right)e^{i\phi}\end{pmatrix}, \chi_- = \begin{pmatrix}-\sin\left(\frac{\theta}{2}\right)e^{-i\phi}\\ \cos\left(\frac{\theta}{2}\right)\end{pmatrix}$ in spherical coordinates of the momentum $\boldsymbol{p} = p(\sin(\theta)\cos(\phi), \sin(\theta), \sin(\phi), \cos(\theta))$.

We begin by taking the ansatz $\psi(\boldsymbol{x},t) = u(\boldsymbol{p})e^{\frac{i(\boldsymbol{p}\cdot\boldsymbol{x}-Et)}{\hbar}}$ with the spinor $u(\boldsymbol{p}) = \begin{pmatrix}u_1(\boldsymbol{p})\\ u_2(\boldsymbol{p})\end{pmatrix}$ and substituting it in the given equation $\widehat{H}\psi(\boldsymbol{x},t) = E\psi(\boldsymbol{x},t)$,

$$\begin{pmatrix}mc^2\sigma_0 & c\boldsymbol{\sigma}\cdot\widehat{\boldsymbol{p}}\\ c\boldsymbol{\sigma}\cdot\widehat{\boldsymbol{p}} & -mc^2\sigma_0\end{pmatrix}\begin{pmatrix}u_1(\boldsymbol{p})\\ u_2(\boldsymbol{p})\end{pmatrix} = E\begin{pmatrix}u_1(\boldsymbol{p})\\ u_2(\boldsymbol{p})\end{pmatrix} \rightarrow \begin{cases}c\boldsymbol{\sigma}\cdot\boldsymbol{p}u_2(\boldsymbol{p}) = (E - mc^2)u_1(\boldsymbol{p})\\ c\boldsymbol{\sigma}\cdot\boldsymbol{p}u_1(\boldsymbol{p}) = (E + mc^2)u_2(\boldsymbol{p})\end{cases}.$$

Then, it is easy to find

$$u_2(\boldsymbol{p}) = \frac{c\boldsymbol{\sigma}\cdot\boldsymbol{p}}{E + mc^2}u_1(\boldsymbol{p}),$$
$$u_1(\boldsymbol{p}) = \frac{c\boldsymbol{\sigma}\cdot\boldsymbol{p}}{E - mc^2}u_2(\boldsymbol{p}).$$

Inserting the first equation into the second, and using $(\boldsymbol{\sigma}\cdot\boldsymbol{a})(\boldsymbol{\sigma}\cdot\boldsymbol{b}) = \boldsymbol{a}\cdot\boldsymbol{b} + \boldsymbol{\sigma}\cdot(\boldsymbol{a}\times\boldsymbol{b})$, we further obtain

$$u_1(\boldsymbol{p}) = \frac{c\boldsymbol{\sigma}\cdot\boldsymbol{p}}{E - mc^2}\frac{c\boldsymbol{\sigma}\cdot\boldsymbol{p}}{E + mc^2}u_1(\boldsymbol{p}) = \frac{c^2\boldsymbol{p}\cdot\boldsymbol{p}}{E^2 - (mc^2)^2}u_1(\boldsymbol{p}),$$

from where positive and negative eigenenergies are obtained,

$$E_\pm = \pm\sqrt{c^2p^2 + m^2c^4}.$$

Similarly, inserting the second equation into the first leads to the same result for E. Therefore, the eigenenergies are *double degenerate*.

The preceding solutions for positive and negative eigenenergies imply that the eigenspinors of the Dirac equation are

$$u_+(\boldsymbol{p}) = N_+ \begin{pmatrix} 1 \\ \frac{c\boldsymbol{\sigma}\cdot\boldsymbol{p}}{E_+ + mc^2} \end{pmatrix},$$

$$u_-(\boldsymbol{p}) = N_- \begin{pmatrix} \frac{c\boldsymbol{\sigma}\cdot\boldsymbol{p}}{E_- - mc^2} \\ 1 \end{pmatrix}.$$

Finally, we take the solution to $(\boldsymbol{\sigma}\cdot\boldsymbol{p})\chi = \lambda\chi$ (where $\chi = \frac{1}{\sqrt{|a|^2+|b|^2}}\begin{pmatrix} a \\ b \end{pmatrix}$) as a representative basis for $u_\pm(\boldsymbol{p})$. From

$$\boldsymbol{\sigma}\cdot\boldsymbol{p} = \begin{pmatrix} p_3 & p_1 - ip_2 \\ p_1 + ip_2 & -p_3 \end{pmatrix} = p\begin{pmatrix} \cos(\theta) & \sin(\theta)e^{-i\phi} \\ \sin(\theta)e^{i\phi} & -\cos(\theta) \end{pmatrix},$$

one finds the eigenvalues $\lambda_\pm = \pm p$ and they are associated with the following eigenvectors:

$$\chi_+ = \begin{pmatrix} \cos\left(\frac{\theta}{2}\right) \\ \sin\left(\frac{\theta}{2}\right)e^{i\phi} \end{pmatrix}, \quad \chi_- = \begin{pmatrix} -\sin\left(\frac{\theta}{2}\right)e^{-i\phi} \\ \cos\left(\frac{\theta}{2}\right) \end{pmatrix}.$$

Since $\langle\chi_\mu|\chi_\nu\rangle = \delta_{\mu\nu}$, the four eigenvectors of the Dirac equation are

$$u_{+,\eta}(\boldsymbol{p}) = N_+ \begin{pmatrix} \chi_\eta \\ \frac{c\boldsymbol{\sigma}\cdot\boldsymbol{p}}{E_+ + mc^2}\chi_\eta \end{pmatrix} = N_+ \begin{pmatrix} \chi_\eta \\ \frac{\eta cp}{E_+ + mc^2}\chi_\eta \end{pmatrix},$$

$$u_{-,\eta}(\boldsymbol{p}) = N_- \begin{pmatrix} \frac{c\boldsymbol{\sigma}\cdot\boldsymbol{p}}{E_- - mc^2}\chi_\eta \\ \chi_\eta \end{pmatrix} = N_- \begin{pmatrix} \frac{\eta cp}{E_- - mc^2}\chi_\eta \\ \chi_\eta \end{pmatrix},$$

where $\eta = \pm$. After imposing the normalization conditions $1 = \langle u_{\pm,\eta}(\boldsymbol{p})|u_{\pm,\eta}(\boldsymbol{p})\rangle$ and using that $\langle\chi_\eta|\chi_\eta\rangle = 1$, we obtain

$$N_+ = \sqrt{\frac{E_+ + mc^2}{2E_+}}, \quad N_- = \sqrt{\frac{E_- - mc^2}{2E_-}}.$$

The results for the four orthonormalized eigenspinors are expressed as

$$u_{+,+}(p) = \frac{1}{\sqrt{2E_+}}\begin{pmatrix} \sqrt{E_+ + mc^2}\chi_+ \\ \frac{\eta cp}{\sqrt{E_+ + mc^2}}\chi_+ \end{pmatrix}, \quad u_{+,-}(p) = \frac{1}{\sqrt{2E_+}}\begin{pmatrix} \sqrt{E_+ + mc^2}\chi_- \\ \frac{\eta cp}{\sqrt{E_+ + mc^2}}\chi_- \end{pmatrix}$$

with $E_+ = \sqrt{m^2c^4 + c^2p^2}$, and

$$u_{-,+}(p) = \frac{1}{\sqrt{2E_-}}\begin{pmatrix} \frac{\eta cp}{\sqrt{E_- - mc^2}}\chi_+ \\ \sqrt{E_- - mc^2}\chi_+ \end{pmatrix}, \quad u_{-,-}(p) = \frac{1}{\sqrt{2E_-}}\begin{pmatrix} \frac{\eta cp}{\sqrt{E_- - mc^2}}\chi_- \\ \sqrt{E_- - mc^2}\chi_- \end{pmatrix}$$

with $E_- = -\sqrt{m^2c^4 + c^2p^2}$.

Food for thought: How would the eigenenergies and eigenspinors change in the limit of $m \to 0$? Show that, in the zeroth mass limit, the helicity operator $\widehat{\Sigma} = \widehat{\boldsymbol{S}} \cdot \widehat{\boldsymbol{p}} = \frac{s\hbar}{|p|}\widehat{\boldsymbol{\sigma}} \cdot \widehat{\boldsymbol{u}}_p$ is conserved and that the Dirac equation can be split into two independent parts, corresponding to the two possible helicities of the Dirac fermion.

Problem 6.6

a) Is the angular momentum $\widehat{\boldsymbol{L}} = \widehat{\boldsymbol{r}} \times \widehat{\boldsymbol{p}}$ conserved in the Dirac equation $\widehat{H} = c\boldsymbol{\alpha} \cdot \widehat{\boldsymbol{p}} + mc^2\alpha^4 + \alpha^0 V(\widehat{\boldsymbol{r}})$?

b) Is the total angular momentum defined as $\widehat{\boldsymbol{J}} = \widehat{\boldsymbol{L}} + \frac{\hbar}{2}\widehat{\boldsymbol{\Sigma}}$, where $\boldsymbol{\Sigma} = \begin{pmatrix} \sigma & 0 \\ 0 & \sigma \end{pmatrix}$ conserved in the Dirac equation? (Here $\boldsymbol{\sigma} = (\sigma^1, \sigma^2, \sigma^3)$).

c) The parity operator for the Dirac equation is defined as $\widehat{\Pi} = \eta\alpha^4\widehat{\pi}$, where $\widehat{\pi}\widehat{\boldsymbol{r}}\widehat{\pi}^{-1} = -\widehat{\boldsymbol{r}}$ is the parity operator for the Schrödinger equation, as discussed in Chapter 1, and η is an arbitrary phase. Is the operator $\widehat{\Pi}$ conserved in the Dirac equation?

---

a) Let us consider the commutator of the Dirac Hamiltonian with the ith component of the angular momentum operator ($\widehat{L}_i = \varepsilon_{ijk}\widehat{r}^j\widehat{p}^k$) and utilize the property $\left[\widehat{A}\widehat{B}, \widehat{C}\right] = \widehat{A}\left[\widehat{B}, \widehat{C}\right] + \left[\widehat{A}, \widehat{C}\right]\widehat{B}$,

$$\begin{aligned}
\left[\widehat{H}, \widehat{L}_i\right] &= \left[(c\boldsymbol{\alpha} \cdot \widehat{\boldsymbol{p}} + mc^2\alpha^4 + \alpha^0 V(\widehat{\boldsymbol{r}}),\ \varepsilon_{ijk}\widehat{r}^j\widehat{p}^k\right] \\
&= c\,\varepsilon_{ijk}\delta_{\ell m}\left[\alpha^\ell\widehat{p}^m, \widehat{r}^j\widehat{p}^k\right] + mc^2\varepsilon_{ijk}\left[\alpha^4, \widehat{r}^j\widehat{p}^k\right] + \varepsilon_{ijk}\left[\alpha^0 V(\widehat{\boldsymbol{r}}), \widehat{r}^j\widehat{p}^k\right] \\
&= c\,\varepsilon_{ijk}\delta_{\ell m}(\widehat{r}^j\alpha^\ell\left[\widehat{p}^m, \widehat{p}^k\right] + \widehat{r}^j\left[\widehat{\alpha}^\ell, \widehat{p}^k\right]\widehat{p}^m + \alpha^\ell\left[\widehat{p}^m, \widehat{r}^j\right]\widehat{p}^k + \left[\alpha^\ell, \widehat{r}^j\right]\widehat{p}^m\widehat{p}^k) \\
&\quad + mc^2\varepsilon_{ijk}\left(\widehat{r}^j\left[\alpha^4, \widehat{p}^k\right] + \left[\alpha^4, \widehat{r}^j\right]\widehat{p}^k\right) + \varepsilon_{ijk}\alpha^0(\widehat{r}^j\left[V(\widehat{\boldsymbol{r}}), \widehat{p}^k\right] + \left[V(\widehat{\boldsymbol{r}}), \widehat{r}^j\right]\widehat{p}^k).
\end{aligned}$$

Taking into account that all commutators with $\boldsymbol{\alpha}$ matrices are zero together with the relations of conjugate variables $\left[\widehat{r}^j, \widehat{p}^m\right] = i\hbar\delta^{jm}$ and $\left[\widehat{p}^m, \widehat{p}^k\right] = \left[\widehat{r}^m, \widehat{r}^k\right] = 0$, we find (see Problem 3.19)

$$\begin{aligned}
\left[\widehat{H}, \widehat{L}_i\right] &= c\,\varepsilon_{ijk}\delta_{\ell m}(-\alpha^\ell i\hbar\delta^{jm}\widehat{p}^k) + \alpha^0\varepsilon_{ijk}\widehat{r}_j\frac{\partial V(\widehat{\boldsymbol{r}})}{\partial\widehat{r}^\ell}i\hbar\delta^{\ell k} \\
&= -i\hbar c\,\varepsilon_{ijk}\delta^j_\ell\alpha^\ell\widehat{p}^k + i\hbar\alpha^0\varepsilon_{ijk}r^j\partial^k V(\widehat{\boldsymbol{r}}) = -i\hbar\varepsilon_{ijk}(c\,\alpha^j\widehat{p}^k - \alpha^0\widehat{r}^j\partial^k V(\widehat{\boldsymbol{r}})) \neq 0.
\end{aligned}$$

The angular momentum is not a conserved quantity for the Dirac equation.

b) For the total angular momentum, we have to evaluate the commutator of the Dirac Hamiltonian with the components of the spin operator,

$$\begin{aligned}
\left[\widehat{H}, \widehat{\Sigma}^i\right] &= \left[(c\boldsymbol{\alpha} \cdot \widehat{\boldsymbol{p}} + mc^2\alpha^4 + \alpha^0 V(\widehat{\boldsymbol{r}})), \widehat{\Sigma}^i\right] \\
&= c\delta_{\ell m}\left(\alpha^\ell\left[\widehat{p}^m, \widehat{\Sigma}^i\right] + \left[\alpha^\ell, \widehat{\Sigma}^i\right]\widehat{p}^m\right) + mc^2\left[\alpha^4, \widehat{\Sigma}^i\right] + \left[\alpha^0 V(\widehat{\boldsymbol{r}}), \widehat{\Sigma}^i\right].
\end{aligned}$$

The only nonzero contribution comes from the second term in the preceding equation, for which it is easy to find that

$$\left[\alpha^\ell,\widehat{\Sigma}^i\right]=\left[\begin{pmatrix}0&\sigma^\ell\\\sigma^\ell&0\end{pmatrix},\begin{pmatrix}\sigma^i&0\\0&\sigma^i\end{pmatrix}\right]$$
$$=\begin{pmatrix}0&\sigma^\ell\\\sigma^\ell&0\end{pmatrix}\begin{pmatrix}\sigma^i&0\\0&\sigma^i\end{pmatrix}-\begin{pmatrix}\sigma^i&0\\0&\sigma^i\end{pmatrix}\begin{pmatrix}0&\sigma^\ell\\\sigma^\ell&0\end{pmatrix}$$
$$=[\sigma^\ell,\sigma^i]\otimes\begin{pmatrix}0&1\\1&0\end{pmatrix}=2\varepsilon^{\ell ik}\sigma_k\otimes\begin{pmatrix}0&1\\1&0\end{pmatrix}=2\varepsilon^{\ell ik}\begin{pmatrix}0&\sigma_k\\\sigma_k&0\end{pmatrix}=2i\varepsilon^{\ell ik}\alpha_k.$$

Combining with the results for the angular momentum, we obtain

$$[\widehat{H},\widehat{J}^i]=\left[\widehat{H},\widehat{L}^i+\frac{\hbar}{2}\widehat{\Sigma}^i\right]=i\hbar\varepsilon^{ijk}\left(c\alpha_j\widehat{p}_k-\alpha^0\widehat{r}_j\partial_kV(\widehat{\boldsymbol{r}})\right)+\frac{\hbar}{2}2ic\varepsilon^{\ell ik}\alpha_k\widehat{p}_\ell$$
$$=i\hbar\alpha^0\varepsilon^{ijk}\widehat{r}_j\partial_kV(\widehat{\boldsymbol{r}})\neq0.$$

It appears that the total angular momentum is conserved for a Dirac equation whose potential is a constant. In general, however, $\widehat{\boldsymbol{J}}$ is not conserved.

c) Let's consider how the Dirac Hamiltonian transforms under parity ($\widehat{\Pi}=\eta\alpha^4\widehat{\pi}$),

$$\widehat{H}_\Pi=\widehat{\Pi}\widehat{H}\widehat{\Pi}^{-1}=\widehat{\Pi}\left(c\boldsymbol{\alpha}\cdot\widehat{\boldsymbol{p}}+mc^2\alpha^4+\alpha^0V(\widehat{\boldsymbol{r}})\right)\widehat{\Pi}^{-1}$$
$$=\widehat{\Pi}c\alpha^i\widehat{p}_i\widehat{\Pi}^{-1}+\widehat{\Pi}mc^2\alpha^4\widehat{\Pi}^{-1}+\widehat{\Pi}\alpha^0V(\widehat{\boldsymbol{r}})\widehat{\Pi}^{-1}$$
$$=c\eta(\alpha^4\alpha^i(\alpha^4)^{-1})\left(\widehat{\pi}\widehat{p}_i\widehat{\pi}^{-1}\right)\eta^{-1}+mc^2(\alpha^4\alpha^4(\alpha^4)^{-1})+\eta(\widehat{\pi}V(\widehat{\boldsymbol{r}})\widehat{\pi}^{-1})\left(\alpha^4\alpha^0\left(\alpha^4\right)^{-1}\right)\eta^{-1}$$

Using $\widehat{\pi}\widehat{\boldsymbol{r}}\widehat{\pi}^{-1}=-\widehat{\boldsymbol{r}}$, $\widehat{\pi}\widehat{\boldsymbol{p}}\widehat{\pi}^{-1}=-\widehat{\boldsymbol{p}}$, $\left[\alpha^4,\alpha^0\right]=0$ and $\{\alpha^4,\alpha^i\}=0$, we obtain

$$\widehat{H}_\Pi=c(-\alpha^i)(-\widehat{p}_i)+mc^2(\alpha^4)+(\alpha^0)V(-\widehat{\boldsymbol{r}})=c\alpha^i\widehat{p}_i+mc^2\alpha^4+\alpha^0V(-\widehat{\boldsymbol{r}}).$$

The Dirac Hamiltonian is invariant under parity transformations only if $V(\widehat{\boldsymbol{r}})=V(-\widehat{\boldsymbol{r}})$.

Problem 6.7

Consider the Dirac equation in 1D with no interacting potential, $\widehat{h}=v_F\,\widehat{p}_x\sigma_x+m(\widehat{x})\sigma_z$, where the "mass" of the medium is $m(x)=\begin{cases}m_-, & x<0\\-m_+, & x>0\end{cases}$. Here $m_\pm>0$.

Find the bound solution and its energy and eigenstate.

Using $\widehat{p}_j=-i\hbar\partial_j$, the Hamiltonian is written in a matrix form,

$$\widehat{h}=\begin{pmatrix}m(\widehat{x})&v_F\,\widehat{p}_x\\v_F\,\widehat{p}_x&-m(\widehat{x})\end{pmatrix}=\begin{pmatrix}m(x)&-i\hbar v_F\partial_x\\-i\hbar v_F\partial_x&-m(x)\end{pmatrix}.$$

Due to the form of the mass term, we can solve the problem independently in each region and impose an appropriate boundary condition at the boundary $x=0$. As we are looking for a bound state, the solution $\psi(x)=\begin{pmatrix}u_1(x)\\u_2(x)\end{pmatrix}$ must vanish at $x=\pm\infty$.

Therefore, we propose the following ansatz for each region: $\begin{pmatrix} u_1(x) \\ u_2(x) \end{pmatrix} = u^- e^{k_- x} = \begin{pmatrix} u_1^- \\ u_2^- \end{pmatrix} e^{k_- x}$ for $x < 0$ and $\begin{pmatrix} u_1(x) \\ u_2(x) \end{pmatrix} = u^+ e^{-k_+ x} = \begin{pmatrix} u_1^+ \\ u_2^+ \end{pmatrix} e^{-k_+ x}$ for $x > 0$ with $k_\pm > 0$, that automatically fulfils the boundary conditions at $x \to \pm\infty$. The eigenvalue problem for the two regions becomes

$$\begin{pmatrix} \mp m_\pm & \pm i\hbar v_F k_+ \\ \pm i\hbar v_F k_+ & \pm m_\pm \end{pmatrix} \begin{pmatrix} u_1^\pm \\ u_2^\pm \end{pmatrix} = E \begin{pmatrix} u_1^\pm \\ u_2^\pm \end{pmatrix}.$$

The constant $k_\pm$ can be determined from the eigenvalue equations,

$$\begin{vmatrix} \mp m_\pm - E & \pm i\hbar v_F k_\pm \\ \pm i\hbar v_F k_\pm & \pm m_\pm - E \end{vmatrix} = 0 \to k_\pm = \frac{\sqrt{m_\pm^2 - E^2}}{\hbar v_F}.$$

The eigenvectors can also be found as $u^\pm = C_\pm \begin{pmatrix} i\frac{m_\pm \mp E}{\hbar v_F k_\pm} \\ 1 \end{pmatrix}$. At the boundary $x = 0$, the wave function must be continuous $\lim_{x \to 0^-} \psi(x) = \lim_{x \to 0^+} \psi(x)$, which imposes that

$$\left. \begin{array}{r} iC_- \frac{m_- + E}{\hbar v_F k_-} = iC_+ \frac{m_+ - E}{\hbar v_F k_+} \\ C_- = C_+ \end{array} \right\} \to \frac{m_- + E}{\hbar v_F k_-} = \frac{m_+ - E}{\hbar v_F k_+}.$$

After substituting $k_\pm$ and some algebra, we obtain

$$2E\,(m_+ + m_-) = 0.$$

Given that $m_\pm > 0$, the preceding relation can only be satisfied when $E = 0$, which further shows that $\hbar v_F k_\pm = m_\pm$. Thus, the orthonormalized eigenstates reduce to

$$u^\pm = \frac{1}{\sqrt{2}} \begin{pmatrix} i \\ 1 \end{pmatrix}.$$

The full solution of the wave function is further written as

$$\psi(x) = N \left[u^- e^{k_- x} \Theta(-x) + u^+ e^{-k_+ x} \Theta(x) \right],$$

where the normalization constant N can be found from $\langle \psi(x) | \psi(x) \rangle = 1$. The complete solution to this eigenvalue problem is finally written as

$$\begin{aligned} E = 0; \quad \psi(x) &= \eta \sqrt{\frac{1}{\hbar v_F} \frac{m_+ m_-}{m_+ + m_-}} \begin{pmatrix} i \\ 1 \end{pmatrix} \left[e^{k_- x} \Theta(-x) + e^{-k_+ x} \Theta(x) \right] \\ &= \eta \sqrt{\frac{1}{\hbar v_F} \frac{m_+ m_-}{m_+ + m_-}} \begin{pmatrix} i \\ 1 \end{pmatrix} e^{\frac{m(x)}{\hbar v_F} x}. \end{aligned}$$

Food for thought: We note that the solution exists even if $m_+ \to \infty$, in which case $\psi(x) \to 0$ when $x > 0$, but it is finite at the boundary. This problem was first considered by Jackiw and Rebbi (1976) in the context of solitons with spin $1/2$. It is also considered as the mathematical basis for the existence of topological excitations in 1D systems and their edge states at an interface (Shen, 2012).

Problem 6.8

Consider the 1D Dirac equation with velocity v (instead of the speed of light c) but with an added quadratic correction

$$\widehat{h} = v\widehat{p}_x\sigma_x + \left(mv^2 - B\widehat{p}_x^2\right)\sigma_z,$$

where $B > 0$ is a real constant. We assume that this type of equation applies to the region $x > 0$ and the eigenenergy is $E = 0$.

Find the corresponding eigenstate for this equation at the surface $x = 0$.

We need to solve the 1D eigenvalue problem

$$\left[v\widehat{p}_x\sigma_x + \left(mv^2 - B\widehat{p}_x^2\right)\sigma_z\right]\Psi(p_x) = 0,$$

where the two-component spinor $\Psi(x)$ must vanish at $x \le 0$. Let us multiply the preceding equation on the left by σ_x and use the property $\sigma_x\sigma_z = -i\sigma_y$:

$$\left[v\widehat{p}_x\sigma_x^2 + \left(mv^2 - B\widehat{p}_x^2\right)\sigma_x\sigma_z\right]\Psi(p_x) = 0 \to \left[v\widehat{p}_x\sigma_0 - \left(mv^2 - B\widehat{p}_x^2\right)i\sigma_y\right]\Psi(p_x) = 0.$$

This suggests using $\sigma_y\chi_\eta = \eta\chi_\eta$, whose eigenspinor is $\chi_\eta = \frac{1}{\sqrt{2}}\begin{pmatrix}\eta \\ i\end{pmatrix}$ with its eigenvalues $\eta = \pm 1$. The spinor is represented as $\Psi(p_x) = \chi_\eta\psi(p_x)$. From here, we find the following equation for $\psi(p_x)$,

$$\left[v\widehat{p}_x - i\left(mv^2 - B\widehat{p}_x^2\right)\eta\right]\chi_\eta\psi(p_x) = 0.$$

Multiplying on the left by η and using explicitly $\widehat{p}_x = -i\hbar\partial_x$, we further find

$$B\hbar^2\partial_x^2\psi(x) + \eta\hbar v\partial_x\psi(x) + mv^2\psi(x) = 0.$$

Here we assume that $\psi(x) \sim e^{-kx}$, with $k > 0$, which ensures that the wave function vanishes at $x \to \infty$. This yields a quadratic equation for k, whose roots can be found easily from

$$B\hbar^2k^2 - \eta\hbar vk + mv^2 = 0,$$

$$k_\pm = \eta\frac{v}{2B\hbar}\left[-1 \pm \sqrt{1 - 4Bm}\right].$$

Further imposing that the wave function vanishes at $x = 0$ yields

$$\Psi(x) = \frac{C}{\sqrt{2}}\begin{pmatrix}1 \\ i\end{pmatrix}\left(e^{\frac{v\sqrt{1-4Bm}}{2B\hbar}x} - e^{\frac{-v\sqrt{1-4Bm}}{2B\hbar}x}\right)e^{\frac{-v}{2B\hbar}x},$$

where the normalization constant can be found as

$$\langle\Psi|\Psi\rangle = 1 = \int_0^\infty \left[\frac{C^*}{\sqrt{2}}(\eta, -i)\psi^*(x)\right]\left[\frac{C}{\sqrt{2}}\begin{pmatrix}\eta\\ i\end{pmatrix}\psi(x)\right]dx = |C|^2\int_0^\infty |\psi(x)|^2 dx$$

$$= |C|^2\int_0^\infty \left(e^{-k_+x} - e^{-k_-x}\right)^2 dx = |C|^2\left(\frac{1}{k_+} + \frac{1}{k_-} - \frac{4}{k_+ + k_-}\right),$$

$$C = \frac{\sqrt{2}\sqrt{k_+k_-(k_+ + k_-)}}{|k_+ - k_-|} = \sqrt{\frac{v}{B\hbar}}\sqrt{\frac{v^2}{1-4Bm} - 1}.$$

Food for thought: It is interesting to graphically analyze $|\Psi(x)|^2$ versus x. We note that the wave function probability is mainly localized around the edge $x=0$ and the spread of this probability distribution is dictated by the parameters in k_+ and k_-. Can you plot $|\Psi(x)|^2$ versus x for some parameter choices to illustrate this point?

Problem 6.9

Consider the Dirac equation from the previous problem, but in 2D with velocity v (instead of c) with quadratic corrections to the mass term $B\left(\hat{p}_x^2 + \hat{p}_y^2\right) = B\hat{\boldsymbol{p}}^2, (B > 0)$. We assume that this type of equation applies to the semi-infinite region $x > 0$ and the eigenenergy is $E = v\hbar k_y$, where k_y is the wave vector in the y-direction.

Find the corresponding eigenstate for this equation at the surface $x = 0$.

The Dirac Hamiltonian is now two-dimensional, and it is written as

$$\hat{H} = v\hat{p}_x\sigma_x + v\hat{p}_y\sigma_y + \left(mv^2 - B\hat{p}_x^2 - B\hat{p}_y^2\right)\sigma_z.$$

As the problem is invariant under displacements in the y-direction, we realize that $\hat{p}_y = \hbar k_y$ is a good quantum number. Also note that, for $k_y = 0$, the problem is very similar to the 1D problem already solved earlier. Then, we use again the solution of the eigenproblem $\sigma_y\chi_\eta = \eta\chi_\eta$ whose eigenspinor is $\chi_\eta = \frac{1}{\sqrt{2}}\begin{pmatrix}\eta\\ i\end{pmatrix}$, with eigenvalues $\eta = \pm 1$, and represent $\Psi(\boldsymbol{x}) = \chi_\eta\psi(\boldsymbol{x})$, such that

$$\hat{H}\chi_\eta\psi(\boldsymbol{x}) = v\hbar k_y\chi_\eta\psi(\boldsymbol{x}),$$

$$\left[v\hat{p}_x\sigma_x + v\hat{p}_y\sigma_y + \left(mv^2 - B\hat{p}_x^2 - B\hat{p}_y^2\right)\sigma_z\right]\chi_\eta e^{-kx}e^{ik_yy} = v\hbar k_y\chi_\eta e^{-kx}e^{ik_yy}.$$

Note that it was assumed that $\psi(\boldsymbol{x}) \sim e^{-kx}e^{ik_yy}$, which ensures that the wave function vanishes at $x \to \infty$ and that $\hbar k_y$ is a good quantum number $\left(\hat{p}_ye^{ik_yy} = \hbar k_ye^{ik_yy}\right)$. After left-multiplying both sides of the preceding equation by σ_x and using that $\sigma_x\sigma_z = -i\sigma_y$,

$$\left[v\hat{p}_x - i\eta\left(mv^2 - B\hat{p}_x^2 - B\hat{p}_y^2\right)\right]\chi_\eta e^{-kx}e^{ik_yy} = v\left(\hbar k_y - p_y\eta\right)\sigma_x\chi_\eta e^{-kx}e^{ik_yy},$$

$$\left[v\hat{p}_x - i\eta\left(mv^2 - B\hat{p}_x^2 - B\hbar^2k_y^2\right)\right]\chi_+ e^{-kx}e^{ik_yy} = 0,$$

which is very similar to the 1D Dirac equation from the previous problem. Going back to the position space by using $\widehat{p}_x = -i\hbar\partial_x$, we obtain that $\psi(\boldsymbol{x})$ must fulfil

$$\left[v(-i\hbar\partial_x) - i\eta\left(mv^2 - B(-i\hbar\partial_x)^2 - B\hbar^2k_y^2\right)\right]\chi_+ e^{-kx}e^{ik_yy} = 0,$$
$$\left[v(i\hbar k) - i\eta\left(mv^2 + B\hbar^2k^2 - B\hbar^2k_y^2\right)\right] = 0,$$
$$B\hbar^2k^2 - \eta v\hbar k + \left(mv^2 - B\hbar^2k_y^2\right) = 0.$$

The roots to this quadratic equation are

$$k_\pm = \eta\frac{v}{2B\hbar}\left[-1 \pm \sqrt{1 - 4Bm + \left(\frac{2B\hbar k_y}{v}\right)^2}\right].$$

In order to ensure that the wave function vanishes at $x \to \infty$, $\eta = 1$ must be imposed, which is also consistent with the equation for the energy. Therefore, the eigenstates corresponding to the 2D Dirac Hamiltonian defined in the boundary of the semi-infinite space $x > 0$ are

$$E = v\hbar k_y, \quad \Psi(x,y) = \frac{C}{\sqrt{2}}\begin{pmatrix} 1 \\ i \end{pmatrix}\left(e^{-k_+x} - e^{-k_-x}\right)e^{ik_yy},$$
$$C = \frac{\sqrt{2}\sqrt{k_+k_-(k_+ + k_-)}}{|k_+ - k_-|}.$$

Food for thought: Suppose we are looking for the eigenstates corresponding to the preceding Hamiltonian and boundary condition but with eigenenergy $E = -v\hbar k_y$. What changes in this case and how is the solution modified?
These two eigenstates with $E = \pm v\hbar k_y$ energies correspond to the so-called helical edge state whose characteristic length λ is found to be dependent on k_y (Zhou et al., 2008).

Problem 6.10

In two dimensions for a Hamiltonian of the form $\widehat{H} = \boldsymbol{d}(\widehat{\boldsymbol{p}})\cdot\boldsymbol{\sigma}$, where $\widehat{\boldsymbol{p}} = (\widehat{p}_x, \widehat{p}_y)$, one can define the Chern number $C = -\frac{1}{4\pi}\int d\boldsymbol{p}\frac{\boldsymbol{d}\cdot(\partial_{p_x}\boldsymbol{d}\times\partial_{p_y}\boldsymbol{d})}{d^3}$. For Chern numbers $C \neq 0$, the system for which this Hamiltonian applies is topologically nontrivial.

The Hamiltonian from the previous problem is precisely of this type,

$$\widehat{H} = v\widehat{p}_x\sigma_x + v\widehat{p}_y\sigma_y + \left(mv^2 - B\widehat{p}_x^2 - B\widehat{p}_y^2\right)\sigma_z,$$

where $\eta = \pm 1$ and B, m can be positive or negative. Under what conditions is the system topologically nontrivial?

From the definition of the Chern number, using $\boldsymbol{d} = (vp_x, \eta vp_y, (mv^2 - Bp_x^2 - Bp_y^2))$, we obtain

$$C = \frac{-1}{4\pi}\int_{\mathbb{R}^2} dp \frac{\eta v^2 (mv^2 + Bp_x^2 + Bp_y^2)}{\left(v^2p_x^2 + v^2p_y^2 + (mv^2 - Bp_x^2 - Bp_y^2)^2\right)^{\frac{3}{2}}}.$$

Due to the rotational symmetry around the origin, the integral is written in polar coordinates $\boldsymbol{p} = (p_x, p_y) = p(\cos(\theta), \sin(\theta))$,

$$C = \frac{-\eta v^2}{4\pi}\int_0^{2\pi} d\theta \int_0^{\infty} \frac{mv^2 + Bp^2}{\left(v^2p^2 + (mv^2 - Bp^2)^2\right)^{\frac{3}{2}}} pdp$$

$$= \frac{-\eta v^2}{4}\int_0^{\infty} \frac{mv^2 + Bq}{\left(v^2q + (mv^2 - Bq)^2\right)^{\frac{3}{2}}} dq,$$

where $q = p^2$. The integral over q can be carried out by applying the change of variables $q = u + \frac{v^2}{2B^2}(2Bm - 1)$, which leads to

$$C = \frac{-\eta v^2}{4}\int_{\frac{v^2}{2B^2}(1-2Bm)}^{\infty} \frac{Bu + 2mv^2 - \frac{v^2}{2B}}{\left(\frac{4B^4u^2 + v^4(4Bm-1)}{B^2}\right)^{\frac{3}{2}}} du$$

$$= \frac{-\eta v^2}{4}\frac{1}{(4B^2)^{\frac{3}{2}}}\int_{\frac{v^2}{2B^2}(1-2Bm)}^{\infty} \frac{Bu + \left(2mv^2 - \frac{v^2}{2B}\right)}{\left(u^2 + \frac{v^4(4Bm-1)}{4B^4}\right)^{\frac{3}{2}}} du.$$

Using that

$$\int \frac{1}{(u^2 + A)^{\frac{3}{2}}} du = \frac{-1}{\sqrt{u^2 + A}}; \qquad \int \frac{u}{(u^2 + A)^{\frac{3}{2}}} du = \frac{u}{A\sqrt{u^2 + A}},$$

it is easy to obtain

$$C = \frac{-\eta}{2}(sign(m) + sign(B)).$$

Thus, it appears that the system is topologically nontrivial when the parameters B, m have the same sign, in which case $C = \pm 1$. However, when B, m have different signs, $C = 0$ and the system is topologically trivial.

> *Food for thought:* The fact that topologically nontrivial solutions exist when $mB > 0$ reflects the *bulk-edge correspondence* consistent with the quantum Hall effect (Hatsugai, 1993; Shen, 2012).

Problem 6.11

Let's now consider the Dirac Hamiltonian in the 3D half-space located above the plane $x = 0$,

$$\widehat{H} = v\widehat{p}_x\sigma_x + \eta v\left(\widehat{p}_y\sigma_y + \widehat{p}_z\sigma_z\right) + \left(mv^2 - B\widehat{\boldsymbol{p}}^2\right)\sigma_z,$$

where $\widehat{\boldsymbol{p}}^2 = \widehat{p}_x^2 + \widehat{p}_y^2 + \widehat{p}_z^2$. The Hamiltonian vanishes at $x = 0$ and the eigenenergies in this case are $E = \pm v\sqrt{p_y^2 + p_z^2}$. Find the corresponding edge eigenstates.

As the problem is invariant under displacements in the y and z directions, we realize that $\widehat{p}_y = \hbar k_y$ and $\widehat{p}_z = \hbar k_z$ are good quantum numbers. Therefore, the eigenspinor is $\Psi(x,y,z) \sim e^{ik_y y} e^{ik_z z}$.

We further note that this problem is quite similar to the 1D and 2D edge problems we solved earlier, thus we use a similar strategy by seeking a solution in terms of the solution of $v\left(p_y\sigma_y + p_z\sigma_z\right)\chi_\pm = \pm e\chi_\pm$, whose eigenvalues $e_\pm$ and eigenspinors are

$$e_\pm = \pm e = \pm v\sqrt{p_y^2 + p_z^2},\ \chi_\pm = \frac{1}{\sqrt{2e(e \pm p_z)}}\begin{pmatrix} p_z \pm e \\ ip_y \end{pmatrix}.$$

Thus, the spinor for the given Dirac Hamiltonian is $\Psi(\boldsymbol{x}) = Ce^{-kx}e^{ik_y y}e^{ik_z z}\chi_\pm$, where e^{-kx} with $k > 0$ ensures that the solution vanishes at $x \to \infty$. After substitution in the equation $\widehat{H}\Psi(\boldsymbol{x}) = E\Psi(\boldsymbol{x})$, we obtain

$$\left[v\widehat{p}_x\sigma_x + \left(mv^2 - B\widehat{p}_x^2 - B\widehat{p}_y^2 - B\widehat{p}_z^2\right)\sigma_z\right]\chi_\eta e^{-kx}e^{ik_y y}e^{ik_z z} = (E - \eta e)\chi_\eta e^{-kx}e^{ik_y y}e^{ik_z z}.$$

Therefore, imposing $E = \eta e = \pm v\sqrt{p_y^2 + p_z^2}$, we have that the boundary state must fulfil

$$\left[v\widehat{p}_x\sigma_x + \left(mv^2 - B\widehat{p}_x^2 - B\widehat{p}_y^2 - B\widehat{p}_z^2\right)\sigma_z\right]\chi_\eta e^{-kx}e^{ik_y y}e^{ik_z z} = 0,$$

which is very similar to the 1D Dirac equation as in the previous problems. Taking into account that $\widehat{p}_i = -i\hbar\partial_i$, we further obtain a secular equation for the constant k,

$$\begin{vmatrix} \left(mv^2 + B\hbar^2k^2 - B\hbar^2k_y^2 - B\hbar^2k_z^2\right) & i\hbar vk \\ i\hbar vk & -\left(mv^2 + B\hbar^2k^2 - B\hbar^2k_y^2 - B\hbar^2k_z^2\right) \end{vmatrix} = 0.$$

We find two roots for which $k > 0$,

$$k_\pm = \frac{v}{2B\hbar}\left[1 \pm \sqrt{1 - 4Bm + \left(\frac{2B\hbar}{v}\right)^2\left(k_y^2 + k_z^2\right)}\right].$$

Therefore, the eigenstates corresponding to the 3D Dirac Hamiltonian at the boundary of the semi-infinite space $x > 0$ with energy $E = \pm v\sqrt{p_y^2 + p_z^2}$ are

$$\Psi_\pm(x,y) = \frac{C}{\sqrt{2e}\sqrt{e \pm p_z}}\begin{pmatrix} p_z \pm e \\ ip_y \end{pmatrix}\left(e^{-k_+x} - e^{-k_-x}\right)e^{ik_y y}e^{ik_z z},$$

$$C = \sqrt{\frac{2k_+k_-(k_+ + k_-)}{|k_+ - k_-|}},$$

where C is the normalization constant, and we have ensured that $\Psi_\pm(x,y)$ vanishes at $x = 0$, as required by the problem.

Problem 6.12

Consider the following effective Dirac Hamiltonian,

$$\widehat{H} = \left(C - D_1\left(\widehat{p}_x^2 + \widehat{p}_y^2\right) - D_3\widehat{p}_z^2\right)\alpha^0 + v(\boldsymbol{\alpha}\cdot\widehat{\boldsymbol{p}}) + \left(M - B_i^2\widehat{p}_i^2\right)\alpha^4,$$

where $D_{1,3}$, B_i, M, v are constants.

Show if this Hamiltonian is invariant under the operations of (a) time reversal, (b) parity, and (c) particle-hole symmetry.

---

a) For spinors, the *anti-Hermitian time-reversal* operator is $\widehat{\Theta} = \eta\alpha^1\alpha^3\widehat{K}$, where η is an arbitrary phase and K is complex conjugation. Therefore,

$$\begin{aligned}\widehat{H}_\Theta &= \widehat{\Theta}\widehat{H}\widehat{\Theta}^{-1} = \widehat{\Theta}\left[\left(C - D_1\left(\widehat{p}_x^2 + \widehat{p}_y^2\right) - D_3\widehat{p}_z^2\right)\alpha^0 + v\alpha^i\widehat{p}_i + \left(M - B_i^2\widehat{p}_i^2\right)\alpha^4\right]\widehat{\Theta}^{-1}\\ &= \eta\Big[\widehat{K}\left(C - D_1\left(\widehat{p}_x^2 + \widehat{p}_y^2\right) - D_3\widehat{p}_z^2\right)\widehat{K}^{-1}\left(\alpha^1\alpha^3\alpha^0(\alpha^1\alpha^3)^{-1}\right)\\ &\quad + v(\alpha^1\alpha^3(\alpha^i)^*(\alpha^1\alpha^3)^{-1})\left(\widehat{K}\widehat{p}_i\widehat{K}^{-1}\right) + \widehat{K}\left(M - B_i^2\widehat{p}_i^2\right)\widehat{K}^{-1}(\alpha^1\alpha^3(\alpha^4)^*(\alpha^1\alpha^3)^{-1})\Big]\eta^{-1}.\end{aligned}$$

To further reduce the preceding expression, we use that $\widehat{K}\widehat{\boldsymbol{p}}\widehat{K}^{-1} = \widehat{\boldsymbol{p}}^* = -\widehat{\boldsymbol{p}}$. Also, the facts that all α matrices are real except $(\alpha^2)^* = -\alpha^2$ further give

$$\begin{aligned}\widehat{H}_\Theta &= \left(C - (-1)^2D_1\left(\widehat{p}_x^2 + \widehat{p}_y^2\right) - (-1)^2D_3\widehat{p}_z^2\right)\alpha^0 + v\left(\alpha^1\alpha^3\alpha^1\left(\alpha^1\alpha^3\right)^{-1}\right)(-\widehat{p}_1)\\ &\quad - v\left(\alpha^1\alpha^3\alpha^2\left(\alpha^1\alpha^3\right)^{-1}\right)(-\widehat{p}_2) + v\left(-\alpha^1\alpha^3\alpha^3\left(\alpha^1\alpha^3\right)^{-1}\right)(-\widehat{p}_3)\\ &\quad + \left(M - (-1)^2B_i^2\widehat{p}_i^2\right)\left(\alpha^1\alpha^3\alpha^4\left(\alpha^1\alpha^3\right)^{-1}\right)\\ &= \left(C - D_1\left(\widehat{p}_x^2 + \widehat{p}_y^2\right) - D_3\widehat{p}_z^2\right)\alpha^0 + v(-\alpha^1)(-\widehat{p}_1) + v(-\alpha^2)(-\widehat{p}_2)\\ &\quad + v(-\alpha^3)(-\widehat{p}_3) + \left(M - B_i^2\widehat{p}_i^2\right)\alpha^4\\ &= \left(C - D_1\left(\widehat{p}_x^2 + \widehat{p}_y^2\right) - D_3\widehat{p}_z^2\right)\alpha^0 + v\alpha^i\widehat{p}_i + \left(M - B_i^2\widehat{p}_i^2\right)\alpha^4 = \widehat{H}.\end{aligned}$$

Therefore, the given Dirac Hamiltonian is invariant under time-reversal.

b) For spinors, the *parity operator* is $\widehat{\Pi} = \eta\alpha^4\widehat{\pi}$, where $\widehat{\pi}\widehat{\boldsymbol{r}}\widehat{\pi}^{-1} = -\widehat{\boldsymbol{r}}$. Therefore,

$$\begin{aligned}\widehat{H}_\Pi &= \widehat{\Pi}\widehat{H}\widehat{\Pi}^{-1} = \widehat{\Pi}\left[\left(C - D_1\left(\widehat{p}_x^2 + \widehat{p}_y^2\right) - D_3\widehat{p}_z^2\right)\alpha^0 + v\alpha^i\widehat{p}_i + \left(M - B_i^2\widehat{p}_i^2\right)\alpha^4\right]\widehat{\Pi}^{-1}\\ &= \eta\left[\widehat{\pi}\left(C - D_1\left(\widehat{p}_x^2 + \widehat{p}_y^2\right) - D_3\widehat{p}_z^2\right)\widehat{\pi}^{-1}\left(\alpha^4\alpha^0(\alpha^4)^{-1}\right) + v\left(\alpha^4\alpha^i(\alpha^4)^{-1}\right)\left(\widehat{\pi}\widehat{p}_i\widehat{\pi}^{-1}\right)\right.\\ &\quad \left. + \widehat{\pi}\left(M - B_i^2\widehat{p}_i^2\right)\widehat{\pi}^{-1}\left(\alpha^4\alpha^4(\alpha^4)^{-1}\right)\right]\eta^{-1}.\end{aligned}$$

Since $\widehat{\pi}\widehat{\boldsymbol{p}}\widehat{\pi}^{-1} = -\widehat{\boldsymbol{p}}$ and $\{\alpha^4, \alpha^i\} = 0$, we obtain

$$\begin{aligned}\widehat{H}_\Pi &= \left(C - D_1(-1)^2\left(\widehat{p}_x^2 + \widehat{p}_y^2\right) - (-1)^2D_3\widehat{p}_z^2\right)\alpha^0 + v(-\alpha^i)(-\widehat{p}_i) + \left(M - (-1)^2B_i^2\widehat{p}_i^2\right)\alpha^4\\ &= \left(C - D_1\left(\widehat{p}_x^2 + \widehat{p}_y^2\right) - D_3\widehat{p}_z^2\right)\alpha^0 + v\alpha^i\widehat{p}_i + \left(M - B_i^2\widehat{p}_i^2\right)\alpha^4 = \widehat{H}.\end{aligned}$$

The given Hamiltonian is also invariant under parity.

c) For spinors, the *particle-hole symmetry operator* is $\widehat{C} = \eta\alpha^2\alpha^4\widehat{K}$. By taking into account that $(\alpha^\mu)^{-1} = \alpha^\mu$, $\widehat{p}_i^* = -\widehat{p}_i$, and that the constants C, D_1, D_3, M, v, and B_i are real numbers, the transformed Hamiltonian is

$$\begin{aligned}\widehat{H}_C &= \widehat{C}\widehat{H}\widehat{C}^{-1} = \eta\alpha^2\alpha^4\widehat{K}\left[\left(C - D_1\left(\widehat{p}_x^2 + \widehat{p}_y^2\right) - D_3\widehat{p}_z^2\right)\alpha^0 + v\alpha^i\widehat{p}_i + \left(M - B_i^2\widehat{p}_i^2\right)\alpha^4\right] \\ &\quad\times \widehat{K}^{-1}\left(\alpha^4\right)^{-1}\left(\alpha^2\right)^{-1}\eta^{-1} \\ &= \eta\left[\left(C - D_1\left(\widehat{p}_x^{*2} + \widehat{p}_y^{*2}\right) - D_3\widehat{p}_z^{*2}\right)\left(\alpha^2\alpha^4\alpha^{0*}\alpha^4\alpha^2\right)\right. \\ &\quad\left. + v\left(\alpha^2\alpha^4\alpha^{i*}\alpha^4\alpha^2\right)\widehat{p}_i^* + \left(M - B_i^2\widehat{p}_i^{*2}\right)\left(\alpha^2\alpha^4\alpha^{4*}\alpha^4\alpha^2\right)\right]\eta^{-1}.\end{aligned}$$

Since $\widehat{K}\widehat{\boldsymbol{p}}\widehat{K}^{-1} = \widehat{\boldsymbol{p}}^* = -\widehat{\boldsymbol{p}}$, $\left[\alpha^0, \alpha^\mu\right] = 0$, $\{\alpha^\mu, \alpha^\nu\} = 2\delta^{\mu\nu}\alpha^0$ for $\mu = 1,2,3,4$, and all α matrices are real except $(\alpha^2)^* = -\alpha^2$, after a little algebra we get

$$\begin{aligned}\widehat{H}_C = \widehat{C}\widehat{H}\widehat{C}^{-1} &= \left[\left(C - D_1\left(\widehat{p}_x^2 + \widehat{p}_y^2\right) - D_3\widehat{p}_z^2\right)\alpha^0 - v\alpha^i\widehat{p}_i - \left(M - B_i^2\widehat{p}_i^2\right)\alpha^4\right] \\ &= 2\left(C - D_1\left(\widehat{p}_x^2 + \widehat{p}_y^2\right) - D_3\widehat{p}_z^2\right)\alpha^0 - \widehat{H}.\end{aligned}$$

The Hamiltonian is invariant under the particle-hole symmetry operation when $\widehat{H}_C = \widehat{C}\widehat{H}\widehat{C}^{-1} = -\widehat{H}$ (see Problem 6.19), therefore, the given Hamiltonian is invariant under the particle-hole symmetry operation only if $C = D_1 = D_3 = 0$.

Problem 6.13

Let us consider the Dirac Hamiltonian at $z \geq 0$ when $p_x = p_y = 0$. In this situation, an effective surface Dirac Hamiltonian can be written as

$$\widehat{H} = \left(C + D_3\hbar^2\partial_z^2\right)\alpha^0 - iv\hbar\alpha^3\partial_z + \left(M + B_3\hbar^2\partial_z^2\right)\alpha^4.$$

Given that the energy is E, find the eigenstates corresponding to $D_3 < B_3$, assuming a Dirichlet boundary condition at $z = 0$. This condition ensures the existence of a topologically nontrivial surface state associated with the topological insulator phase, obtain this state.

Let us write the preceding Hamiltonian in an explicit matrix form using $\widehat{p}_z = -i\hbar\partial_z$,

$$\widehat{H} = \begin{pmatrix} C + M + (D_3 + B_3)\hbar^2\partial_z^2 & 0 & -iv\hbar\partial_z & 0 \\ 0 & C + M + (D_3 + B_3)\hbar^2\partial_z^2 & 0 & iv\hbar\partial_z \\ -iv\hbar\partial_z & 0 & C - M + (D_3 - B_3)\hbar^2\partial_z^2 & 0 \\ 0 & iv\hbar\partial_z & 0 & C - M + (D_3 - B_3)\hbar^2\partial_z^2 \end{pmatrix}.$$

We observe right away that the eigenspinors of this $\widehat{H}$ can be decoupled in the following way:

$$\Psi_1 = \begin{pmatrix} a_1 \\ 0 \\ b_1 \\ 0 \end{pmatrix} e^{\xi z}, \quad \Psi_2 = \begin{pmatrix} 0 \\ a_2 \\ 0 \\ b_2 \end{pmatrix} e^{\xi z}.$$

This ansatz gives two independent eigenvalue equations:

$$\begin{pmatrix} C + M + (D_3 + B_3)\hbar^2\xi^2 & -iv\hbar\xi \\ -iv\hbar\xi & C - M + (D_3 - B_3)\hbar^2\xi^2 \end{pmatrix}\begin{pmatrix} a_1 \\ b_1 \end{pmatrix} = E\begin{pmatrix} a_1 \\ b_1 \end{pmatrix},$$

$$\begin{pmatrix} C + M + (D_3 + B_3)\hbar^2\xi^2 & iv\hbar\xi \\ iv\hbar\xi & C - M + (D_3 - B_3)\hbar^2\xi^2 \end{pmatrix}\begin{pmatrix} a_2 \\ b_2 \end{pmatrix} = E\begin{pmatrix} a_2 \\ b_2 \end{pmatrix}.$$

From here, we obtain the following secular equations for the eigenvalues,

$$\begin{vmatrix} (C+M)+(D_3+B_3)\hbar^2\xi^2-E & -iv\hbar\xi \\ -iv\hbar\xi & (C-M)+(D_3-B_3)\hbar^2\xi^2-E \end{vmatrix}=0,$$

$$\begin{vmatrix} (C+M)+(D_3+B_3)\hbar^2\xi^2-E & iv\hbar\xi \\ iv\hbar\xi & (C-M)+(D_3-B_3)\hbar^2\xi^2-E \end{vmatrix}=0,$$

which lead to the same two roots for the parameter ξ,

$$\xi=\frac{\pm V}{\sqrt{2}\hbar\sqrt{B_3^2-D_3^2}}\sqrt{1\pm\sqrt{1+\frac{4((E-C)^2-M^2)\hbar^2\left(B_3^2-D_3^2\right)}{V^2}}},$$

where $V^2=\hbar^2v^2-2\hbar\left[(E-C)D_3+MB_3\right]$.

We further note that $\Psi_{1,2}(z)$ must vanish at $z=0$ and $z\to\infty$. Therefore, the meaningful solutions are

$$\xi_\pm=\frac{-V}{\sqrt{2}\hbar\sqrt{B_3^2-D_3^2}}\sqrt{1\pm\sqrt{1+\frac{4((E-C)^2-M^2)\hbar^2\left(B_3^2-D_3^2\right)}{V^2}}},$$

$$\Psi_1(z)=\begin{pmatrix} a_1^+ \\ 0 \\ b_1^+ \\ 0 \end{pmatrix}e^{\xi_+z}-\alpha_1\begin{pmatrix} a_1^- \\ 0 \\ b_1^- \\ 0 \end{pmatrix}e^{\xi_-z},\quad \Psi_2(z)=\begin{pmatrix} 0 \\ a_2^+ \\ 0 \\ b_2^+ \end{pmatrix}e^{\xi_+z}-\alpha_2\begin{pmatrix} 0 \\ a_2^- \\ 0 \\ b_2^- \end{pmatrix}e^{\xi_-z}.$$

Imposing the Dirichlet Boundary Condition at the surface $z=0$, we have $\lim_{z\to 0}\Psi_{1,2}(z)=0$, from which we have $\begin{pmatrix} a_1^+ \\ b_1^+ \end{pmatrix}=\alpha_1\begin{pmatrix} a_1^- \\ b_1^- \end{pmatrix}$ and $\begin{pmatrix} a_2^+ \\ b_2^+ \end{pmatrix}=\alpha_2\begin{pmatrix} a_2^- \\ b_2^- \end{pmatrix}$.
Rewriting the reduced spinors further gives

$$\psi_{1,2}^\pm=\begin{pmatrix} a_{1,2}^\pm \\ b_{1,2}^\pm \end{pmatrix}=\frac{1}{N_{1,2}^\pm}\begin{pmatrix} 1 \\ \Lambda_{1,2}^\pm \end{pmatrix},$$

$$\Lambda_n^\pm=\frac{(-1)^n i\hbar v\xi_\pm}{(M-C)+(D_3-B_3)\hbar^2\xi_\pm^2+E};\ N_n^\pm=\sqrt{1+(\Lambda_n^\pm)^2}.$$

The condition $\frac{1}{N_{1,2}^+}\begin{pmatrix} 1 \\ \Lambda_{1,2}^+ \end{pmatrix}=\alpha_{1,2}\frac{1}{N_{1,2}^-}\begin{pmatrix} 1 \\ \Lambda_{1,2}^- \end{pmatrix}$ implies that $\alpha_{1,2}=1$ and $\Lambda_{1,2}^+=\Lambda_{1,2}^-$ are consistent with Dirichlet boundary conditions.

The relation $\Lambda_{1,2}^+=\Lambda_{1,2}^-$ gives a transcendental equation to find the energy E, which can be obtained numerically. The numerical solution is left for the reader.

Problem 6.14

An effective surface Dirac Hamiltonian, when a constant magnetic field $\boldsymbol{B}=B\hat{\boldsymbol{u}}_z$ is applied along the z direction with unit vector $\hat{\boldsymbol{u}}_z$, can be written as

$$\widehat{H}=v(\widehat{\boldsymbol{p}}\times\boldsymbol{\sigma})\cdot\hat{\boldsymbol{u}}_z+\Delta\sigma_z,$$

where Δ is a positive constant. What are the eigenstates and eigenenergies of $\widehat{H}$?

After expressing the given Hamiltonian in the form $\widehat{H} = \boldsymbol{d}(\boldsymbol{k}) \cdot \boldsymbol{\sigma}$, calculate its Hall conductivity, $\sigma_{xy} = \frac{e^2}{2\hbar} \int \frac{dk_x dk_y}{(2\pi)^2} \frac{\boldsymbol{d}(\boldsymbol{k}) \cdot (\partial_{k_x} \boldsymbol{d}(\boldsymbol{k}) \times \partial_{k_y} \boldsymbol{d}(\boldsymbol{k}))}{|\boldsymbol{d}(\boldsymbol{k})|^3}$.

The Hamiltonian can be written as

$$\widehat{H} = v\left(\widehat{p}_x \sigma_y - \widehat{p}_y \sigma_x\right) + \Delta \sigma_z.$$

Thus, using $\widehat{p}_i = \hbar k_i$, one writes the eigenvalue equation $\widehat{H}\chi_\pm = E\chi_\pm$ explicitly,

$$\begin{pmatrix} \Delta & -\hbar v\left(k_y + ik_x\right) \\ -\hbar v\left(k_y - ik_x\right) & -\Delta \end{pmatrix} \chi_\pm = E\chi_\pm.$$

From here, the eigenenergies are calculated as $E_\pm = \pm\sqrt{\hbar^2 v^2 k^2 + \Delta^2}$, where $k^2 = k_x^2 + k_y^2$. The corresponding eigenstates can also be found,

$$\chi_\pm = \frac{1}{\sqrt{2\varepsilon(\varepsilon \pm \Delta)}} \begin{pmatrix} \Delta \pm \varepsilon \\ \hbar v\left(ik_x - k_y\right) \end{pmatrix},$$

where $\varepsilon = \sqrt{\hbar^2 v^2 k^2 + \Delta^2}$.

To calculate the Hall conductivity, we use that $\boldsymbol{d}(\boldsymbol{k}) = \left(-\hbar v k_y, \hbar v k_x, \Delta\right)$; then we have

$$\sigma_{xy} = \frac{e^2}{2\hbar} \int \frac{dk_x dk_y}{(2\pi)^2} \frac{\hbar^2 v^2 \Delta}{(\hbar^2 v^2 k^2 + \Delta^2)^{\frac{3}{2}}} = \frac{e^2}{2\hbar} \frac{\Delta}{(2\pi)^2} \int_0^{2\pi} d\theta \int_0^\infty dk \frac{\hbar^2 v^2 k}{(\hbar^2 v^2 k^2 + \Delta^2)^{\frac{3}{2}}}$$

$$= \sigma_{xy} = \frac{e^2}{\hbar} \int_{|\Delta|}^\infty \frac{d\varepsilon}{\varepsilon^2} = \frac{e^2}{4\pi\hbar} \frac{\Delta}{|\Delta|} = \frac{e^2}{4\pi\hbar} sign(\Delta) = \frac{\alpha c}{4\pi} sign(\Delta),$$

where we have used that $\alpha = \frac{e^2}{\hbar c}$ is the fine structure constant.

> *Food for thought:* Making a connection with Chapter 2 and previous problems, we see that there is an intimate connection between the topology of the surface Dirac Hamiltonian and its Hall conductivity,
>
> $$\sigma_{xy} = \frac{e^2}{2\hbar} \int \frac{dk_x dk_y}{(2\pi)^2} \frac{\boldsymbol{d} \cdot \left(\partial_{k_x} \boldsymbol{d} \times \partial_{k_y} \boldsymbol{d}\right)}{|\boldsymbol{d}|^3}$$
>
> $$= \frac{\alpha c}{2\pi} \left[\frac{1}{4\pi} \int dk_x dk_y \frac{\boldsymbol{d} \cdot \left(\partial_{k_x} \boldsymbol{d} \times \partial_{k_y} \boldsymbol{d}\right)}{|\boldsymbol{d}|^3}\right] = \frac{\alpha c}{2\pi} C,$$
>
> where we have used the Chern number definition and $\alpha = \frac{e^2}{\hbar c}$.

Problem 6.15

Here we would like to consider the surface states described by a Dirac Hamiltonian when a strong magnetic field $\boldsymbol{B} = (0, 0, B)$ is applied. Working in the Landau gauge with the vector potential $\boldsymbol{A} = (-By, 0, 0)$, the effective Dirac Hamiltonian given in position space is

$$\widehat{H} = v\left[\left(\hbar k_x - eBy\right) \sigma_y + i\hbar \partial_y \sigma_x\right] + \Delta \sigma_z.$$

Obtain the eigenenergies and eigenstates. For the calculations, define the dimensionless parameter $M = \frac{\Delta}{v\sqrt{2eB\hbar}}$ and rewrite the effective Hamiltonian in terms of harmonic oscillator auxiliary annihilation and creation operators, such that $\widehat{A} = i\sqrt{\frac{\hbar}{2eB}}\left(\partial_y + \frac{eB}{\hbar}\left(y - \frac{\hbar}{eB}k_x\right)\right)$ for which $\left[\widehat{A}, \widehat{A}^+\right] = 1$.

Let's express the Hamiltonian as suggested in the problem,

$$\begin{aligned}\widehat{H} &= v\left[(\hbar k_x - eBy)\sigma_y + i\hbar\partial_y\sigma_x\right] + \Delta\sigma_z \\ &= \begin{pmatrix} \Delta & v\left[-p_y - i(p_x - eBy)\right] \\ v\left[-p_y + i(p_x - eBy)\right] & -\Delta \end{pmatrix} \\ &= v\sqrt{2eB\hbar}\begin{pmatrix} \frac{\Delta}{v\sqrt{2eB\hbar}} & \widehat{A} \\ \widehat{A}^+ & -\frac{\Delta}{v\sqrt{2eB\hbar}} \end{pmatrix} = v\sqrt{2eB\hbar}\begin{pmatrix} M & \widehat{A} \\ \widehat{A}^+ & -M \end{pmatrix}.\end{aligned}$$

Thus, we arrive at the following eigenvalue equation:

$$v\sqrt{2eB\hbar}\begin{pmatrix} M & \widehat{A} \\ \widehat{A}^+ & -M \end{pmatrix}\begin{pmatrix} \psi_1 \\ \psi_2 \end{pmatrix} = E\begin{pmatrix} \psi_1 \\ \psi_2 \end{pmatrix}.$$

The operators $\widehat{A}$ and $\widehat{A}^+$ are effective annihilation and creation operators, as implied by their property $\left[\widehat{A}, \widehat{A}^+\right] = 1$. Thus, they can be treated as the ladder operators of harmonic oscillator levels $|n\rangle$ already seen in other chapters, and then

$$\widehat{A}|n\rangle = \sqrt{n}|n-1\rangle, \quad \widehat{A}^+|n\rangle = \sqrt{n+1}|n+1\rangle.$$

From the eigenvalue equation, we obtain

$$\begin{cases} M\psi_1 + \widehat{A}\psi_2 = \frac{E}{v\sqrt{2eB\hbar}}\psi_1 \\ \widehat{A}^+\psi_1 - M\psi_2 = \frac{E}{v\sqrt{2eB\hbar}}\psi_2 \end{cases}.$$

This implies that if $\psi_2 \propto |n\rangle$, then $\psi_1 \propto |n-1\rangle$. By defining $E_n = v\sqrt{2eB\hbar}\varepsilon_n$, we propose the following ansatz for the orthonormalized solution

$$\Psi = \begin{pmatrix} \psi_1 \\ \psi_2 \end{pmatrix} = \begin{pmatrix} \alpha_n|n-1\rangle \\ \beta_n|n\rangle \end{pmatrix},$$

for which the normalization condition of the wave function is $\alpha_n^2 + \beta_n^2 = 1$. Inserting the ansatz into the eigenvalue problem, we obtain the following set of relations:

$$\begin{cases} M\alpha_n|n-1\rangle + \sqrt{n}\beta_n|n-1\rangle = \varepsilon_n\alpha_n|n-1\rangle \\ \sqrt{n}\alpha_n|n\rangle - M\beta_n|n\rangle = \varepsilon_n\beta_n|n\rangle \end{cases} \rightarrow \begin{cases} M\alpha_n + \sqrt{n}\beta_n = \varepsilon_n\alpha_n \\ \sqrt{n}\alpha_n - M\beta_n = \varepsilon_n\beta_n \end{cases}.$$

This leads to $\varepsilon_{n,\pm} = \pm\varepsilon_n = \pm\sqrt{M^2+n}$. From the normalization condition $\alpha_n^2+\beta_n^2=1$ and the preceding system of equations, we also find solutions for the $\alpha_{n,\pm}, \beta_{n,\pm}$ corresponding to $\varepsilon_{n,\pm}$.

In summary,

$$E_{n,\pm} = \pm v\varepsilon_n\sqrt{2eB\hbar}, \quad \Psi_{n,\pm} = \begin{pmatrix} \alpha_{n,\pm}|n-1\rangle \\ \beta_{n,\pm}|n\rangle \end{pmatrix},$$

$$\alpha_{n,\pm} = \sqrt{\frac{n}{2\varepsilon_n(\varepsilon_n \mp M)}}, \quad \beta_{n,\pm} = \pm\sqrt{\frac{\varepsilon_n \mp M}{2\varepsilon_n}}.$$

> *Food for thought:* For practice, demonstrate explicitly the following commutation relations: $\left[\widehat{A},\widehat{A}^+\right] = 1$, $\left[\widehat{A},\widehat{A}\right] = \left[\widehat{A}^+,\widehat{A}^+\right] = 0$. These commutators ensure that the $\widehat{A}, \widehat{A}^+$ operators are effective ladder operators for the quantum mechanical oscillator states.
> These $E_{n,\pm}$ are the Landau energy levels of a Dirac particle. Note that from here we can obtain the Landau levels for a massless Dirac particle by taking $M=0$, found as $E_n = \pm v\sqrt{2e\hbar Bn}$. It appears that this behavior is different when the particle is nonrelativistic, for which the Landau levels are equally separated by the cyclotron frequency $E_n = \hbar\omega_c\left(n+\frac{1}{2}\right)$.

Problem 6.16

Here we would like to examine a relativistic electron with velocity v moving in 1D. The electron with energy E_e is scattered from a step-function potential $V(x) = \begin{cases} 0, & x \le 0 \\ V, & x > 0 \end{cases}$, as proposed in Thomson and McKellar (1991). Determine the transmission and reflection coefficients of the relativistic electron for the following cases:

a) $E_e > V + mv^2$;
b) $\left(V - mv^2\right) < E_e < \left(V + mv^2\right)$;
c) $mv^2 < E_e < V - mv^2$.

For this problem we need to solve the equation

$$\widehat{H}\Psi = E_\eta\Psi$$

for regions $x<0$ and $x>0$, where $\widehat{H} = v\widehat{p}_x\sigma_x + mv^2\sigma_z + V\sigma_0$ is the Dirac equation in 1D. Since this is an elastic scattering problem with a given energy E_η, the wave function $\Psi_\eta(x) \sim \begin{pmatrix} a \\ b \end{pmatrix} e^{ikx}$ is considered in momentum space. Here we take advantage of the fact that $p_x = \hbar k$ is a good quantum number for the momentum operator $\widehat{p}_x = -i\hbar\partial_x$. The index $\eta = \{+1, -1\} = \{e, h\}$ is taken to denote scattering of a relativistic electron or a hole particle, respectively.

The general eigenvalue problem becomes

$$\begin{pmatrix} V+mv^2 & \hbar v k_x \\ \hbar v k_x & V-mv^2 \end{pmatrix}\begin{pmatrix} a \\ b \end{pmatrix} = E_\eta \begin{pmatrix} a \\ b \end{pmatrix},$$

which yields $E_\eta = V + \eta\sqrt{(\hbar v k_{\eta,V})^2 + m^2v^4}$. Considering only traveling solutions, we find

$$k_{\eta,\pm} = \pm k_{\eta,V} = \pm\frac{\sqrt{(E_\eta - V)^2 - m^2v^4}}{\hbar v},$$

$$\Psi_\eta = \frac{1}{\sqrt{|C_+|^2+|C_-|^2}}\left(C_+\Psi_{\eta,+}e^{ik_{\eta,V}x} + C_-\Psi_{\eta,-}e^{-ik_{\eta,V}x}\right),$$

$$\Psi_{\eta,\pm} = \frac{1}{\sqrt{(\hbar v k_{\eta,V})^2 + (mv^2+E_\eta - V)^2}}\begin{pmatrix} mv^2+E_\eta-V \\ \pm\hbar v k_{\eta,V}\end{pmatrix} = \frac{1}{N_{\eta,V}}\begin{pmatrix} 1 \\ \pm\Lambda_{\eta,V}\end{pmatrix},$$

with $\Lambda_{\eta,V} = \frac{\hbar v k_{\eta,V}}{mv^2+E_\eta-V}$ and $N_{\eta,V} = \sqrt{1+|\Lambda_{\eta,V}|^2}$.

This compact solution can now be used to obtain the reflection and transmission coefficient for the electron and hole particles. The group velocity for the relativistic electron is

$$v_g = \nabla_k E_\eta = \frac{\partial\left(V+\eta v\sqrt{m^2v^2+\hbar^2k^2}\right)}{\partial k} = \frac{\hbar^2v^2k}{E_\eta - V}.$$

Assuming the particle moves from left to right, v_g is parallel to the wave vector k in the $x<0$ region and antiparallel to k in the $x>0$ region.

This general solution for the eigenstates now enables us to consider the specific cases as specified in the problem.

Case a) In the case of an electron with energy $E_e > V + mv^2$ (see the diagram in Figure 6.1), the wave function is

$$\Psi_e(x) = \begin{cases} \Psi_{e,L}(x) = \Psi_{e,+}e^{ik_{e,0}x} + r\Psi_{e,-}e^{-ik_{e,0}x}, & x<0 \\ \Psi_{e,R}(x) = t\Psi_{e,+}e^{ik_{e,V}x}, & x>0 \end{cases}.$$

From the boundary conditions of continuity of the wave function $\lim_{x\to 0^-}\Psi_L(x) = \lim_{x\to 0^+}\Psi_R(x)$, we find

$$\begin{cases} \dfrac{1+r}{N_{e,0}} = \dfrac{t}{N_{e,V}} \\ \dfrac{\Lambda_{e,0} - r\Lambda_{e,0}}{N_{e,0}} = t\dfrac{\Lambda_{e,V}}{N_{e,V}} \end{cases}.$$

Using that $\hbar v k_{e,V} = \sqrt{(E_e - V)^2 - m^2v^4}$ and the expressions for $\Lambda_{e,V}$ and $N_{e,V}$, we obtain

$$r = \frac{k_{e,0}\left(mv^2+E_e-V\right) - k_{e,V}\left(mv^2+E_e\right)}{k_{e,0}\left(mv^2+E_e-V\right) + k_{e,V}\left(mv^2+E_e\right)},$$

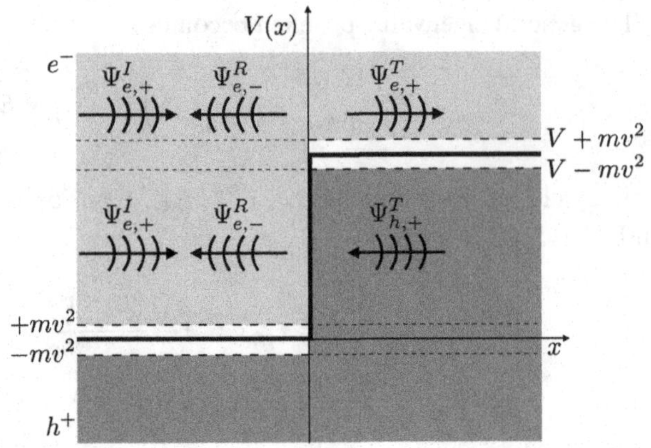

Figure 6.1 A schematic representation of the incident (I), reflected (R), and transmitted (T) electron and hole wave function is given. The potential $V(x)$ is represented as a thick line, the electronic part of the spectrum as light gray, the hole part of the spectrum as a dark gray and the forbidden energies at white. The arrows show the direction of the group velocity.

$$t = \sqrt{\frac{(mv^2+E_e-V)^2+(\hbar v k_{e,V})^2}{(mv^2+E_e)^2+(\hbar v k_{e,0})^2}}\frac{2k_{e,0}\left(mv^2+E_e\right)}{k_{e,0}\left(mv^2+E_e-V\right)+k_{e,V}\left(mv^2+E_e\right)}.$$

Case b) In the case of an electron with energy $(V-mv^2)<E_e<(V+mv^2)$, there is no possibility of transmission of a traveling wave, since such energy does not belong to the given spectrum (see the diagram in Figure 6.1). However, there is a generation of an evanescent wave for $x>0$. In principle, we have

$$\Psi(x)=\begin{cases}\Psi_{e,L}(x)=\Psi_{e,+}e^{ik_{e,0}x}+r\Psi_{e,-}e^{-ik_{e,0}x}, & x<0\\ \Psi_R(x)=t\Psi_{\eta,+}e^{-\kappa_{\eta,V}x}, & x>0\end{cases},$$

where $-mv^2<(E_e-V)<mv^2$. Therefore, the associated momentum of the evanescent wave is

$$k_{\eta,+}=+k_{\eta,V}=\frac{\sqrt{\left(E_\eta-V\right)^2-m^2v^4}}{\hbar v}=i\frac{\sqrt{m^2v^4-\left(E_\eta-V\right)^2}}{\hbar v}=i\kappa_{\eta,V},$$

where $\kappa_{\eta,V}$ is a positive real quantity. Therefore, we have $\Psi_{\eta,+}=\psi_{\eta,+}e^{-\kappa_{\eta,V}x}$, with

$$\psi_{\eta,+}=\frac{1}{\sqrt{(mv^2+E_e-V)^2+\left(\hbar v\kappa_{\eta,V}\right)^2}}\begin{pmatrix}mv^2+E_e-V\\ \hbar v i\kappa_{\eta,V}\end{pmatrix}=\frac{1}{N_{\eta,V}}\begin{pmatrix}1\\ i\left|\Lambda_{\eta,V}\right|\end{pmatrix},$$

$$\Lambda_{\eta,V}=i\frac{\hbar v\kappa_{\eta,V}}{mv^2+E_e-V},\ N_{\eta,V}=\sqrt{1+\left|\Lambda_{\eta,V}\right|^2}.$$

From the boundary conditions of continuity of the wave function $\lim_{x\to 0^-}\Psi_L(x)=\lim_{x\to 0^+}\Psi_R(x)$, we find

$$\begin{cases} \dfrac{1+r}{N_{e,0}}=\dfrac{t}{N_{\eta,V}} \\ \dfrac{\Lambda_{e,0}-r\Lambda_{e,0}}{N_{e,0}}=\dfrac{t\Lambda_{\eta,V}}{N_{\eta,V}} \end{cases}.$$

Thus,

$$r=\frac{\Lambda_{e,0}-\Lambda_{\eta,V}}{\Lambda_{e,0}+\Lambda_{\eta,V}}=\frac{k_{e,0}\left(mv^2+E_e-V\right)-i\kappa_{\eta,V}\left(mv^2+E_e\right)}{k_{e,0}\left(mv^2+E_e-V\right)+i\kappa_{\eta,V}\left(mv^2+E_e\right)},$$

$$t=\frac{N_{\eta,V}}{N_{e,0}}\frac{2\Lambda_{e,0}}{\Lambda_{e,0}+\Lambda_{\eta,V}}$$

$$=\sqrt{\frac{\left(mv^2+E_e-V\right)^2+\left(\hbar v\kappa_{\eta,V}\right)^2}{\left(mv^2+E_e\right)^2+\left(\hbar vk_{e,0}\right)^2}}\frac{2k_{e,0}\left(mv^2+E_e\right)}{k_{e,0}\left(mv^2+E_e-V\right)+i\kappa_{\eta,V}\left(mv^2+E_e\right)}.$$

Case c) For the incident energy $mv^2<E_e<V-mv^2$ (see the diagram in Figure 6.1), the transmitted particle is a relativistic hole, as its energy will be $E_h=E_e$, such that $E_h-V<-mv^2<0$, whose group velocity is positive. The solution can be written as

$$\Psi(x)=\begin{cases}\Psi_{e,L}(x)=\Psi_{e,+}e^{ik_{e,0}x}+r\Psi_{e,-}e^{-ik_{e,0}x}, & x<0\\ \Psi_{h,R}(x)=t\Psi_{h,-}e^{-ik_{h,V}x}, & x>0\end{cases}.$$

From the boundary conditions of continuity of the wave function $\lim_{x\to 0^-}\Psi_L(x)=\lim_{x\to 0^+}\Psi_R(x)$, we find

$$r=\frac{k_{e,0}\left(V-E_h-mv^2\right)-k_{h,V}\left(mv^2+E_e\right)}{k_{e,0}\left(V-E_h-mv^2\right)+k_{h,V}\left(mv^2+E_e\right)},$$

$$t=\sqrt{\frac{\left(V-E_h-mv^2\right)^2+\left(\hbar vk_{h,V}\right)^2}{\left(mv^2+E_e\right)^2+\left(\hbar vk_{e,0}\right)^2}}\frac{2k_{e,0}\left(mv^2+E_e\right)}{k_{e,0}\left(V-E_h-mv^2\right)+k_{h,V}\left(mv^2+E_e\right)}.$$

Food for thought: Further insight can be gained by comparing the scattering of a nonrelativistic particle from the same potential obeying the Schrödinger equation $\left(\frac{\widehat{p}_x^2}{2m}+V\right)\phi=E\phi$. Using that $p_x=\hbar k$ is a good quantum number for the momentum operator $\widehat{p}_x=-i\hbar\partial_x$, we find that

$$k_\pm=\pm k_V=\pm\frac{\sqrt{2m(E-V)}}{\hbar},$$

$$\phi=\frac{1}{\sqrt{|C_+|^2+|C_-|^2}}\left(C_+e^{ik_Vx}+C_-e^{-ik_Vx}\right).$$

The wave function for the entire region can then be given as

$$\Phi(x) = \begin{cases} \phi_L(x) = e^{ik_0 x} + re^{-ik_0 x}, & \text{for } x < 0 \\ \phi_R(x) = te^{ik_V x}, & \text{for } x > 0 \end{cases}.$$

From the continuity of the wave function and its first derivative across $x = 0$, we find that

$$r = \frac{k_0 - k_V}{k_0 + k_V}, \quad t = \frac{2k_0}{k_0 + k_V}.$$

Problem 6.17

Now let us consider again the reflection and transmission of the 1D relativistic electron with energy E_e being scattered from the $V(x) = \begin{cases} 0, & x < 0 \\ V, & x > 0 \end{cases}$ step-potential and understand the behavior of the various currents. What is the *current density* of the particle in both regions as specified by the potential for the following:

a) $E_e > V + mv^2$;
b) $V - mv^2 < E_e < V + mv^2$;
c) $mv^2 < E_e < V - mv^2$.
d) Check the current's continuity condition at the boundary $x = 0$ and comment on the reflected and transmitted probability conservation relation. Consider the limit of $V \to \infty$ explicitly and compare with the nonrelativistic scattering.

The particle currents on the left and right sides of the potential are given by

$$j_L(x) = \Psi_L^+(x)\sigma_x \Psi_L(x); \; j_R(x) = \Psi_R^+(x)\sigma_x \Psi_R(x),$$

and they must fulfil the boundary condition $j_L(0) = j_R(0)$. From the previous problem for relativistic electrons and holes, we have

$$k_{\eta,\pm} = \pm k_{\eta,V} = \pm \frac{\sqrt{(E_\eta - V)^2 - m^2 v^4}}{\hbar v};$$

$$\Psi_{\eta,\pm}(E) = \frac{1}{\sqrt{(\hbar v k_{\eta,V})^2 + (mv^2 + E_\eta - V)^2}} \begin{pmatrix} mv^2 + E_\eta - V \\ \pm \hbar v k_{\eta,V} \end{pmatrix} = \frac{1}{N_{\eta,V}} \begin{pmatrix} 1 \\ \pm \Lambda_{\eta,V} \end{pmatrix};$$

$$\Lambda_{\eta,V} = \frac{\hbar v k_{\eta,V}}{mv^2 + E_\eta - V}; \; N_{\eta,V} = \sqrt{1 + |\Lambda_{\eta,V}|^2}.$$

Case a) In the case of an electron with energy $E_e > V + mv^2$ (see the diagram in Figure 6.1), the wave function is

$$\Psi(x) = \begin{cases} \Psi_{e,L}(x) = \Psi_{e,+} e^{ik_{e,0}x} + r\Psi_{e,-} e^{-ik_{e,0}x}, & \text{for } x < 0 \\ \Psi_{e,R}(x) = t\Psi_{e,+} e^{ik_{e,V}x}, & \text{for } x > 0 \end{cases}.$$

At the potential step $x=0$, the current obtained from each side of the wave function is

$$j_L(0) = \Psi_L^{+}(0)\sigma_x\Psi_L(0) = \left[\frac{1}{N_{e,0}}(1,\Lambda_{e,0}) + \frac{r^*}{N_{e,0}}(1,-\Lambda_{e,0})\right]$$
$$\times\,\sigma_x\left[\frac{1}{N_{e,0}}\begin{pmatrix}1\\ \Lambda_{e,0}\end{pmatrix} + \frac{r}{N_{e,0}}\begin{pmatrix}1\\ -\Lambda_{e,0}\end{pmatrix}\right]$$
$$= \frac{1}{N_{e,0}^2}\left[(1,\Lambda_{e,0})\begin{pmatrix}\Lambda_{e,0}\\ 1\end{pmatrix} + r(1,\Lambda_{e,0})\begin{pmatrix}-\Lambda_{e,0}\\ 1\end{pmatrix}\right.$$
$$\left. + r^*(1,-\Lambda_{e,0})\begin{pmatrix}\Lambda_{e,0}\\ 1\end{pmatrix} + |r|^2(1,-\Lambda_{e,0})\begin{pmatrix}-\Lambda_{e,0}\\ 1\end{pmatrix}\right]$$
$$= \frac{2\Lambda_{e,0}}{N_{e,0}^2}\left(1-|r|^2\right),$$

$$j_R(0) = \Psi_R^{+}(0)\sigma_x\Psi_R(0) = \left[\frac{t^*}{N_{e,V}}(1,\Lambda_{e,V})\right]\sigma_x\left[\frac{t}{N_{e,V}}\begin{pmatrix}1\\ \Lambda_{e,V}\end{pmatrix}\right]$$
$$= \frac{|t|^2}{N_{e,V}^2}\left[(1,\Lambda_{e,V})\begin{pmatrix}\Lambda_{e,V}\\ 1\end{pmatrix}\right] = \frac{2\Lambda_{e,V}}{N_{e,V}^2}|t|^2,$$

$$j_O = \left[\frac{1}{N_{e,0}}(1,\Lambda_{e,0})\right]\sigma_x\left[\frac{1}{N_{e,0}}\begin{pmatrix}1\\ \Lambda_{e,0}\end{pmatrix}\right] = \frac{2\Lambda_{e,0}}{N_{e,0}^2},$$

where j_O is the current density when $V=0$. Thus, the reflected and transmitted currents are, respectively,

$$j_r = \left[\frac{r^*}{N_{e,0}}(1,-\Lambda_{e,0})\right]\sigma_x\left[\frac{r}{N_{e,0}}\begin{pmatrix}1\\ -\Lambda_{e,0}\end{pmatrix}\right] = -\frac{2\Lambda_{e,0}}{N_{e,0}^2}|r|^2,$$
$$j_t = j_R(0) = \frac{2\Lambda_{e,V}}{N_{e,V}^2}|t|^2.$$

Therefore, from $j_L(0) = j_O + j_r$ and $j_R(0) = j_t$, and we have

$$R = \frac{|j_r|}{|j_O|} = \frac{\left|-\frac{2\Lambda_{e,0}}{N_{e,0}^2}|r|^2\right|}{\left|\frac{2\Lambda_{e,0}}{N_{e,0}^2}\right|} = |r|^2, \quad r = \frac{\Lambda_{e,0}-\Lambda_{e,V}}{\Lambda_{e,0}+\Lambda_{e,V}},$$

$$T = \frac{|j_t|}{|j_O|} = \frac{\left|\frac{2\Lambda_{e,V}}{N_{e,V}^2}|t|^2\right|}{\left|\frac{2\Lambda_{e,0}}{N_{e,0}^2}\right|} = \frac{\Lambda_{e,V}N_{e,0}^2}{\Lambda_{e,0}N_{e,V}^2}|t|^2, \quad t = \frac{N_{e,V}}{N_{e,0}}\frac{2\Lambda_{e,0}}{\Lambda_{e,0}+\Lambda_{e,V}},$$

$$R+T = |r|^2 + \frac{\Lambda_{e,V}N_{e,0}^2}{\Lambda_{e,0}N_{e,V}^2}|t|^2 = 1,$$

which is a well-known result from scattering theory.

Case b) In the case of an electron with energy $(V - mv^2) < E_e < (V + mv^2)$ (see the diagram in Figure 6.1), there is a transmission of an evanescent wave, for which

$$\Psi(x) = \begin{cases} \Psi_{e,L}(x) = \Psi_{e,+}e^{ik_{e,0}x} + r\Psi_{e,-}e^{-ik_{e,0}x}, & x < 0 \\ \Psi_R(x) = t\Psi_{\eta,+}e^{-\kappa_{\eta,V}x}, & x > 0, \end{cases}$$

where $\kappa_{\eta,V}$ is a positive real quantity. In the potential step at $x = 0$, the current obtained from each side of the wave function is $j_L(0) = j_0 + j_r$ and $j_R(0) = j_t$; then we have

$$j_0 = \left[\frac{1}{N_{e,0}}(1, \Lambda_{e,0})\right]\sigma_x\left[\frac{1}{N_{e,0}}\begin{pmatrix}1\\ \Lambda_{e,0}\end{pmatrix}\right] = \frac{2\Lambda_{e,0}}{N_{e,0}^2},$$

$$j_r = \left[\frac{r^*}{N_{e,0}}(1, -\Lambda_{e,0})\right]\sigma_x\left[\frac{r}{N_{e,0}}\begin{pmatrix}1\\ -\Lambda_{e,0}\end{pmatrix}\right] = -\frac{2\Lambda_{e,0}}{N_{e,0}^2}|r|^2.$$

It is important to note that, for the evanescent wave, $\Lambda_{\eta,V}$ becomes a purely imaginary quantity (see the preceding problem); then, $\Psi_{\eta,+} = \psi_{\eta,+}e^{-\kappa_{\eta,V}x}$ with $\psi_{\eta,+} = \frac{1}{N_{\eta,V}}\begin{pmatrix}1\\ i|\Lambda_{\eta,V}|\end{pmatrix}$, resulting in

$$\begin{aligned} j_t &= \left[\frac{t^*}{N_{\eta,V}}(1, -i|\Lambda_{\eta,V}|)\right]\sigma_x\left[\frac{t}{N_{\eta,V}}\begin{pmatrix}1\\ i|\Lambda_{\eta,V}|\end{pmatrix}\right] \\ &= \frac{|t|^2}{N_{h,V}^2}\left[(1, -i|\Lambda_{\eta,V}|)\begin{pmatrix}i|\Lambda_{\eta,V}|\\ 1\end{pmatrix}\right] = 0. \end{aligned}$$

The reflection and transmission coefficients become

$$R = \frac{|j_r|}{|j_0|} = \frac{\left|-\frac{2\Lambda_{e,0}}{N_{e,0}^2}|r|^2\right|}{\left|\frac{2\Lambda_{e,0}}{N_{e,0}^2}\right|} = |r|^2, \qquad r = \frac{\Lambda_{e,0} - i|\Lambda_{\eta,V}|}{\Lambda_{e,0} + i|\Lambda_{\eta,V}|},$$

$$T = \frac{|j_t|}{|j_0|} = 0, \qquad t = \frac{N_{\eta,V}}{N_{e,0}}\frac{2\Lambda_{e,0}}{\Lambda_{e,0} + \Lambda_{\eta,V}}.$$

Then, we obtain

$$|r|^2 = r^*r = \frac{\Lambda_{e,0} + i|\Lambda_{\eta,V}|}{\Lambda_{e,0} - i|\Lambda_{\eta,V}|}\frac{\Lambda_{e,0} - i|\Lambda_{\eta,V}|}{\Lambda_{e,0} + i|\Lambda_{\eta,V}|} = 1,$$

showing that the known relation $R + T = 1$ is fulfilled in this case as well.
Case c) In the case of an electron with incident energy $mv^2 < E_e < V - mv^2$, which leads to a reflected electron and a transmitted hole whose energy is $E_h - V < -mv^2 < 0$ (see the diagram in Figure 6.1), we recall that the solution is

$$\Psi(x) = \begin{cases} \Psi_{e,L}(x) = \Psi_{e,+}e^{ik_{e,0}x} + r\Psi_{e,-}e^{-ik_{e,0}x}, & \text{for } x < 0 \\ \Psi_{h,R}(x) = t\Psi_{h,-}e^{-ik_{h,V}x}, & \text{for } x > 0 \end{cases}.$$

At the potential step $x = 0$, the current obtained from each side of the wave function is $j_L(0) = j_0 + j_r$ and $j_R(0) = j_t$; then we have

$$j_0 = \left[\frac{1}{N_{e,0}}(1, \Lambda_{e,0})\right]\sigma_x\left[\frac{1}{N_{e,0}}\begin{pmatrix}1\\ \Lambda_{e,0}\end{pmatrix}\right] = \frac{2\Lambda_{e,0}}{N_{e,0}^2},$$

$$j_r = \left[\frac{r^*}{N_{e,0}}(1, -\Lambda_{e,0})\right]\sigma_x\left[\frac{r}{N_{e,0}}\begin{pmatrix}1\\ -\Lambda_{e,0}\end{pmatrix}\right] = -\frac{2\Lambda_{e,0}}{N_{e,0}^2}|r|^2,$$

$$j_t = \left[\frac{t^*}{N_{h,V}}(1, -\Lambda_{h,V})\right]\sigma_x\left[\frac{t}{N_{h,V}}\begin{pmatrix}1\\ -\Lambda_{h,V}\end{pmatrix}\right]$$
$$= \frac{|t|^2}{N_{h,V}^2}\left[(1, -\Lambda_{h,V})\begin{pmatrix}-\Lambda_{h,V}\\ 1\end{pmatrix}\right] = -\frac{2\Lambda_{h,V}}{N_{h,V}^2}|t|^2.$$

Then, the reflection and transmission coefficients are

$$R = \frac{|j_r|}{|j_0|} = \frac{\left|-\frac{2\Lambda_{e,0}}{N_{e,0}^2}|r|^2\right|}{\left|\frac{2\Lambda_{e,0}}{N_{e,0}^2}\right|} = |r|^2, \qquad r = \frac{\Lambda_{e,0} + \Lambda_{h,V}}{\Lambda_{e,0} - \Lambda_{h,V}},$$

$$T = \frac{|j_t|}{|j_0|} = \frac{\left|-\frac{2\Lambda_{h,V}}{N_{h,V}^2}|t|^2\right|}{\left|\frac{2\Lambda_{e,0}}{N_{e,0}^2}\right|} = \left|\frac{\Lambda_{h,V}N_{e,0}^2}{\Lambda_{e,0}N_{h,V}^2}\right||t|^2, \qquad t = \frac{N_{h,V}}{N_{e,0}}\frac{2\Lambda_{e,0}}{\Lambda_{e,0} - \Lambda_{h,V}}.$$

Further realizing that $\Lambda_{h,V} = \frac{\hbar v k_{h,V}}{mv^2 + E_h - V} = -\sqrt{\frac{V - E_h + mv^2}{V - E_h - mv^2}} < 0$ and $\Lambda_{e,0} = \frac{\hbar v k_{e,0}}{mv^2 + E_e} = \sqrt{\frac{E_e - mv^2}{E_e + mv^2}} > 0$, then

$$R = \left|\frac{\Lambda_{e,0} - |\Lambda_{h,V}|}{\Lambda_{e,0} + |\Lambda_{h,V}|}\right|^2, \qquad T = \frac{4\Lambda_{e,0}|\Lambda_{h,V}|}{(\Lambda_{e,0} + |\Lambda_{h,V}|)^2} \to R + T = 1.$$

d) Let's see now what happens at the $V \to \infty$ limit for $E_h - V < -mv^2 < 0$ using that $\lim_{V\to\infty} \Lambda_{h,V} = -1$,

$$\lim_{V\to\infty} R = \lim_{V\to\infty}\left|\frac{\Lambda_{e,0} - |\Lambda_{h,V}|}{\Lambda_{e,0} + |\Lambda_{h,V}|}\right|^2 = \left|\frac{\Lambda_{e,0} - 1}{\Lambda_{e,0} + 1}\right|^2 = \frac{E_e - \sqrt{E_e^2 - m^2v^4}}{E_e + \sqrt{E_e^2 - m^2v^4}} \neq 1,$$

$$\lim_{V_0\to\infty} T = \lim_{V\to\infty}\frac{4\Lambda_{e,0}|\Lambda_{h,V}|}{(\Lambda_{e,0} + |\Lambda_{h,V}|)^2} = \frac{4\Lambda_{e,0}}{(\Lambda_{e,0} + 1)^2} = \frac{2\sqrt{E_e^2 - m^2v^4}}{E_e + \sqrt{E_e^2 - m^2v^4}} \neq 0.$$

Note that, even when the potential step becomes infinite, it cannot block the transmission of flux of fermions (even though we have a transmitted flux of holes instead of electrons). This is known as the *Klein paradox* for relativistic particle scattering (Calogeracos & Dombey, 1999).

Let us compare with the nonrelativistic particle scattering, which was also discussed in the previous problem. The motion is subject to the Schrödinger equation. The particle current is $\boldsymbol{j}(x) = \frac{\hbar}{2mi}[\phi^*(x)\boldsymbol{\nabla}\phi(x) - \phi(x)\boldsymbol{\nabla}\phi^*(x)]$ and the continuity equation $j_L(0) = j_0 + j_r$ and $j_R(0) = j_t$ still holds.

We have to distinguish two different possible cases here:

$$\begin{cases} E > V \to & k_V = \sqrt{\frac{2m(E-V)}{\hbar}} > 0, \\ E < V \to & k_V = i\frac{\sqrt{2m(V-E)}}{\hbar} = i\kappa_V. \end{cases}$$

For both cases we have the same wave function,

$$\Phi(x) = \begin{cases} \phi_L(x) = e^{ik_0x} + re^{-ik_0x}, & \text{for } x < 0 \\ \phi_R(x) = te^{ik_Vx}, & \text{for } x > 0 \end{cases},$$

and the same result for r and t,

$$r = \frac{k_0 - k_V}{k_0 + k_V}, \quad t = \frac{2k_0}{k_0 + k_V}.$$

However, the $E > V$ case corresponds to a propagating transmitted wave, while the $E < V$ case corresponds to an evanescent transmitted wave.

The incoming, reflected, and transmitted currents are found as

$$j_0 = \frac{\hbar k_0}{m}, \qquad j_r = -|r|^2\frac{\hbar k_0}{m},$$

$$j_t = |t|^2\frac{\hbar}{2mi}\left[e^{-ik_Vx}\partial_x e^{ik_Vx} - e^{ik_Vx}\partial_x e^{-ik_Vx}\right] = |t|^2\frac{\hbar k_V}{m}.$$

For the propagating transmission, we arrive at

$$R = \frac{|j_r|}{|j_0|} = |r|^2, \quad r = \frac{k_0 - k_V}{k_0 + k_V} \in [0,1],$$

$$T = \frac{|j_t|}{|j_0|} = \frac{k_V}{k_0}|t|^2 > 0, \qquad t = \frac{2k_0}{k_0 + k_V} \in [1,2].$$

For the evanescent transmission, we have

$$j_t(x) = |t|^2\frac{\hbar}{2mi}\left[e^{-\kappa_Vx}\partial_x e^{-\kappa_Vx} - e^{-\kappa_Vx}\partial_x e^{-\kappa_Vx}\right] = 0,$$

$$R = \frac{|j_r|}{|j_0|} = |r|^2 = 1, \quad r = \frac{k_0 - i\kappa_V}{k_0 + i\kappa_V},$$

$$T = \frac{|j_t|}{|j_0|} = 0, \qquad t = \frac{2k_0}{k_0 + i\kappa_V}.$$

The Klein paradox appears precisely for the evanescent transmission corresponding to the creation of holes due to the scattering of the incoming relativistic electron.

Problem 6.18

Consider a Dirac particle with mass m moving in 1D with velocity v and energy E being scattered by a potential $V_0\delta(x)$. The particle also experiences a quadratic correction Bp_x^2 to the mass term with the constant $B > 0$. Assuming an incoming particle on the left side of the potential barrier, obtain the scattered and reflected wave functions.

We begin with the Dirac equation of this 1D massive particle:

$$\left[v\widehat{p}_x\sigma_x + \left(mv^2 - B\widehat{p}_x^2\right)\sigma_z + V_0\delta(\widehat{x})\alpha^0\right]\Psi(x) = E\Psi(x).$$

The potential separates the space region into two parts and the corresponding wave functions of the two regions will be connected by some boundary conditions at $x = 0$.

For the regions to the left and right of the potential barrier, the solution to the eigenvalue problem $\widehat{H}\Psi_\eta(x) = E_\eta\Psi_\eta(x)$ for $\widehat{p}_x = \hbar k$ is

$$\Psi_{s,\eta}(x) = u_\eta(k)e^{ikx} = \frac{1}{N_\eta}\begin{pmatrix} 1 \\ s\Lambda_\eta \end{pmatrix} e^{ik_{s,\eta}x},$$

$$E_\eta = \eta\sqrt{\left(mv^2 - B\hbar^2k^2\right)^2 + (\hbar vk)^2},$$

where $\eta = \pm 1$, $N_\eta = \sqrt{1+\Lambda_\eta^2}$, $\Lambda_\eta = \frac{\hbar vk}{mv^2 - B\hbar^2k^2 + E_\eta}$. Similarly, there are two possible momenta ($s = \pm$) for each η:

$$k_{\pm,\eta} = \pm\frac{v}{\sqrt{2}\hbar B}\sqrt{(2mB-1) + \eta\sqrt{(1-4mB) + \frac{4E_\eta^2B^2}{v^4}}}.$$

The general form of the spinor for a given energy $E = E_\eta$ can then be written as

$$\Psi_\eta(x) = C_{+,\eta}u_\eta\left(k_{+,\eta}\right)e^{ik_{+,\eta}x} + C_{-,\eta}u_\eta\left(k_{-,\eta}\right)e^{ik_{-,\eta}x}.$$

The so-obtained wave function is valid everywhere except at the $x = 0$ point. At that point, we must ensure the continuity of the Dirac equation due to the inhomogeneity from the Dirac delta function, by considering $\widehat{H}\left(\widehat{p}_x = -i\hbar\partial_x\right)\Psi(x) = E\Psi(x)$. Using that $\Psi(x) = \begin{pmatrix} a(x) \\ b(x) \end{pmatrix}$, we arrive at

$$\begin{pmatrix} \left(mv^2 + B\hbar^2\partial_x^2\right) + V_0\delta(x) - E & -i\hbar\partial_x \\ -i\hbar\partial_x & -\left(mv^2 + B\hbar^2\partial_x^2\right) + V_0\delta(x) - E \end{pmatrix}\begin{pmatrix} a(x) \\ b(x) \end{pmatrix} = \begin{pmatrix} 0 \\ 0 \end{pmatrix}.$$

This is a linear system of two second-order ordinary differential equations. For our purposes, it is useful to define $\alpha(x) = \partial_x a(x)$ and $\beta(x) = \partial_x b(x)$, which transforms the system to first-order ordinary differential equations,

$$\begin{cases} \alpha(x) = \partial_x a(x) \\ \beta(x) = \partial_x b(x) \\ \left(mv^2 + V_0\delta(x) - E\right)a(x) + B\hbar^2\partial_x\alpha(x) - i\hbar\beta(x) = 0 \\ -i\hbar\alpha(x) + \left(-mv^2 + V_0\delta(x) - E\right)b(x) - B\hbar^2\partial_x\beta(x) = 0 \end{cases}.$$

Collectively, one writes

$$B\hbar^2\partial_x\begin{pmatrix} a(x) \\ b(x) \\ \alpha(x) \\ \beta(x) \end{pmatrix} = \begin{pmatrix} 0 & 0 & B\hbar^2 & 0 \\ 0 & 0 & 0 & B\hbar^2 \\ -\left(V_0\delta(x) + mv^2 - E\right) & 0 & 0 & i\hbar \\ 0 & \left(V_0\delta(x) - mv^2 - E\right) & -i\hbar & 0 \end{pmatrix}\begin{pmatrix} a(x) \\ b(x) \\ \alpha(x) \\ \beta(x) \end{pmatrix},$$

where each equation from the preceding system is of the form $\partial_x\Phi(x) = \frac{1}{B\hbar^2}M(x)\Phi(x)$. Using the Neumann solution ansatz $\Phi(x) = \mathcal{P}_x e^{\frac{1}{B\hbar^2}\int_{x_0}^{x} dxM(x)}\Phi(x_0)$, where $\mathcal{P}_x$ is the Dyson ordering operator, one finds

$$\lim_{\varepsilon\to 0}\int_{-\varepsilon}^{\varepsilon} dxM(x)$$

$$= \lim_{\varepsilon\to 0}\int_{-\varepsilon}^{\varepsilon}\begin{pmatrix} 0 & 0 & B\hbar^2 & 0 \\ 0 & 0 & 0 & B\hbar^2 \\ -\left(V_0\delta(x)+mv^2-E\right) & 0 & 0 & i\hbar \\ 0 & \left(V_0\delta(x)-mv^2-E\right) & -i\hbar & 0 \end{pmatrix} dx$$

$$= \lim_{\varepsilon\to 0}\begin{pmatrix} 0 & 0 & 2\varepsilon B\hbar^2 & 0 \\ 0 & 0 & 0 & 2\varepsilon B\hbar^2 \\ -V_0-2\varepsilon\left(mv^2-E\right) & 0 & 0 & 2\varepsilon i\hbar \\ 0 & -V_0-2\varepsilon\left(mv^2+E\right) & -2\varepsilon i\hbar & 0 \end{pmatrix}$$

$$= V_0\begin{pmatrix} 0 & 0 & 0 & 0 \\ 0 & 0 & 0 & 0 \\ -1 & 0 & 0 & 0 \\ 0 & 1 & 0 & 0 \end{pmatrix}.$$

This solution leads to the following boundary condition:

$$\lim_{\varepsilon\to 0}\Psi(\varepsilon) = \lim_{\varepsilon\to 0}\mathcal{P}_x e^{\frac{1}{B\hbar^2}\int_{-\varepsilon}^{\varepsilon} dxM(x)}\Psi(-\varepsilon) = \begin{pmatrix} 1 & 0 & 0 & 0 \\ 0 & 1 & 0 & 0 \\ -\frac{V_0}{B\hbar^2} & 0 & 1 & 0 \\ 0 & \frac{V_0}{B\hbar^2} & 0 & 1 \end{pmatrix}\Psi(-\varepsilon),$$

$$\begin{cases} a(\varepsilon) = a(-\varepsilon) \\ b(\varepsilon) = b(-\varepsilon) \\ \alpha(\varepsilon) = \alpha(-\varepsilon) - \dfrac{V_0}{B\hbar^2}a(-\varepsilon) \\ \beta(\varepsilon) = \beta(-\varepsilon) - \dfrac{V_0}{B\hbar^2}b(-\varepsilon) \end{cases}.$$

Thus, for $\Psi(x) = \begin{pmatrix} a(x) \\ b(x) \end{pmatrix}$ we have the following boundary conditions,

$$\begin{pmatrix} a(0^+) \\ b(0^+) \end{pmatrix} = \begin{pmatrix} a(0^-) \\ b(0^-) \end{pmatrix} \to \Psi(0^+) = \Psi(0^-),$$

$$\partial_x\begin{pmatrix} a(0^+) \\ b(0^+) \end{pmatrix} = \partial_x\begin{pmatrix} a(0^-) \\ b(0^-) \end{pmatrix} - \frac{V_0}{B\hbar^2}\begin{pmatrix} 1 & 0 \\ 0 & -1 \end{pmatrix}\begin{pmatrix} a(0^-) \\ b(0^-) \end{pmatrix}$$

$$\to \partial_x\Psi(0^+) = \partial_x\Psi(0^-) - \lambda\sigma_3\Psi(0^-),$$

where $\lambda = \frac{V_0}{B\hbar^2}$. These boundary conditions are now used to study the wave function scattering (see the scheme in Figure 6.2),

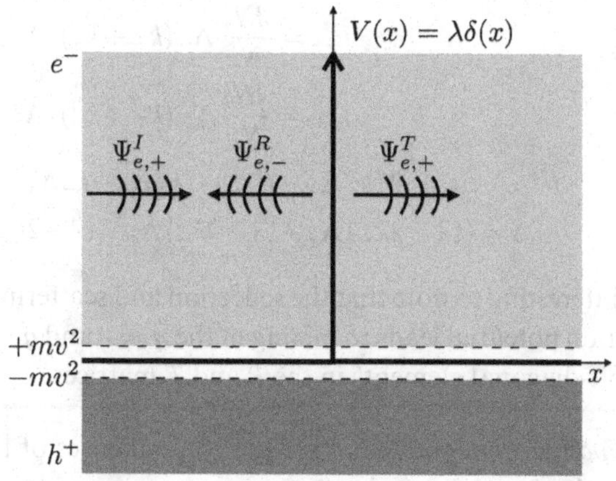

Figure 6.2 A schematic representation of the potential and the two regions for which the Dirac equation must be solved is given. The incident (I), reflected (R), and transmitted (T) electron wave function are denoted.

$$\Psi(x) = \begin{cases} \Psi_L(x) = \sum_{\eta=\pm1} \left(A_\eta u_\eta(k_\eta) e^{ik_\eta x} + R_\eta u_\eta(-k_\eta) e^{-ik_\eta x}\right) & \text{for } x < 0 \\ \Psi_R(x) = \sum_{\eta=\pm1} \left(T_\eta u_\eta(k_\eta) e^{ik_\eta x}\right) & \text{for } x > 0 \end{cases},$$

where the R_η and T_η terms denote reflected and transmitted waves, respectively. We find

$$\begin{aligned} &T_- u_-(k_-) + T_+ u_+(k_+) \\ &\quad = A_- u_-(k_-) + A_+ u_+(k_+) + R_- u_-(-k_-) + R_+ u_+(-k_+), \\ &ik_- T_- u_-(k_-) + ik_+ T_+ u_+(k_+) \\ &\quad = ik_- A_- u_-(k_-) + ik_+ A_+ u_+(k_+) - ik_- R_- u_-(-k_-) - ik_+ R_+ u_+(-k_+) \\ &\quad - \lambda\sigma_3\left(A_- u_-(k_-) + A_+ u_+(k_+) + R_- u_-(-k_-) + R_+ u_+(-k_+)\right), \end{aligned}$$

which can be written in a matrix form:

$$\begin{aligned} R &= \begin{pmatrix} R_+ \\ R_- \end{pmatrix} = \frac{\lambda}{\Delta}\begin{pmatrix} r_{++} & r_{+-} \\ r_{-+} & r_{--} \end{pmatrix}\begin{pmatrix} A_+ \\ A_- \end{pmatrix}, \\ T &= \begin{pmatrix} T_+ \\ T_- \end{pmatrix} = \frac{2}{\Delta}\begin{pmatrix} t_{++} & t_{+-} \\ t_{-+} & t_{--} \end{pmatrix}\begin{pmatrix} A_+ \\ A_- \end{pmatrix}, \\ r_{++} &= 2k_-\left(\Lambda_+^2 + \Lambda_-^2\right) - 4k_+\Lambda_+\Lambda_- + \lambda\left(\Lambda_+^2 - \Lambda_-^2\right), \\ r_{+-} &= \frac{2\Lambda_- N_+}{N_-}\left((k_- - k_+)(\Lambda_+ + \Lambda_-) + \lambda(\Lambda_+ - \Lambda_-)\right), \\ r_{-+} &= -\frac{2\Lambda_+ N_-}{N_+}\left((k_- - k_+)(\Lambda_+ + \Lambda_-) + \lambda(\Lambda_+ - \Lambda_-)\right), \\ r_{--} &= 2k_+\left(\Lambda_-^2 + \Lambda_+^2\right) - 4k_-\Lambda_+\Lambda_- + \lambda\left(\Lambda_-^2 - \Lambda_+^2\right), \\ t_{++} &= 2\left(k_+\Lambda_+ - k_-\Lambda_-\right)\left(k_+\Lambda_- - k_-\Lambda_+\right) + \lambda k_+\left(\Lambda_-^2 - \Lambda_+^2\right), \end{aligned}$$

$$t_{+-} = \frac{\lambda N_+}{N_-}\Lambda_-(k_+ + k_-)(\Lambda_- - \Lambda_+),$$
$$t_{-+} = \frac{\lambda N_-}{N_+}\Lambda_+(k_+ + k_-)(\Lambda_- - \Lambda_+),$$
$$t_{--} = 2(k_+\Lambda_+ - k_-\Lambda_-)(k_+\Lambda_- - k_-\Lambda_+) - \lambda k_-(\Lambda_-^2 - \Lambda_+^2),$$
$$\Delta = ((\lambda + 2k_+)\Lambda_- - (\lambda + 2k_-)\Lambda_+)((\lambda - 2k_-)\Lambda_- - (\lambda - 2k_+)\Lambda_+).$$

It is interesting to note that the reflection and scattering of a Dirac particle from a delta function potential leads to mixing of the $\eta = +$ and $\eta = -$ polarizations, as evident from the off-diagonal elements in the R and T matrices.

Food for thought: What are the consequences of the $\lambda \to 0$ and $\lambda \to \pm\infty$ limits in the R and T matrices?

Problem 6.19

Consider the Dirac equation $\widehat{H} = c\boldsymbol{\alpha}\cdot\widehat{\boldsymbol{p}} + mc^2\alpha^4$.

Obtain the (a) parity, (b) time-reversal, and (c) charge conjugation under which this Dirac Hamiltonian is invariant.

We recall that the invariance of a given Hamiltonian $\widehat{H}$ under a symmetry transformation g of a given symmetry group G implies

$$\widehat{H}(gt, g\widehat{\boldsymbol{r}}, g\widehat{\boldsymbol{p}}) = U(g)\widehat{H}(t, \widehat{\boldsymbol{r}}, \widehat{\boldsymbol{p}})U^{-1}(g),$$

or, similarly

$$\widehat{H}(t, \widehat{\boldsymbol{r}}, \widehat{\boldsymbol{p}}) = U^{-1}(g)\widehat{H}(gt, g\widehat{\boldsymbol{r}}, g\widehat{\boldsymbol{p}})U(g),$$

where $U(g)$ is a unitary transformation that is a representation of the symmetry transformation g in the Hilbert space of the system.

Then, the eigenfunction is transformed as $|\psi(gt, g\boldsymbol{x}, g\boldsymbol{p})\rangle = U(g)|\psi(t, \boldsymbol{x}, \boldsymbol{p})\rangle$. Invariant observables $\widehat{A}$, under the symmetry operation are given as

$$\begin{aligned}
\langle A_g\rangle_\psi &= \left\langle \psi(gt, g\boldsymbol{x}, g\boldsymbol{p})|\widehat{A}(gt, g\widehat{\boldsymbol{r}}, g\widehat{\boldsymbol{p}})\,|\psi(gt, g\boldsymbol{x}, g\boldsymbol{p})\right\rangle \\
&= \left\langle U(g)\,\psi(t, \boldsymbol{x}, \boldsymbol{p})\,|U(g)\widehat{A}(t, \widehat{\boldsymbol{r}}, \widehat{\boldsymbol{p}})U^{-1}(g)\,|U(g)\psi(t, \boldsymbol{x}, \boldsymbol{p})\right\rangle \\
&= \left\langle \psi(t, \boldsymbol{x}, \boldsymbol{p})\,|U^{-1}(g)\,U(g)\widehat{A}(t, \widehat{\boldsymbol{r}}, \widehat{\boldsymbol{p}})\,U^{-1}(g)\,U(g)\,|\psi(t, \boldsymbol{x}, \boldsymbol{p})\right\rangle \\
&= \left\langle \psi(t, \boldsymbol{x}, \boldsymbol{p})|\widehat{A}(t, \widehat{\boldsymbol{r}}, \widehat{\boldsymbol{p}})\,|\psi(t, \boldsymbol{x}, \boldsymbol{p})\right\rangle = \langle A\rangle_\psi.
\end{aligned}$$

The Hamiltonian is invariant under the symmetry transformation g if

$$\widehat{H}(t, \widehat{\boldsymbol{r}}, \widehat{\boldsymbol{p}}) = \widehat{H}(gt, g\widehat{\boldsymbol{r}}, g\widehat{\boldsymbol{p}}) = U(g)\widehat{H}(t, \widehat{\boldsymbol{r}}, \widehat{\boldsymbol{p}})U^{-1}(g).$$

a) We recall that the **parity operator** $\widehat{\pi}$ reverses the position $\pi\widehat{r}\pi^{-1} = -\widehat{r}$; consequently, it also reverses momentum $\widehat{\pi}\widehat{p}\widehat{\pi}^{-1} = -\widehat{p}$. Also, $\widehat{\pi}^2\widehat{r} = -\widehat{\pi}\widehat{r} = (-1)^2\widehat{r}$, therefore, $\widehat{\pi}^{-1} = \widehat{\pi}^+$.

For the Hamiltonian of the Dirac equation, we use the equality

$$
\begin{aligned}
\widehat{H}_\Pi(t,\widehat{r},\widehat{p}) &= U_\pi^{-1}\widehat{\pi}\widehat{H}(t,\widehat{r},\widehat{p})\widehat{\pi}^{-1}U_\pi = U_\pi^{-1}\widehat{H}\left(\widehat{\pi}t,\widehat{\pi}\widehat{r}\widehat{\pi}^{-1},\widehat{\pi}\widehat{p}\widehat{\pi}^{-1}\right)U_\pi \\
&= U_\pi^{-1}\widehat{H}(t,-\widehat{r},-\widehat{p})U_\pi, \\
U_\pi^{-1}\widehat{\pi}\widehat{H}(t,\widehat{r},\widehat{p})\widehat{\pi}^{-1}U_\pi &= U_\pi^{-1}\widehat{\pi}\left(c\boldsymbol{\alpha}\cdot\widehat{\boldsymbol{p}}+mc^2\alpha^4\right)\widehat{\pi}^{-1}U_\pi \\
&= c\left(U_\pi^{-1}\alpha^iU_\pi\right)\left(\widehat{\pi}\widehat{p}_i\widehat{\pi}^{-1}\right)+mc^2\left(U_\pi^{-1}\alpha^4U_\pi\right) \\
&= c\left(-U_\pi^{-1}\alpha^iU_\pi\right)\widehat{p}_i+mc^2\left(U_\pi^{-1}\alpha^4U_\pi\right) \\
&= H(t,\widehat{r},\widehat{p}) = c\alpha^i\widehat{p}_i+mc^2\alpha^4.
\end{aligned}
$$

This means that the unitary operator U_π must fulfil

$$
\begin{cases}
-U_\pi^{-1}\alpha^iU_\pi = \alpha^i \\
U_\pi^{-1}\alpha^4U_\pi = \alpha^4
\end{cases}.
$$

Equivalently, $\{\alpha^i, U_\pi\} = 0$ and $[\alpha^4, U_\pi] = 0$. Therefore, by using the anticommutation relations of the Dirac matrices $\{\alpha^i, \alpha^4\} = 0$, it is very easy to obtain that $U_\pi = \alpha^4 = U_\pi^{-1}$. Therefore, the parity operator can be represented as $\widehat{\Pi} = \eta\alpha^4\widehat{\pi}$, and $\widehat{\Pi}|\psi(t,\boldsymbol{x},\boldsymbol{p})\rangle = \eta\alpha^4\widehat{\pi}|\psi(t,\boldsymbol{x},\boldsymbol{p})\rangle = \eta\alpha^4|\psi(t,-\boldsymbol{x},-\boldsymbol{p})\rangle$.

b) The **time-reversal** operator reverses time $\widehat{\Theta}t\widehat{\Theta}^{-1} = -t$, but not position $\widehat{\Theta}\widehat{r}\widehat{\Theta}^{-1} = +\widehat{r}$; consequently, this operator must be odd for the momentum $\widehat{\Theta}\widehat{p}\widehat{\Theta}^{-1} = -\widehat{p}$. Using $\widehat{p} = -i\hbar\nabla$ and applying the time-reversal operator, we obtain

$$
\begin{aligned}
\widehat{\Theta}\widehat{p}\widehat{\Theta}^{-1} &= \widehat{\Theta}(-i\hbar\nabla)\widehat{\Theta}^{-1} = -\hbar\left(\widehat{\Theta}i\widehat{\Theta}^{-1}\right)\left(\widehat{\Theta}\nabla\widehat{\Theta}^{-1}\right) \\
&= -\hbar\left(\widehat{\Theta}i\widehat{\Theta}^{-1}\right)\nabla = -\widehat{p} = -(-i\hbar\nabla) = i\hbar\nabla.
\end{aligned}
$$

To obtain the antiunitary operator that represents the time-reversal transformations for the Dirac equation, the following equality must be fulfilled

$$
\widehat{H}\left(\widehat{\Theta}t\widehat{\Theta}^{-1},\widehat{\Theta}\widehat{r}\widehat{\Theta}^{-1},\widehat{\Theta}\widehat{p}\widehat{\Theta}^{-1}\right) = \widehat{\Theta}\widehat{H}(t,\widehat{r},\widehat{p})\widehat{\Theta}^{-1} = \widehat{H}(t,\widehat{r},\widehat{p}).
$$

Representing the time-reversal operator as $\widehat{\Theta} = U_\Theta\widehat{K}$ with $\widehat{K}$ being the complex conjugation operator, we obtain

$$
\begin{aligned}
\widehat{\Theta}\widehat{H}(t,\widehat{r},\widehat{p})\widehat{\Theta}^{-1} &= U_\Theta\widehat{K}\left[c\alpha^i\widehat{p}_i+mc^2\alpha^4\right]\widehat{K}^{-1}U_\Theta^{-1} = U_\Theta\left[c\alpha^{i*}\widehat{p}_i^*+mc^2\alpha^{4*}\right]U_\Theta^{-1} \\
&= c\left(U_\Theta\alpha^{i*}U_\Theta^{-1}\right)(-\widehat{p}_i)+mc^2\left(U_\Theta\alpha^{4*}U_\Theta^{-1}\right) = \widehat{H}(t,\widehat{r},\widehat{p}) = c\alpha^i\widehat{p}_i+mc^2\alpha^4,
\end{aligned}
$$

where we have used that c and m are real. This means that the symmetry transformation matrix U_Θ must fulfil

$$
\begin{cases}
U_\Theta\alpha^{i*}U_\Theta^{-1} = -\alpha^i \\
U_\Theta\alpha^{4*}U_\Theta^{-1} = \alpha^4
\end{cases}.
$$

As all α^μ matrices are real except the purely imaginary matrix α^2, the above relations are equivalent to

$$\{\alpha^1, U_\Theta\} = \{\alpha^3, U_\Theta\} = 0 = [\alpha^2, U_\Theta] = [\alpha^4, U_\Theta].$$

It is easy to obtain that $U_\Theta = \alpha^1\alpha^3$, therefore, the time-reversal operator can be represented as $\widehat{\Theta} = \eta\alpha^1\alpha^3\widehat{K}$, and $\widehat{\Theta}|\psi(t,\boldsymbol{x},\boldsymbol{p})\rangle = U_\Theta\widehat{K}|\psi(t,\boldsymbol{x},\boldsymbol{p})\rangle = \eta\alpha^1\alpha^3|\psi^*(-t,\boldsymbol{x},-\boldsymbol{p})\rangle$.

c) The **particle-hole symmetry operator** transforms particles into antiparticles, and vice versa. It is also called the **charge-conjugation operator** because it can be proved that, for consistency, all gauge charges (as the electric charge) change sign under the effect of this operator.

As we are looking for an operator that transforms particles into antiparticles, we need the equation for the particle and for the antiparticle (if it exists). The particle (electron) is represented as the ket ($|\psi\rangle = \psi$, while the antiparticle (hole) will be the bra ($\langle\psi| = \psi^*$), therefore, we have

$$\widehat{H}\psi = \left[c\alpha^i(-i\hbar\partial_i) + mc^2\alpha^4\right]\psi = E_e\psi.$$

Our goal is to transform the particle equation into the antiparticle one. To this end, we conjugate the Dirac equation for the electron

$$(\widehat{H}\psi)^* = \widehat{H}^*\psi^* = \left[c\alpha^i(-i\hbar\partial_i) + mc^2\alpha^4\right]^*\psi^* = \left[c\alpha^{i*}(+i\hbar\partial_i) + mc^2\alpha^{4*}\right]\psi^* = E_e\psi^*.$$

Note that the hole has equal energy to the electron, but with opposite sign: $E_h = -E_e$, a property that the Hamiltonian has to obey.

Now, we multiply the conjugate Dirac equation by (-1) in order to exchange the global sign to obtain the correct energy for the antiparticle

$$-\widehat{H}^*\psi^* = \left[c\alpha^{i*}(-i\hbar\partial_i) - mc^2\alpha^{4*}\right]\psi^* = -E_e\psi^* = E_h\psi^*.$$

Finally, to obtain the unitary representation of the particle-hole symmetry for the Dirac equation, we assume that there is a unitary matrix transformation U_C that transforms the initial Hamiltonian into the antiparticle Hamiltonian,

$$\begin{aligned} U_C\widehat{H}U_C^{-1}U_C\psi &= U_C\left[c\alpha^i(-i\hbar\partial_i) + mc^2\alpha^4\right]U_C^{-1}\psi_C \\ &= \left[c\left(U_C\alpha^iU_C^{-1}\right)(-i\hbar\partial_i) + mc^2\left(U_C\alpha^4U_C^{-1}\right)\right]\psi^* \\ &= -\widehat{H}^*\psi^* = \left[c\alpha^{i*}(-i\hbar\partial_i) - mc^2\alpha^{4*}\right]\psi^* \end{aligned}$$

where we have used that c and m are real and $\psi_C = U_C\psi = \psi^*$. This means that the symmetry transformation matrix U_C must fulfil

$$\begin{cases} U_C\alpha^iU_C^{-1} = \alpha^{i*} \\ U_C\alpha^4U_C^{-1} = -\alpha^{4*} \end{cases}.$$

As all α^μ matrices are real except the purely imaginary matrix α^2, the above relations are equivalent to

$$[\alpha^1, U_C] = [\alpha^3, U_C] = 0 = \{\alpha^2, U_C\} = \{\alpha^4, U_C\}.$$

It is easy to obtain that $U_C = \eta\alpha^2\alpha^4$; therefore, the particle-hole operator can be represented as $\widehat{C} = \eta\alpha^2\alpha^4\widehat{K}$, and $\widehat{C}|\psi(t,\boldsymbol{x},\boldsymbol{p})\rangle = U_C\widehat{K}|\psi(t,\boldsymbol{x},\boldsymbol{p})\rangle = \eta\alpha^2\alpha^4|\psi^*(t,\boldsymbol{x},\boldsymbol{p})\rangle$.

Finally, note that, contrary to the rest of symmetries studied here, a Hamiltonian is invariant under the particle-hole symmetry if, under this transformation, the hamiltonian changes its sign:

$$\widehat{C}\widehat{H}(t,\widehat{\boldsymbol{r}},\widehat{\boldsymbol{p}})\widehat{C}^{-1} = -\widehat{H}(t,\widehat{\boldsymbol{r}},\widehat{\boldsymbol{p}})$$

Food for thought: Starting from the Dirac Hamiltonian in the presence of an electromagnetic field,

$$\widehat{H}_\psi = \left[c\alpha^i\left(-i\hbar\partial_i + qA_i\right) + mc^2\alpha^4 + q\phi\alpha^0\right]\psi = E_e\psi,$$

check that the antiparticles have the opposite charges of the particles.

References

Berry, M. V. 1984. Quantal phase factors accompanying adiabatic changes. *Proceedings of the Royal Society A*, Volume 392, p. 45.

Berry, M. V., and Mondragon, R. J. 1987. Neutrino billiards: time-reversal symmetry-breaking without magnetic fields. *Proceedings of the Royal Society A: Mathematical, Physical and Engineering Sciences*, Volume 412, p. 53.

Bordag, M., Klimchitskaya, G. L., Mohideen, U., and Mostepanenko, V. M. 2009. *Advances in the Casimir Effect*. Oxford University Press.

Brandsen, B. H., and Joachain, C. J. 2000. *Quantum Mechanics*. Second ed. Prentice Hall.

Brown, E. 1964. Bloch electrons in a uniform magnetic field. *Physical Review*, Volume 133, pp. A1038–A1044.

Calogeracos, A., and Dombey, N. 1999. History and physics of the Klein paradox. *Contemporary Physics,* Volume 40, pp. 313–321.

Castro Neto, A. H. et al. 2009. The electronic properties of graphene. *Reviews of Modern Physics*, Volume 81, pp. 109–162.

Chakrabarti, A. 1964. On the coupling of 3 angular momenta. *Annales de l'I. H. P., section A*, Volume 1, pp. 301–327.

Clauser, J. F., Horne, M. A., Shirmony, A., and Holt, R. A. 1969. Proposed experiment to test local hidden-variable theories. *Physical Review Letters*, Volume 23, pp. 880–884.

Coleman, P. 2015. *Introduction to Many Body Physics*. Cambridge University Press.

Griffiths, D. 2004. *Introduction to Quantum Mechanics*. Second ed. Pearson Prentice Hall.

Hatsugai, Y. 1993. Chern number and edge states in the integer quantum Hall effect. *Physical Review Letters*, Volume 71, pp. 3697–3700.

Jackiw, R., and Rebbi, C. 1976. Solitons with fermion number 1/2. *Physical Review D*, Volume 13, pp. 3398–3409.

Johnston, N. 2021. *Advanced Linear and Matrix Algebra*. Springer.

Levine, I. N. 2008. *Physical Chemistry*. Sixth ed. McGraw-Hill.

Lu, H.-Z., and Shen, S.-Q. 2017. Quantum transport in topological semimetals. *Frontiers in Physics*, Volume 12, p. 127201.

Mahan, G. D. 2000. *Many-Particle Physics*. Third ed. Springer Series in Physics of Solids and Liquids.

Nowakowski, M. 1999. The quantum mechanical current of the Pauli equation. *American Journal of Physics*, Volume 67, pp. 916–919.

Sakurai, J. J., and Napolitano, J. 2017. *Modern Quantum Mechanics*. Second ed. Cambridge University Press.

Shen, S.-Q. 2012. *Topological Insulators: Dirac Equation in Condensed Matters*. Springer Series in Solid State Sciences, Volume 174. Springer.

Steeb, W.-H., and Hardy, Y. 2018. *Problems and Solutions in Quantum Computing and Quantum Information*. Fourth ed. World Scientific.

Stoler, D. 1970. Equivalence classes of minimum uncertainty packets. *Physical Review D*, Volume 1, pp. 3217–3219.

Szabo, A., and Ostlund, N. S. 1989. *Modern Quantum Chemistry: Introduction to Advanced Electronic Structure Theory*. Dover.

Tamvakis, K. 2019. *Basic Quantum Mechanics*. Springer.

Taylor, P. L., and Heinonen, O. 2002. *A Quantum Approach to Condensed Matter Physics*. Cambridge University Press.

Thomson, M. J., and McKellar, B. H. J. 1991. The solution of the Dirac equation for a high square barrier. *American Journal of Physics*, Volume 59, pp. 340–346.

Vanderbilt, D. 2018. *Berry Phases in Electronic Structure Theory: Electric Polarization, Orbital Magnetization and Topological Insulators*. Cambridge University Press.

Wikipedia, n.d. *Hermite Polynomials*. [Online] https://en.wikipedia.org/wiki/Hermite_polynomials.

Xiao, D., Chang, M.-C., and Niu, Q. 2010. Berry phase effects on electronic properties. *Reviews of Modern Physics*, Volume 82, pp. 1959–2007.

Xu, C., and Moore, J. E. 2006. Stability of the quantum spin Hall effect: effects of interactions, disorder, and Z2 topology. *Physical Review B*, Volume 73, p. 045322.

Zak, J. 1964. Magnetic translation group. *Physical Review*, Volume 134, pp. A1602–A1606.

Zak, J. 1989. Berry's phase for energy bands in solids. *Physical Review Letters*, Volume 62, pp. 2747–2750.

Zhou, B., Lu, H.-Z., Chu, R.-L., Shen, S.-Q., and Niu, Q. 2008. Finite size effects on helical edge states in a quantum spin-Hall system. *Physical Review Letters*, Volume 101, p. 246807.

Zubarev, D. N. 1960. Double-time Green functions in statistical physics. *Soviet Physics Uspekhi*, Volume 3, p. 320.

Index

Printed in the United States
by Baker & Taylor Publisher Services